Atoms, Stars, and Nebulae

The Harvard Books on Astronomy
edited by Harlow Shapley and
Cecilia Payne-Gaposchkin

Galaxies
Harlow Shapley

The Story of Variable Stars
Leon Campbell and Luigi Jacchia

Earth, Moon, and Planets
Fred L. Whipple

Stars in the Making
Cecilia Payne-Gaposchkin

Our Sun
Donald H. Menzel

Between the Planets
Fletcher G. Watson

The Milky Way
Bart J. Bok and Priscilla F. Bok

Tools of the Astronomer
G. R. Miczaika and
William M. Sinton

The nebula 30 Doradus. (Mount Stromlo Observatory.)

Atoms, Stars, and Nebulae *Revised Edition*

Lawrence H. Aller

Harvard University Press, Cambridge, Massachusetts, 1971

A revision of *Atoms, Stars and Nebulae* by Leo Goldberg and Lawrence H. Aller

Distributed in Great Britain by Oxford University Press
Library of Congress Catalog Card Number 76–134951
SBN 674–05264–1
Printed in the United States of America

To Leo Goldberg

Preface

In the generation that has elapsed since Leo Goldberg and I wrote the first version of *Atoms, Stars, and Nebulae,* the nature of astrophysics has been wondrously transformed. At that time, quantitative chemical analyses of stars from their spectra were being successfully carried out. Plausible subatomic sources of stellar energy had been identified and there occurred the first groping toward a comprehensive picture of the life history of a star. The interstellar medium was becoming recognized as the material from which stars and solar systems were emerging—perhaps under our very eyes.

Subsequent developments have greatly extended these early efforts. A vast range of astronomical phenomena—from chemical composition differences between stars to the details of the spectrum luminosity diagram—have been embraced in a ubiquitous theory of element building and stellar structure. Bits and pieces of seemingly unrelated data now often emerge as essential connecting links in a steadily sharpening picture.

All the physical processes involved in our interpretations of observed features of stars and nebulae had one thing in common: Everywhere matter radiated because it was hot, reflected and scattered light, or fluoresced. Then, a quarter-century ago, the new technique of radio astronomy appeared and a brand new window was opened on the universe. Through that window the outer world looked strangely different. Copious amounts of power were radiated by streams of charged particles moving with nearly the velocity of light in vast magnetized clouds in the deep recesses of space. Yet other windows, opened in the infrared, the ultraviolet, and the x-ray region, expanded and embellished this picture. The mysterious cosmic rays, long a province worked by a small band of devoted physicists, were revealed as an integral part of the expanding scene. Radio galaxies and quasars suggested powerhouses of unbelievably high wattage radiating away in space, while the pulsars made sense only in terms of stars of incredible density where the very nuclei of the atoms themselves were crushed beyond redemption.

In the following pages I have attempted to summarize the state of the art in what might now be called the classical astrophysics and have but briefly peeked into that mysterious magnetized realm of high-energy particles and bizarre power sources where the fragmentary facts sometimes seem to belie any reasonable interpretation.

At the end of the thirties, many astronomers felt that we had made the great breakthrough and that an understanding of the essential processes of stellar origin, development, and decline was close at hand. In spite of what appears to be steady progress toward achieving these particular goals, we contemplate more confusing puzzles than ever before. They might be likened to a creator's Zen koans, loaded with apparent contradictions, but carrying in their solution a fuller enlightenment as to the nature of the enchanting universe in which we live.

Composing even a broad-brush, quick sketch of contemporary astrophysics is a frustrating assignment. The vast expansion of the field constitutes only one problem. So many facets of new knowledge are disclosed that it is very difficult to select those that may lead to the most exciting future progress.

My greatest regret is that heavy administrative and other responsibilities have prevented Dr. Leo Goldberg, my former colleague and coauthor, from participating in this revision.

I am deeply indebted, however, to colleagues in the U.S.A. and abroad who have generously read much of the text and have supplied illustrations and photographs. Without their help the task could not have been accomplished.

L. H. A.

Contents

Appendices

Atoms, Stars, and Nebulae

1 Introducing the Stars and Nebulae

To men of ancient times the universe was a stable, if not always secure, place, created, so it seemed, for the sole convenience of humanity. That man's abode, the earth, should occupy the dominant central position could scarcely be doubted, while the sun's justification for existence was to provide mankind with light and life-sustaining energy. The gleaming stars, fixed in the revolving celestial sphere, were regarded as bits of a cosmic mural designed to beautify the night.

It was only natural, too, that the details of the celestial scenery should have become identified with heroes and objects of mythology, identifications that remain in current use as names of star groups, or constellations. Thus the unexcelled constellation of the winter sky is Orion, the mighty hunter, whose club is upraised against the charging bull Taurus (Fig. 1). Three bright equally spaced stars represent his belt; a misty group of stars forms his sword. Behind Orion his two dogs pursue Lepus, the fleeing hare. Marking the eye of the greater dog is sparkling Sirius, the Dog Star. To the ancient Egyptians Sirius was the popular Nile star, whose rising just before the sun foretold the impending flooding of the Nile. Sirius was distasteful to the Greeks, however, for they believed that the blending of its rays with those of the August sun produced hot summer weather. In Persian mythology, Sirius was Tishtrya, the Great Rain

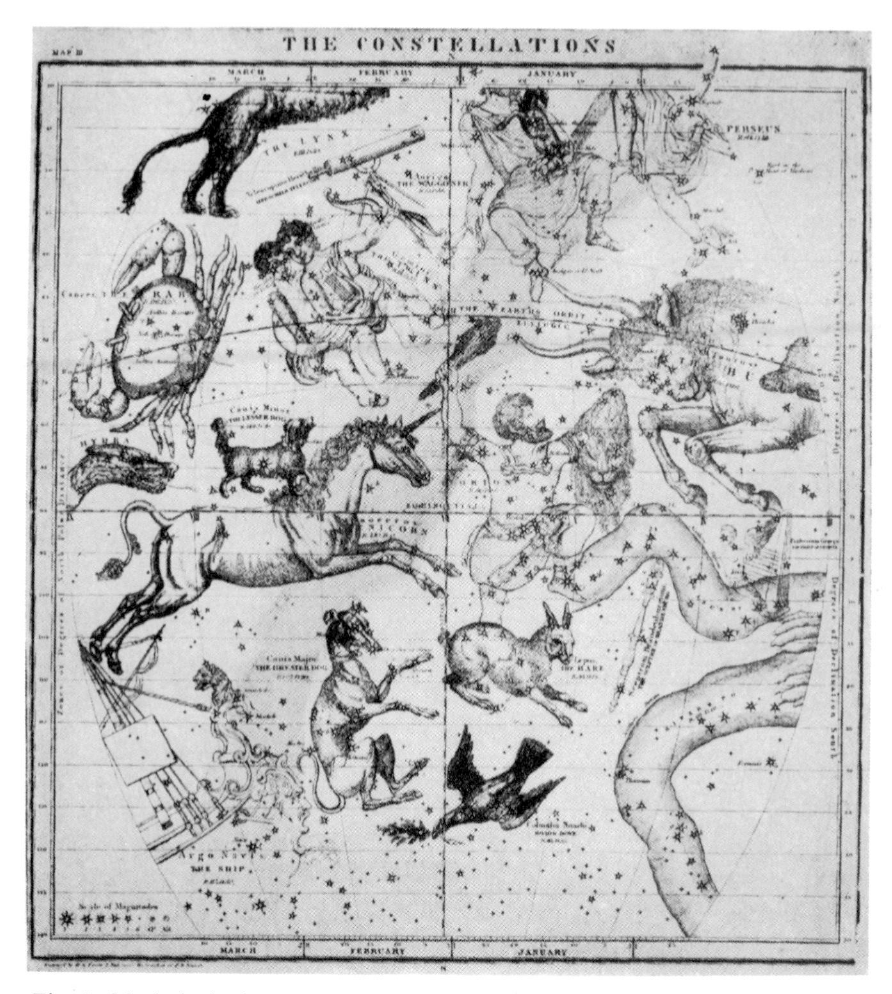

Fig. 1. Mythological map of the sky in the neighborhood of Orion.

Star, who battled Apaosha, the demon of drought. Immortalized in the constellations are Hercules, the Nemean Lion, Hydra, Perseus and Andromeda, and the equipment of gods and heroes—Jason's ship Argo, the Harp of Orpheus, and the arrow from the bow of Chiron.

With the passage of time, these legends, which represented man's earliest attempts to relate himself to his surroundings, became replaced by objective studies of the stars. The astronomical explorer has found the universe a treasure-house of existing discoveries wherein, to add zest to the chase, each great addition to knowledge has brought forth scores of fresh unsolved problems. Mysteries will continue to appear as long as there are men to ponder them.

In this book we embark on a journey of astronomical exploration in which the reader may himself sample a little of the thrill of discovery. During the course of our journey we shall probe the seething atmospheres of the stars and even dig into the interiors themselves. We shall encounter all kinds of curious objects, not only single stars, multiple stars, dwarf stars, giant stars, pulsating stars, and some whose surface layers are occasionally wrenched away in cataclysmic stellar explosions, but also clouds of gas and diffuse matter, the bizarre pulsars, and mysterious emitters of huge amounts of energy whose nature is not understood at the present time.

Our course among the stars has already been charted, for, in broad outlines at least, the geography of the local regions of the universe is known. The earth is but one of a family of planets, satellites, comets, planetoids, and meteoric particles that revolve periodically about the sun. The sun in turn is one of a vast host of stars, about a hundred thousand million, which are grouped together in the form of a thin lenslike system. This stellar system, which contains all naked-eye stars as well as millions too faint to appear visually, is known as the Galaxy or the Milky Way System. The sun's position in the Galaxy is at a point approximately two-thirds of the way from the center to the circumference. It is not an easy matter to determine the details or even the broad outlines of the structure of the Galaxy, since we are located inside the system. Our Galaxy, however, is but one of hundreds of millions of widely separated galaxies that collectively make up the observable universe. Many external galaxies are not unlike our own Galaxy in form and in the kinds of stars they contain. One of these is the famous Andromeda galaxy M 31 (that is, number 31 in the catalog of nebulae compiled in 1781 by Charles Messier; Fig. 2); another is the Triangulum spiral M 33 (Fig. 3). Both are sufficiently close that their shape and contents may be examined in detail with the largest telescopes. Studies of M 31 in particular have revealed its similarity in size, form, and stellar content to the Milky Way and have suggested how we should proceed to discover the structure of our own Galaxy. It is now well established, for example, that both galaxies possess pronounced bulges at their centers and that many of the stars in the main disks are arranged in tightly wound spiral arms. We shall see later that there are fundamental differences in the natures of the stars that populate the spiral arms as compared with those found in the surrounding regions and in the central bulge.

Fig. 2. The Andromeda spiral, Messier 31. Note the spiral arms with the dark lines of absorbing material. On either side of the spiral is a companion elliptical galaxy. The larger one, M 32, is almost directly opposite the center of the spiral. The arms of the spiral actually extend out farther than they can be traced on this photograph. (Agassiz Station of Harvard College Observatory.)

Despite the space-penetrating powers of the 200-inch telescope on Mount Palomar, there is yet no indication that we have approached the boundaries of the universe, if any exist. Most of our tour of exploration will be within the confines of our own Galaxy or the local system of galaxies, but we have reason to believe that our sample is more or less typical of the universe as a whole.

The voyages between stars are likely to be smoggy, for interstellar space is strewn with great clouds of gas and solid grains of dust that dim and redden the light from the stars beyond. Like powerful searchlights, bright stars illuminate many of these clouds, revealing them to the astronomical explorer as bright nebulae. This interstellar matter is spread so thinly that, by comparison, the density of gas present in the best laboratory vacuum seems enormous. Yet, despite its extreme tenuity, enough dust is scattered between the stars to hide from view distant regions of our Galaxy. The gas emits radiation in both the optical and the radio-frequency ranges, and much progress has been made by studying the interstellar medium with radio telescopes.

Fig. 3. The Triangulum spiral, Messier 33. This spiral, which is much smaller than M 31, is seen nearly in plan. There are a large number of gaseous nebulosities and aggregates of stars associated with the spiral arms. (Lick Observatory, University of California.)

One of the features of our tour of discovery is that it can be made without the usual perils of exploration. In fact, owing to the magical powers of light rays, x-rays, and radio waves, we are able to explore far corners of the universe without leaving the comfort and security of the earth. Radiations that are absorbed in the earth's atmosphere (x-rays and certain radio waves) can be studied by rockets and satellites flown above the atmosphere.

The astronomer of a century ago mapped the positions of stars upon the sky and designated their locations much as a geographer maps the earth from accurate measurements of latitude and longitude upon its surface. The positions of the stars are found from the directions of the light rays they emit, but direction is only one characteristic of light rays. Starlight also carries a message about the physical nature of the stars, their masses, brightnesses, chemical compositions, surface temperatures, and even the nature of their internal structure. Radio waves from clouds of interstellar gas tell something about their temperature, density, and chemical composition, and reveal the presence of large-scale magnetic fields. Only relatively recently have we learned to read these hieroglyphic messages from the stars and nebulae. Modern physics, which describes how atoms behave and how they can radiate light, has made this analysis possible. The story of the interpretation of stars and nebulae, so highly dependent on the findings of modern physics, is one that we shall emphasize in this book.

Stellar Distances and Brightnesses

Four obvious questions will occur immediately to anyone interested in probing the physical nature of the stars, namely, how distant, how bright, how big, and how heavy they are. To answer these questions we must employ measuring rods and scales that can be applied over large distances. An astronomer uses the same principle to measure the distance of a star that a surveyor uses to measure the distance across a lake. Figure 4(*a*) illustrates the surveyor's problem; Fig. 4(*b*) the astronomer's. The former measures the length of the line AB and the angles ABC and CAB. The determination of two angles and an included side serves to fix the dimensions of the triangle ABC and the side AC or BC may be computed. Analogously, the astronomer uses as his baseline AB the diameter of the earth's orbit around the sun. When the earth is at A the star lies in the direction AC; six months later the earth is at B and the direction of the star is now BC. One-half of the angle of displacement, that is, the

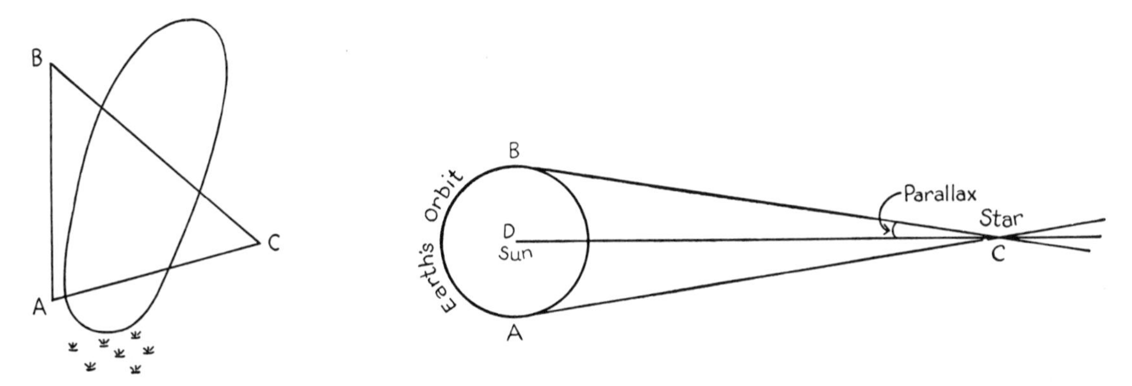

Fig. 4. The measurement of distance to an inaccessible point: (*a*) from one side of a lake to the other; (*b*) from the earth to a star.

angle *BCD* or *ACD*, is called the *parallax* of the star. The amount of the shift clearly depends on the proximity of the star, the more distant ones being the least affected. (Actually the star is moving in a straight line with respect to the sun, and additional observations must be secured to obtain both the parallax and the motion of the star across the line of sight.)

The unit of stellar parallax is the second of arc (1/3600 of a degree), which is roughly the angle subtended by a penny at a distance of 2.5 miles. So small an angle cannot be distinguished with the unaided eye, but modern telescopes enable parallaxes of 0.01 second of arc to be measured with fair accuracy. The parallax of α Centauri, the nearest star, is 0".752, corresponding to a distance of about 25 million million miles. To express distances such as this in miles is more awkward than giving the distance from New York to Bombay in millimeters; hence stellar distances are often expressed in light-years, at least in popular writing. One light-year, the distance traversed in 1 year by a ray of light traveling at 186,000 miles per second, is nearly 6 million million miles. The nearest star is 4.33 light-years away; Sirius, which appears as the brightest star in the sky, is at a distance of 8.7 light-years, and our entire system of stars, the Milky Way, probably measures 100,000 light-years across. On such a scale our solar system seems tiny indeed: if we represent the distance from the earth to the sun by 1 inch, 1 light-year corresponds very nearly to 1 mile.

Two other units of distance, the astronomical unit and the parsec, are also useful in astronomy. For expressing distances intermediate between the mile and the light-year, the radius of the earth's orbit, which is called the astronomical unit (abbreviated A.U.), is commonly used. (This unit should not be confused with the Angstrom unit, abbreviated Å, used to express the wavelength of light.) The parsec is the distance of a star whose parallax is 1 second of arc; it is equal to 206,265 A.U. or 3.26 light-years. Since parallax is inversely proportional to distance, the distance in parsecs is simply the reciprocal of the parallax in seconds of arc. Thus a star 10 parsecs or 32.6 light-years away has a parallax of 0".1, a star 100 parsecs or 326 light-years away has a parallax of 0".01, and so on.

The surveyor's method of measuring parallax is inadequate for any but the very nearest stars; angles smaller than 0".01 cannot be measured with any accuracy. Fortunately, astronomers have devised ways of estimating the distances of remoter stars. One can take advantage of the fact that the stars are in motion, both in regard to one another and with respect to the sun. The actual velocity of motion along the line of sight can be measured by the Doppler effect (see Chapter 2). Then, by measuring the apparent angular motions of selected stars across the sky in different parts of the celestial sphere, one can obtain average or statistical distances much as one could estimate the distance of a lighted speedboat seen on a harbor at night if one knew its actual speed in the water. Other methods, which we shall describe later, are based on the principle that we measure accurately the intrinsic luminosities of certain kinds of stars that we can recognize in distant parts of the Galaxy or even in other stellar systems. Then from the apparent brightness of the star and its known intrinsic brightness we can get its distance, since the brightness of a point source of light diminishes as the square of its distance. If α Centauri were 8.66 instead of 4.33 light-years away it would appear one-fourth as bright.

Conversely, if the distance of a star has been found, we may, knowing its apparent brightness, establish its true brightness. Our current practice of expressing the apparent brightness of a star as seen in the sky in terms of magnitudes was initiated 2000 years ago, when ancient astronomers graded the stars from the first (brightest) to the sixth magnitude, the latter being just barely visible to the naked eye. For the past century, the magnitude scale has been so adjusted that a star of the first magnitude is exactly 100 times as bright as one of the sixth magnitude. The scale goes as a geometrical progression, that is, the brightness ratio corresponding to a one-magnitude step is constant. Thus, a first-magnitude star is 2.512 times as bright as a second-magnitude star, which in turn is 2.512 times as bright as a third-magnitude star, and so on. The original scale of six magnitudes has been extended to include the very faint as well as the very bright stars. Stars as faint as the 23rd or 24th magnitude can be detected photoelectrically or by special photographic techniques with the 200-inch telescope. The brighter stars in the sky, like Aldebaran and Altair, are of the first magnitude. The two very brightest stars in the sky, however, have negative magnitudes; thus, for Canopus, $m = -0.7$, and for Sirius, $m = -1.6$. (On the same scale the apparent magnitude of the full moon is -12.7 and that of the sun is -26.8.)

Stellar magnitudes may be measured with the eye or other light-sensitive devices, such as the photographic plate or photoelectric cell with appropriate filters. By using different filters the color of a star may be measured. The visual magnitudes measured by early observers have been replaced by photoelectric magnitudes measured with a yellow filter—the so-called V-magnitudes. If we wish to express the apparent brightness of a star taking into account all the radiation it emits—infrared, red, green, blue, violet, and ultraviolet (Chapter 2)—we use the bolometric magnitude. The bolometric magnitude is a quantity derived from the observations and the temperature of the star; it will be an observed quantity only when stellar brightnesses can be measured from above the earth's atmosphere. Both very cool and very hot stars are very much brighter bolometrically than visually, since most of their energy is emitted as radiation to which our eyes are not sensitive.

Were all stars equally distant from us, their apparent magnitudes would represent their true relative brightnesses. In practice, we define the intrinsic luminosity of a star by its so-called absolute magnitude, which is the apparent magnitude it would have at a standard distance of 10 parsecs = 32.6 light-years (see Appendix E). The bolometric absolute magnitude of the sun is +4.77. This is the quantity that is important when we want to actually compare the energy outputs of stars. The absolute "photoelectric visual" magnitude of the sun is +4.84 (according to Kron and Stebbins; see Appendix D), which means that at a distance of 10 parsecs it would be comfortably visible on a clear moonless night. Arcturus, whose distance is about 33 light-years, would appear at about its present brightness. Sirius would be about 1/14 as bright as at present and no longer conspicuous. Rigel, in the constellation Orion, which is 50,000 times brighter than the sun, would outshine any object in the present night sky save the moon.

Most of what we know of the universe has been discovered by the detection and measurement of radiation by optical methods, that is, with devices employing ordinary lenses and mirrors. In recent years, however, it has become known that stars, gas clouds, and galaxies also emit radio waves, in addition to light and heat waves. As radio telescopes, the radio astronomer uses large antennas of various shapes, including parabolic dishes that resemble conventional optical telescopes but are usually much larger. A typical radio telescope would be about 85 feet in diameter, and the largest steerable one, at Manchester, England, measures 250 feet across—although the most effective is probably the 210-foot dish at Parkes, Australia. As seen through the eye of a radio telescope, the sky has a totally different "appearance" from that in visible light. Most of the radio radiation comes from gas clouds rather than from individual stars; hence the familiar constellations are not seen in the radio telescope, but are replaced by a variety of radio sources that have quite a different arrangement in the sky.

Weighing the Stars

The motion of the earth about the sun makes possible the determination of stellar distances. Curiously enough, the circling of one star about another permits the determination of stellar masses. Like all planets, and stars too for that matter, the earth is imbued with a wanderlust. Were the restraining influence of the sun's gravitational attraction suddenly to be removed, the earth would fly off in a straight line and eventually lose itself in interstellar space. Just as the earth is kept in its path by the gravitational attraction of the sun, so also are a large number of stars denied a carefree existence by the gravitational attractions of companion stars. Stars so inhibited pursue circular or elliptical orbits about each other. The more massive the two stars, the faster will they move about each other, which we may see from a simple analogy.

Suppose we were in a space ship in interstellar space, where there was no gravitational attraction, so that we floated freely about, and suppose further that it was necessary to measure the mass of a small solid object. Since gravity would not exist inside the space ship, we could not just put the object on a scale and weigh it; some other technique would have to be used. If a spring scale were available, the unknown mass could be found by attaching the object to the scale and swinging them both in a circle at the end of a string. The spring scale would measure the tension in the string, which would depend on the speed of revolution and the mass of the object. The tension would be greater, the greater the mass or the greater the speed of revolution. From the measured tension and the speed of whirling, we could find the mass of the object.

By an analogous procedure the astronomer weighs the stars. The rate of motion of two stars in a double-star system about each other depends on the gravitational force between them. This attractive force, analogous to the tension in the string, is proportional to the masses of the stars (and also to the inverse square of the distance between them), according to Newton's law of gravitation. By observing the time required for the two stars to circle each other and measuring the distance between them, we find the restraining force and hence the masses.

Double-star, or binary, systems are common among the stars. In fact, groups have been found in which three, four, five, and even six stars revolve about one another. Some of these multiple systems merit a brief description.

α Centauri consists of two stars that revolve about each other in 80 years in rather elongated elliptical orbits, so that at times they approach as near as 11 astronomical units (a little more than the distance of Saturn from the sun) and sometimes they recede to 35 astronomical units (nearly the distance of Pluto from the sun). The brighter component is almost a duplicate of the sun, save that it is a little brighter and perhaps a little heavier and a little hotter. The fainter component is cooler and less massive. In 1915, Innes discovered a faint red star 2 degrees away that shares the same motion through space as α Centauri but is 15,000 times fainter than the sun. It is at least 10,000 or 12,000 astronomical units from the brighter pair and must take about a million years to complete its orbit.

Of particular interest is the lesser Dog Star, Procyon, a binary with a period of 40.65 years and a mean separation of 4$''$.55, which corresponds to a distance of 15.8 astronomical units, somewhat less than the separation of the sun and Uranus. The brighter ($V = 0.35$) star has a mass about 1.75 times that of the sun. The companion is a very faint star ($m = 10.8$). It is an aged, superdense star, commonly called a white dwarf (see Chapter 9). Now the orbit of the bright star is known in terms of shape, orientation, and diameter (in seconds of arc). Also the motion of the bright star in the line of sight can be measured from its spectrum (see Chapter 2) and, since the orbit is known, the orbital speed can be found. Then from the period one obtains the actual diameter of the orbit in miles or kilometers, which can be compared with the diameter of the orbit obtained from the separation of the stars in seconds of arc and parallax. The relation is

$$\text{Diameter of orbit} = \frac{\text{Diameter (arcsec)}}{\text{Parallax (arcsec)}} \times 93{,}000{,}000 \text{ miles.}$$

In this way Strand obtained an independent check on the stellar parallax, which he found to be in good agreement with the trigonometric value.

Among multiple stars we mention ζ Cancri and Castor. The brighter component of ζ Cancri is itself a binary consisting of two nearly equally bright components revolving about each other with a period of 59.7 years. The fainter component also comprises two stars, one of which can be detected only by its gravitational effects on the other. They revolve about each other with a period of 17.5 years. This fainter pair revolves about the brighter one with a period of 1150 years. The masses of all four stars are comparable with that of the sun.

As seen in a telescope, Castor is a double star, whose components, separated by a distance of about 80 astronomical units, move about each other with a period of 340 years. Spectroscopic observations (see Chapter 2) show that both stars are actually double, with periods of about 9 and 3 days, respectively. An even more interesting result is that more than 1 minute of arc in the sky away from Castor is a faint star, Castor C, associated physically with the brighter pair. This object itself consists of two faint red stars, smaller and less massive than the sun, separated by about 1,700,000 miles and revolving about each other in less than a day. Thus Castor is a sextuple star whose components appear in pairs.

Although more than 17,000 double stars are listed in Aitken's catalogue, reliable orbits are known for relatively few of them. A hundred years from now our knowledge of binary orbits, and of stellar masses, should have improved greatly. Fortunately, as Russell and Hertzsprung independently showed, it is possible to get good average values of the masses of stars whose parallaxes are known, even if we observe the motion of a star over only a part of its orbit.

The multiple systems are composed of all kinds of stars, large and small, cool and hot; consequently we are able to determine masses for most varieties of stars. When the weighing operation is completed, we find that the most massive stars are between 50 and 100 times heavier than the sun; the lightest stars have probably between one-fifth and one-tenth the solar mass, with the majority weighing a bit less than the sun.

In many instances a star trapped in a double-star system is forced to reveal not only its mass but also its size. Double-star components are frequently so close together that even the most powerful telescope fails to reveal them separately. If, however, the plane of the orbit is so tilted as to appear edgewise in the sky, the passage of one star in front of the other will produce a periodic eclipse, not unlike an eclipse of the sun by the moon. Such double stars are known as eclipsing binary stars. In general, each star masks the other once during a revolution, thus producing two eclipses per cycle. If the two stars are of equal size and brightness, the amount of light received on the earth will be cut in half twice during a revolution. Usually, however, the components of known eclipsing stars are of unequal brightness and size; the pairing of a large, faint star with one that is small and bright is a frequent occurrence. Such a pair of stars is shown diagrammatically in Fig. 5. The passage of the bright star across the faint one produces a partial eclipse of the latter and a resultant dimming of the total light. Half a revolution later, the relative positions of the stars are reversed and, since the bright star is now obscured, the loss of light is much greater. If the observed brightness of the eclipsing binary is plotted against the time, we find a periodic variation in the light as shown in Fig. 6. When the bright star is at positions in the orbit corresponding to *A* and *C* in Fig. 5, the light is undimmed. In *D*, the brighter of the two stars is obscured and the star is said to be at primary minimum. In *B*, the faint star is partially obscured, only a small fraction of the light is lost, and the star is said to be at secondary minimum. It is clear that the duration of each eclipse, which may be learned from the light curve, depends both upon the diameters of the stars and upon

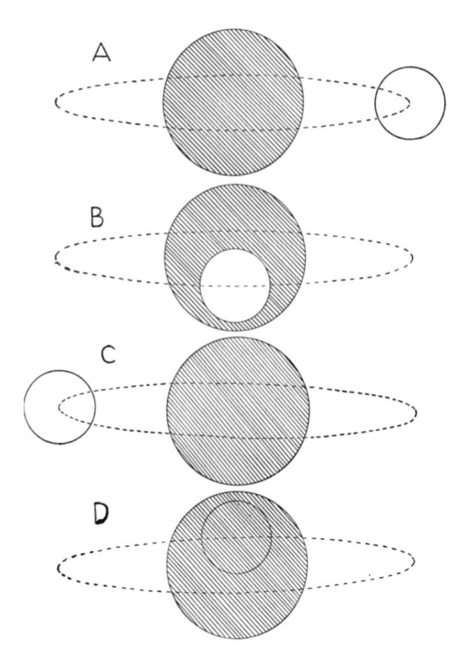

Fig. 5. The relative positions of the stars in an eclipsing-binary system during three-quarters of a period. Here a small bright star revolves around a large dim one. At *B* the bright star is in front of the dim one, whereas at *D* it is behind it.

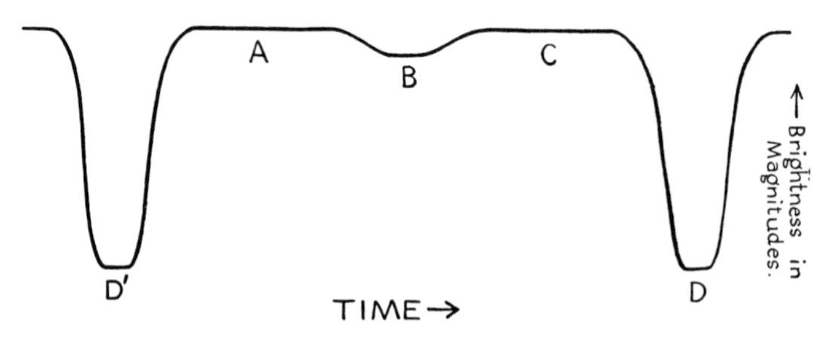

Fig. 6. The light curve of the eclipsing-binary system represented in Fig. 5; the letters correspond to the positions indicated there.

the speed of revolution. Since, as we shall see in the next chapter, we may frequently determine the latter by means of the spectroscope, we can find the stellar diameters.

If, as is usually true, the plane of the orbit does not quite pass through the earth (that is, the inclination is not exactly 90°), the situation will be as depicted in Fig. 5. From an accurately determined light curve, one can determine this tilt or inclination of the orbit, the sizes of the two stars in terms of the diameter of the orbit, and the ratios of the surface brightnesses of the stars (which depends on their surface temperatures).

We can do even more. If we know the orbital speeds in miles (or kilometers) per second and the period, we can find the size of the orbit. From the period and orbital radius we can find the masses of the stars, in terms of that of the sun (see Appendix H). Since the stellar sizes relative to their orbits are known from the light curve, and since the orbital size is found from the spectroscopic measurements, the diameters of the stars are known in miles (or kilometers). With both its diameter and its mass known, the density of a star may be found. In some instances, where the light curve has been followed for several decades, we can even ascertain something about the rate at which the density increases toward the center in the heavier star of the eclipsing system. These studies show that the stars are not homogenous; rather the density increases markedly toward the center (see Chapter 8).

Tables 8 and 9 (Chapter 6) list masses, sizes, periods, and densities for some well-known eclipsing binaries for which data have been obtained. Perhaps the best known of these systems is Algol, the second brightest star in the constellation Perseus, which at intervals of 2.87 days suddenly fades to about one-third its usual brightness. The brighter component has about three times the diameter of the sun, while the larger but much fainter component measures 3.7 solar diameters. Further discussion of these systems is deferred to Chapter 6.

One further point must be mentioned. Although we can learn much from eclipsing binaries, which H. N. Russell referred to as the Royal Road of Eclipses, we must recognize that some of the stars often found in eclipsing systems are abnormal objects in the sense that similar stars are not found singly or in wide binaries. The evolution or life history of a star (Chapter 9) may be modified or distorted if it has a close companion. This situation offers some interesting possibilities for understanding certain remarkable variable stars (see Chapter 11).

Although the separation of stars in binaries ranges from a few times their diameters in systems of the W Ursae Majoris type (where two solar-sized stars swing about each other almost in contact) to thousands of astronomical units, most double-star systems seem to be built on a scale not greatly different from that of the solar system. Frequently it has been suggested that the formation of a solar system and that of a binary system are different aspects of the same fundamental process. Usually the primordial material is assumed to be collected in two or more large masses, forming a binary or multiple system, but sometimes much of it may be lost and a star may be surrounded by a system of planets.

There exists a lower limit to the mass that a body may have and yet shine as a star (see Chapter 8). The faintest known star is the companion to BD + 4° 4048 discovered by Van Biesbroeck (see Appendix A for the meaning of BD designations). It has an absolute magnitude of +19, that is, it is a million times fainter than the sun. If it had a life-sustaining planet, that body would have to revolve around it at a distance less than that of the moon from the earth.

Planetlike companions have been discovered in several binaries by their gravitational attractions upon visible members. Some years ago Strand found a companion to one of the components of the visual binary 61 Cygni for which he determined a mass 1/60 that of the sun, or 16 times that of Jupiter, and a period of 1.89 years. It has been suggested that the fainter of the optical components may also have one (or even two) similar companions.

Even more remarkable is Barnard's star. On the basis of a long series of observations, van de Kamp, of Swarthmore College, has found evidence for two companions. One appears to be slightly more massive than Jupiter, moving with a period of 26 years in an orbit of about the same size as that of Jupiter. The other, whose mass is about 0.8 that of

Jupiter, moves in an orbit that would correspond in size to a position in the asteroid belt in our own solar system. Thus there exists here a solar system with at least two bonafide planets—neither of which, however, is likely to be a fit abode for life.

Dwarf stars, normal stars, giants, supergiants, clouds of dust and gas all go to make up the Milky Way. But the fundamental building blocks for all material structures are tiny atoms, a few million-millionths of an inch in diameter. From atoms and molecules come the light rays that enable us to see and study the stars and nebulae. It is our good fortune that the kind of light emitted by atoms is controlled by their physical environment. Thus the light rays from galactic space carry with them vivid code messages of the climatic conditions in the stars and nebulae. We now turn to the story of how the message of starlight is decoded.

2 Stellar Rainbows

The Spectroscope

The fact that sunlight is composed of a mixture of colors was discovered in 1666 by Sir Isaac Newton. He admitted sunlight into a darkened room through "a small Hole in [his] Window-shuts" and then allowed the light to pass through a triangular glass prism and to fall on the opposite wall of the room (Fig. 7). The original spot of white light was replaced by a brilliant rainbow or *spectrum* of colors, arranged in a band with violet at one end and changing slowly to blue, green, yellow, orange, and finally red at the other. By placing a second prism in reverse orientation behind the first, Newton demonstrated that they could be recombined and white light would be restored. Thus "white" sunlight was proved to be actually a mixture of all the colors of the rainbow.

The glass prism sorts out the separate colored rays by changing their directions by amounts that depend on the color of the light. When a light ray passes from one medium to another its direction usually changes (Fig. 8). (Physically this bending, or refraction, of a ray of light arises from the fact that the velocity of light in the denser medium, such as glass, is lower than in air.) Were all light rays deviated by the same amount in passing through a prism, the emergent light beam would be uncolored.

Fig. 7. Newton's experiment on spectra. (Voltaire, *Elémens de la philosophie de Newton*, Amsterdam, 1738.)

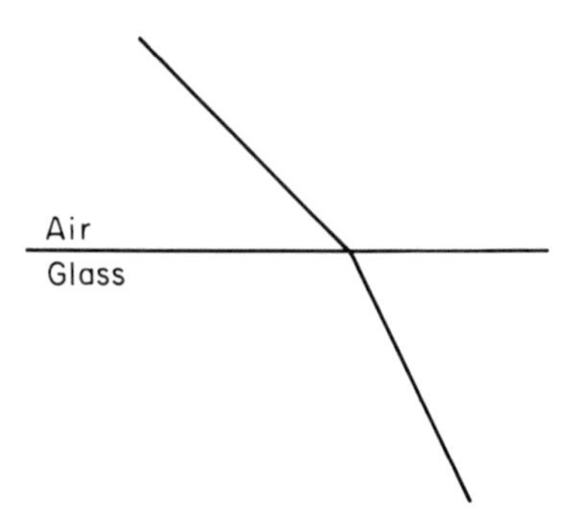

Fig. 8. Refraction of light at a surface between two media.

However, the violet rays are bent more than the blue rays, the blue rays more than the green, the green more than the yellow, with the result that the original white light is spread, or *dispersed,* into its component colors. Similarly, droplets of water in the earth's atmosphere act like tiny prisms and disperse the sunlight to produce the rainbow.

The prism spectroscope (Fig. 9) is essentially patterned after Newton's experimental arrangement. To prevent overlapping of the separate colors, the light source is first focused on a narrow slit, perhaps 0.01 inch wide. After passing through the slit, the diverging beam is collimated, or made parallel, by a lens at *C,* and then directed through a glass prism, *D*. The lens *T* then brings the rays to a focus along the line *PP′*. The spectrum at *PP′* consists of a series of "lines," which are images of the slit, each of a different color; we may examine it with an eyepiece or photograph it upon a plate or film. In an alternative form of the spectroscope, the prism

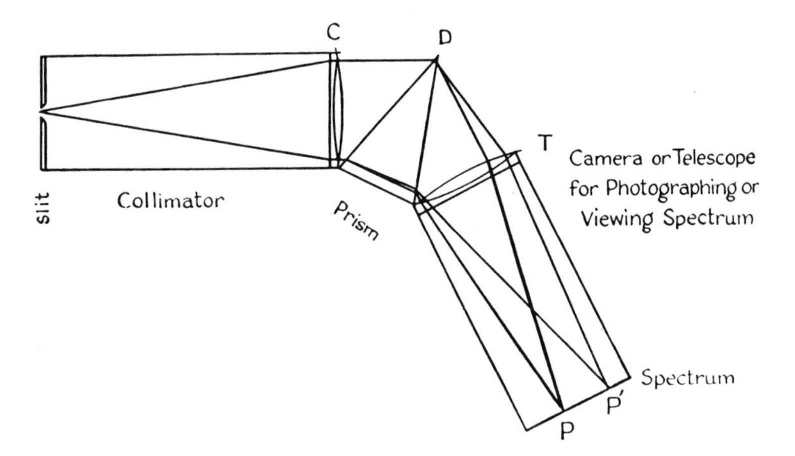

Fig. 9. Schematic diagram of a prism spectroscope.

is replaced by a so-called diffraction grating. In its most widely used form the grating consists of a flat, reflecting surface upon which a series of exceedingly fine parallel grooves are ruled with a microscopically sharp diamond point. The grooves are uniformly and closely spaced, up to 30,000 per inch. When parallel white light falls upon the grooved surface, the component colors are reflected at different angles and are thus dispersed into a spectrum.

In conjunction with a large telescope, the spectroscope in its various forms is the single most important observational device used in astrophysics, and the results obtained with it will be the subject of much of the remainder of this book. Before discussing this instrument further, however, we shall first comment on the physical meaning of color.

The Meaning of Color

Just what is meant by the color of a light ray? The sensation of color is purely subjective, resulting from the response of the retina to some physical property of light. Laboratory experiments have shown that light is propagated in the form of waves, at a speed (in vacuum) of 186,000 miles per second. The distance between successive crests or troughs in the waves is known as the wavelength. The interesting property of light waves is that the phenomenon of color, which is a physiological sensation, is directly related to the wavelength of the light; red light waves are the longest waves visible, the yellow ones are shorter; and the waves of violet light are the shortest that can be seen. The wavelength of red

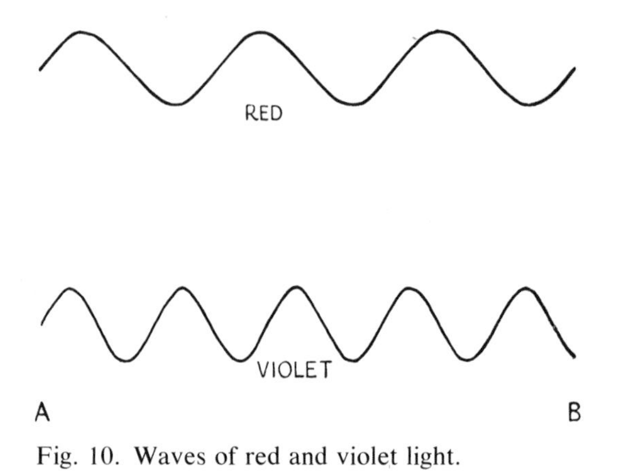

Fig. 10. Waves of red and violet light.

light, for example, is about 25 millionths of an inch, whereas the wavelength of violet light is only about 16 millionths of an inch. Two different waves of light, one red and the other violet, are shown schematically in Fig. 10. Both waves move from A to B in the same time, since the velocity of all light is the same in a vacuum. Since the violet ray has a shorter wavelength than the red ray, it undergoes a greater number of vibrations over the same distance. The number of such vibrations per second, or the *frequency* of the light wave, is equal to the velocity of light divided by the wavelength. Thus the frequency of short-wavelength violet light is 750 million million vibrations per second. (This number is usually abbreviated to read 7.5×10^{14}, where 10^{14} signifies the number 1 followed by 14 zeroes; similarly, the reciprocal of 10^{14} is written 10^{-14}.) The frequency of violet light is $^{25}/_{16}$ times that of the longer-wavelength red light. Wavelengths are usually expressed in angstrom units, named in honor of the Swedish physicist A. J. Ångstrom. One angstrom (abbreviated Å) is equal to 1 hundred-millionth (1×10^{-8}) of a centimeter, or about 4 thousand-millionths (4×10^{-9}) of an inch.

The limited color sensitivity of the human eye confines the visible portion of the spectrum to a strip extending from 4000 Å in the violet region to about 7000 Å in the red.

Various devices for detecting radiant energy, such as the photographic plate, the photoelectric cell, and the thermocouple, show that the radiation spectrum extends far on either side of the visible region. Immediately shortward of the violet region lies the ultraviolet, which can be detected by photographic plates or photoelectric detectors. Solar ultraviolet radiation produces sunburn. Yet farther lies the region of the soft x-rays (10 Å to 100 Å approximately.) High-frequency x-rays fall in the neighborhood of 1 or 2 Å; still higher frequencies are represented by gamma rays, which are emitted by elements like radium or artificially produced radioactive substances.

Longward of the red lies the infrared, which merges continuously through heat rays into the region of the microwaves, "short" radio waves, and ultimately broadcast radio waves hundreds of meters long. Up to 10,000 or 12,000 Å the infrared can be studied by photographic plates. Beyond, devices such as lead sulfide cells, lead telluride cells, and Golay cells must be used. Millimeter and centimeter waves can be detected by ultrahigh-frequency receivers, while meter waves and broadcast waves can be detected by more conventional radio receivers.

Although the stars radiate energy at all wavelengths, most regions of the spectrum cannot be observed from the ground, owing to their absorption by the atoms and molecules in the earth's atmosphere. All parts of the spectrum are at least partially absorbed by the atmosphere, but in certain wavelength regions the absorption is so thorough that no radiation at all can penetrate the atmosphere, even when observations are made from high mountain tops. The inaccessible regions of the spectrum are shown by the shaded areas in Fig. 11, from which it is seen that no radiation of wavelength shorter than about 3000 Å can reach the surface of the earth. The shortest wavelengths, up to 1000 Å, are absorbed by atoms of oxygen and nitrogen at heights of more than 100 kilometers above the ground. From 1000 to 2300 Å, the radiation is blocked by molecules of oxygen and nitrogen, while between 2300 and 3000 Å the absorbing agent is ozone. The spectral regions visible to the human eye are relatively unobscured, but at longer wavelengths large sections of the infrared spectrum are blotted out by molecules of water vapor and carbon dioxide.

Finally the atmosphere becomes totally opaque and remains so until the millimeter region is reached, when it again becomes transparent, this time to radio waves up to about 20 meters long, beyond which the ionosphere cuts out all radiation.

As will be shown in Chapter 7, the radio "window" furnishes an amazing new view of the universe which emphasizes the role of high-energy particles and magnetic fields in interstellar space.

Investigation of remaining spectral regions, however, requires telescopes and detectors flown above the earth's atmosphere in rockets and satellites. Monochromatic pictures of the solar disk as well as far-ultraviolet and x-ray solar spectra show the sun to be a variable star. Stellar ultraviolet spectra reveal extremely important clues to the structure of the

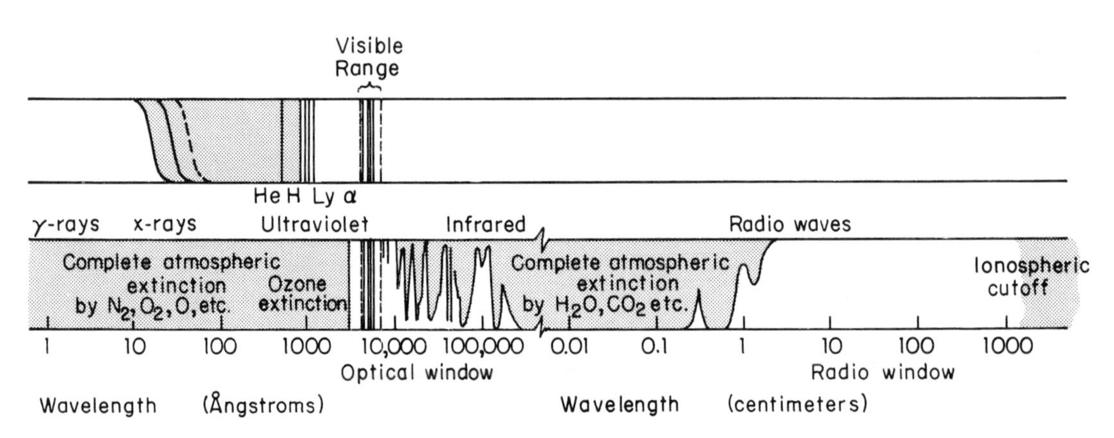

Fig. 11. The spectrum of electromagnetic waves from γ-rays to long radio waves. In the lower strip, the optical and radio "windows" are indicated by white areas and the regions of atmospheric extinction by shaded areas. The upper strip shows the range of radiations accessible to detectors flown in a rocket or satellite above the earth's atmosphere. Absorption by hydrogen, and to some extent by helium, cuts off the light of distant stars in the ultraviolet far beyond the cutoff of the earth's atmosphere. Note, however, that γ-rays, most x-rays, and other regions of the spectrum can be observed without much interference. The narrow range of wavelengths to which the eye is sensitive is indicated. (Adapted from L. H. Aller, *Astrophysics: The Atmospheres of the Sun and Stars* (Ronald Press, New York, ed. 2, 1963), p. 4.)

atmospheres of stars. Investigations by D. C. Morton and by T. D. Stecher and others show that hot stars are ejecting their outer atmospheres at high speeds. X-ray sources have been detected, and some have been identified with catastrophic variable stars such as novae and supernovae (see Chapters 11 and 12).

Atomic Thumbprints

Newton's discovery that a light source like the sun radiates a brilliant spectrum of color, although artistically appealing, is hardly as significant as the fact that different types of light sources are characterized by different types of spectra. The laws of spectrum analysis, generally known as Kirchhoff's laws, are of fundamental importance in our problem. Suppose that, simulating the experiments of Kirchhoff and Bunsen, we were to place the glowing white-hot tungsten filament of an incandescent lamp before the slit of a spectroscope. We would find that the spectrum consists of a bright, continuous band of colors, very similar, in fact, to the rainbow. A piece of iron, or any other solid, heated to red or white heat,

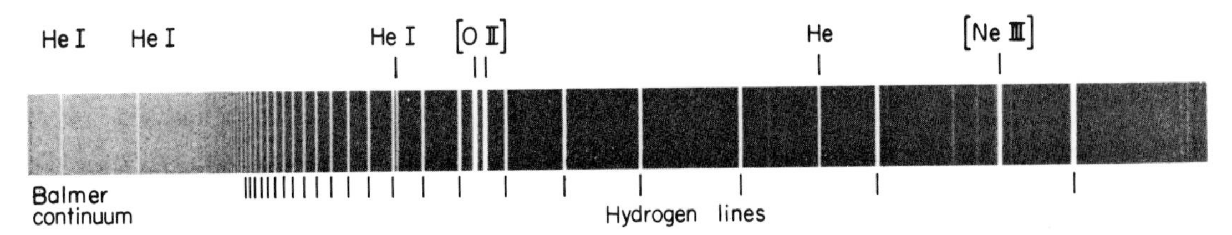

Fig. 12. The emission spectrum of the nebula in Orion, showing the Balmer series of hydrogen. This spectrum of a gaseous nebula shows how the hydrogen lines of the Balmer series converge to a limit, which is followed by a continuum. Note that lines of helium, indicated by He I, the so-called forbidden lines of oxygen [O II], and those of neon [Ne III] are also present. (Lick Observatory, University of California, 120-inch telescope with coudé spectrograph.)

but not vaporized, likewise displays a continuous spectrum. But now if we employ as our light source a glass tube filled with rarefied hydrogen that is carrying an electric current and therefore is luminescent, we observe a spectrum radically different from that of a shining solid. In place of a brilliant continuum there are three bright, colored lines, or slit images, red, blue, and blue-violet, the last just above the limit of visibility at 4102 Å. We note that the spaces between the lines appear black, and also that there is a remarkable regularity in the positions of the lines, with the separations between successive bright images decreasing steadily from red to violet. On the photographic plate, the series continues into the ultraviolet, with the lines crowding together until they terminate near 3650 Å; see Fig. 12, which shows lines of hydrogen and other elements emitted by the Orion nebula. The spectrum of heated sodium vapor likewise shows discrete bright lines, notably a pair close together in the yellow and a series in the ultraviolet. Other glowing gases and vapors radiate bright-line spectra, too, but each element, be it hydrogen, helium, sodium, calcium, iron, lead, or radium, is marked by a different set of radiations, which the spectroscope sorts out as bright lines. Because no two elements display identical spectra, we see that nature has provided us with the means for fingerprinting every element. Once the spectra of the known elements have been recorded in the laboratory, the composition of any mixture may be ascertained, regardless of whether the sample to be analyzed is located on the earth or in a distant star or nebula.

If we now interpose cool sodium vapor between a hot tungsten filament and the slit of the spectroscope, we obtain still a third type of spectrum. In the visible part of the spectrum the brilliant continuum of color from the incandescent lamp appears unchanged except for two dark lines at precisely the same wavelengths at which the bright sodium lines had been

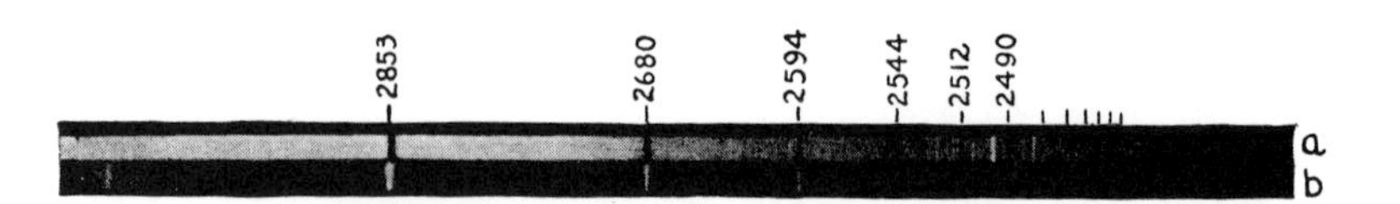

Fig. 13. The ultraviolet spectrum of sodium vapor, illustrating Kirchhoff's second and third laws of spectroscopy. In strip (*a*), light from a carbon arc is passed through a heated tube containing sodium vapor, whose spectral lines appear dark against a (nearly) continuous background; in strip (*b*) the pole pieces of the carbon arc are drilled and filled with sodium carbonate, which then gives the bright-line spectrum of sodium. A few bright lines of the carbon arc are visible in both photographs.

seen before. In Fig. 13 we show the bright-line or emission spectrum of sodium in the invisible ultraviolet, and also the dark-line absorption spectrum with a carbon arc as the source of continuous radiation. The cooler sodium vapor has evidently absorbed light from the bright background, but only in those wavelengths which it is capable of emitting. Similar results are obtained for other vaporized elements; their characteristic spectra appear as dark rather than as bright lines.

Experiments of the sort we have described led to the formulation of Kirchhoff's three laws of spectroscopy: (1) an incandescent, that is, glowing, solid or liquid (or very dense gas) radiates a continuous spectrum; (2) a rarefied glowing gas emits a characteristic bright-line spectrum; (3) the spectrum of a gas placed in front of a hotter source of continuous radiation consists of dark absorption lines at just those wavelengths that the gas emits when it is heated.

In 1802, Wollaston, repeating Newton's experiment, found four dark lines in the spectrum of the sun and interpreted the lines as divisions separating the colors of white light: red, yellow-green, blue, and violet. Spectra of the sun obtained through liquid prisms containing nitric acid, oil of turpentine, oil of sassafras, and Canada balsam were similar, showing that the spectra did not depend on the dispersing medium. About 1815, Fraunhofer mapped 574 lines in the spectrum of the sun; a section of his map is shown in Fig. 14. The suggestion that these lines were caused by absorption in the atmosphere of the earth was disproved when Fraunhofer found that the spectra of several bright stars were quite unlike that of the sun (see Fig. 19). He also noticed agreements between the positions of lines of terrestrial elements and the dark lines of the solar and stellar spectra, but unfortunately attached no significance to the coincidences. In Fig. 15 we reproduce a portion of the solar spectrum in the neighborhood of the blue hydrogen line at 4861 A.

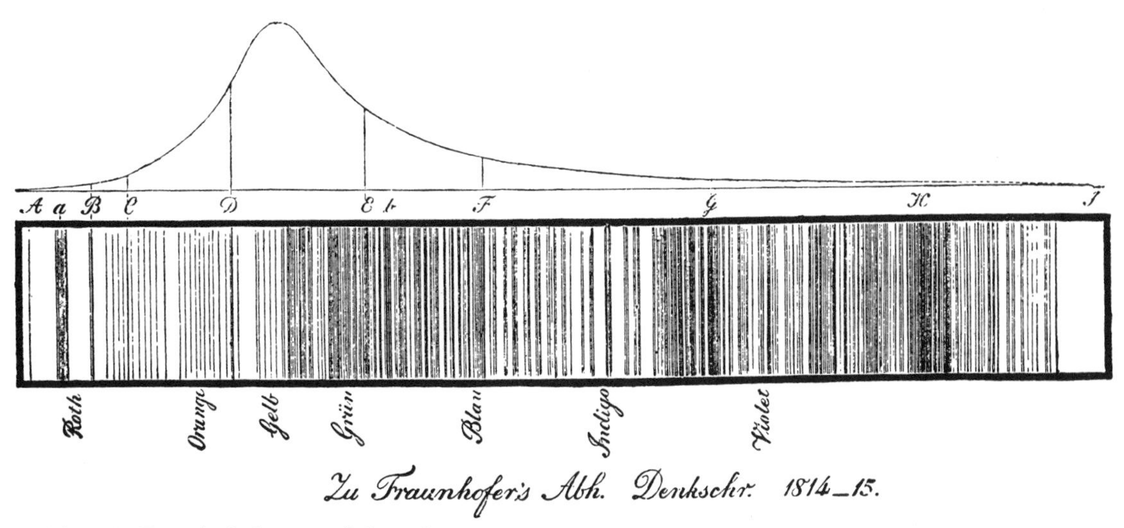

Fig. 14. Fraunhofer's map of the solar spectrum.

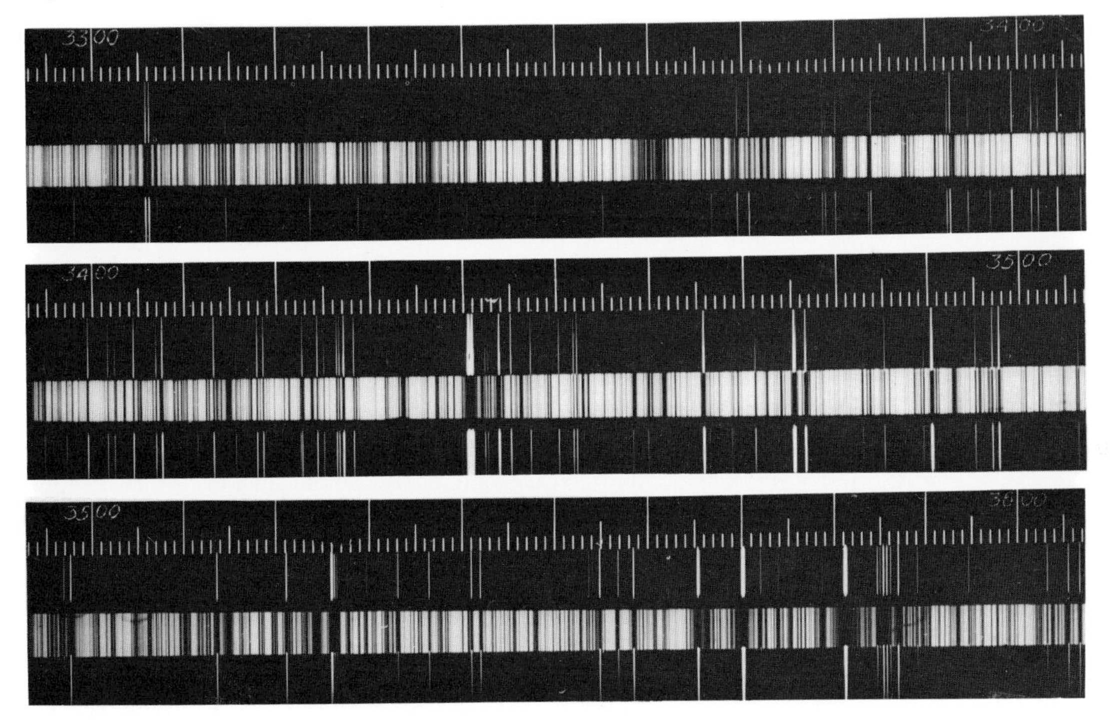

Fig. 15. A portion of the solar spectrum (*center*) with lines of the iron arc above and below it in the wavelength region 3300–3600 Å. The numerous coincidences between lines of the two spectra reveal the presence of iron in the sun. (Mount Wilson Observatory.)

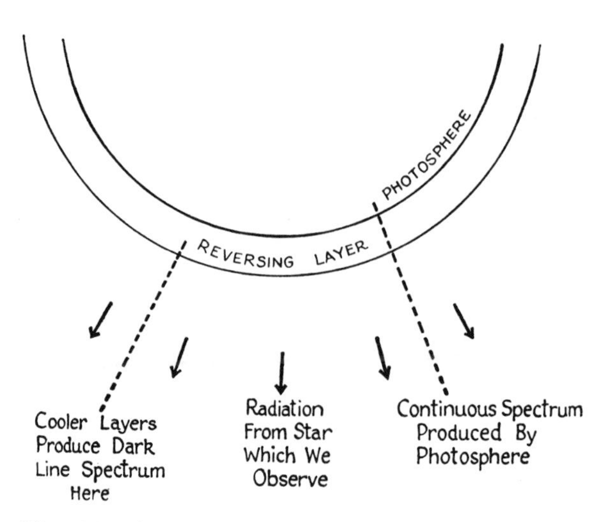

Fig. 16. Kirchhoff's model for the formation of the continuous and dark-line spectra of a star.

Kirchhoff concluded from his observations that the sun and stars must be incandescent hot bodies surrounded by relatively cool thin atmospheres (Fig. 16). He proposed a simple model in which the elements comprising the gaseous atmosphere, or *reversing layer,* of a star absorbed intense, continuous radiation emerging from the lower surface, or *photosphere,* and thus imprint their dark lines upon the spectrum. Kirchhoff's model is useful for visualizing the process of formation of spectral lines, and we shall employ it in some of our later discussions. Nearly 40 years ago, however, Menzel and others showed that this highly stratified model was an oversimplification; both the solar continuous radiation and the absorption lines originate in the same region of the atmosphere, which is still called the photosphere. It is true that on the average the continuous radiation comes from deeper atmospheric layers than do the absorption lines, but there is no sharp demarcation between the photosphere and the reversing layer.

These developments revealed each star as a gigantic laboratory in which matter often could be studied under extreme physical conditions not attainable on earth.

Sir William Huggins, an English amateur astronomer, and independently Sir Norman Lockyer, examined the spectra of a large number of stars, comparing the positions of the dark lines with those of bright lines emitted by elements in the laboratory. They found many coincidences and concluded that matter everywhere in the universe must be alike. The great Orion nebula (Fig. 17), believed by many astronomers to be an

Fig. 17. The region of the great nebula in Orion, showing the confusion of stars and bright and dark nebulosity that make this one of the most fascinating areas of the sky. (Photographed with the Jewett Schmidt telescope of the Harvard College Observatory.)

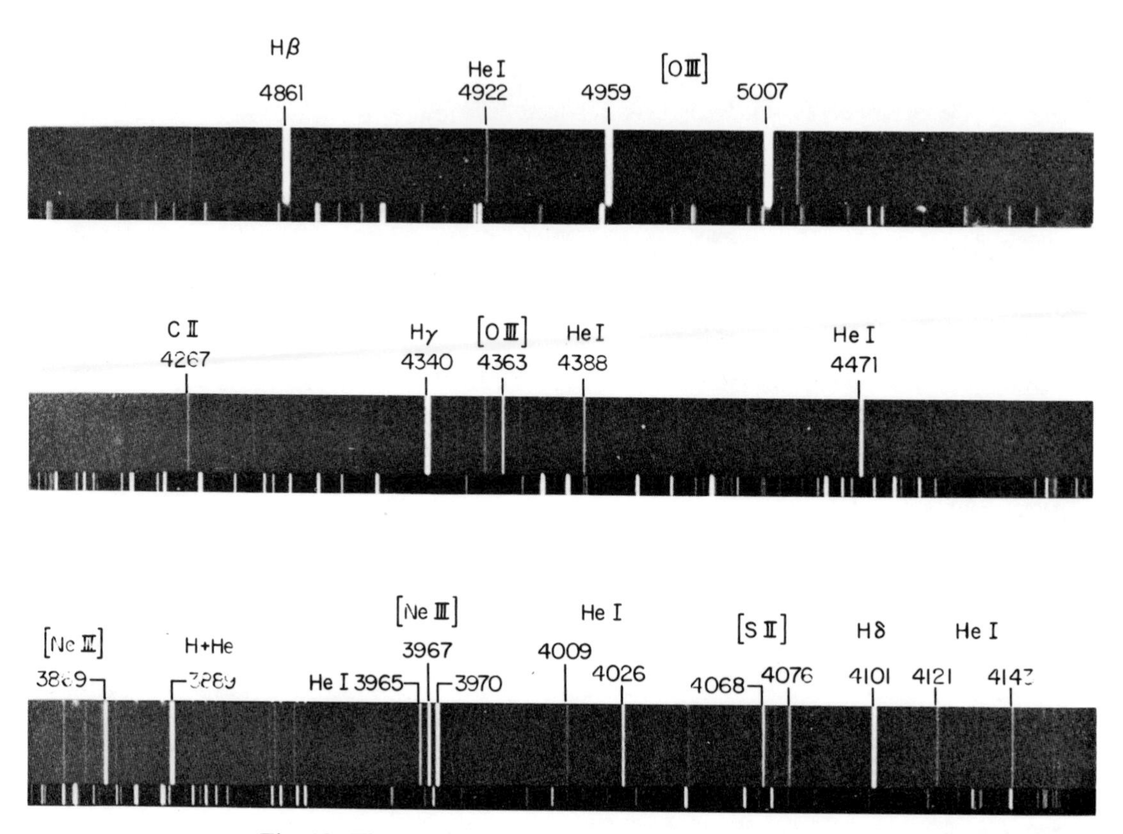

Fig. 18. The spectrum of the great nebula in Orion. A large number of emission lines of permanent gases, notably hydrogen, helium, oxygen, and neon, together with two lines due to sulfur, are indicated (wavelengths in angstroms). Many of these lines, in particular those of oxygen, neon, and sulfur noted here, are of the so-called forbidden type (see Chapter 7). (Lick Observatory, University of California, 120-inch telescope with coudé spectrograph.)

aggregation of stars too far away and too close to one another to be resolved by existing telescopes, was expected to show a continuous spectrum. Huggins found instead, to his astonishment, that the spectrum (Fig. 18) consisted entirely of a few bright lines, some of which could be identified with hydrogen and helium, although certain strong lines defied interpretation. These lines were originally attributed to a hypothetical element called "nebulium," but were later identified as arising from doubly ionized oxygen (see Chapter 7). Later studies also showed a weak continuous spectrum to be present, but there were no dark lines, such as would be produced by stars. The Orion nebula was thus found to be a cloud of tenuous gas rather than a cluster of stars.

Sorting the Stars

While Huggins was interested in the chemical composition of the stars, Father P. A. Secchi, at Rome, was attracted by the diversity in appearance among stellar spectra. Some, like the sun, featured large numbers of lines of metallic elements, notably calcium, sodium, and iron. Others showed only broad lines of hydrogen, while still others, the red stars, exhibited a wealth of complex detail, characterized by dark fluted bands. Secchi found that he could arrange the vast majority of stellar spectra into four distinct types, with all the stars in each group sharing roughly the same spectral features. This contribution was very important, for if the spectrum of a star were related to its physical characteristics, and if all the stars fell into one of four classes of spectra, the detailed study of one star might reveal the characteristics of many more. Secchi found that stars whose brightnesses fluctuated irregularly belonged to the class showing fluted spectra. Stars of Type 1, the blue and white stars, showed some tendency to collect in certain parts of the sky. For example, five of the stars in the Big Dipper, which form a physical cluster of stars moving through space in the same direction and with the same speed, are of this type.

Secchi's achievement was remarkable, considering that his observations were made visually, during long hours at the telescope. With the advent of photography, E. C. Pickering, director of the Harvard College Observatory, embarked on a huge program of spectral classification, with the collaboration of Mrs. Williamina P. S. Fleming, Miss Antonia C. Maury, and Miss Annie J. Cannon. Pickering placed a large glass prism in front of the telescope objective, and used the lens to focus the spectra on the photographic plate. The advantage of the objective-prism technique is that a great many spectra may be photographed on a single plate, whereas the slit spectrograph records only one spectrum at a time.

The aim of the Harvard classification was to group the stars in such a way that the spectral features of one group merged as smoothly as possible into those of the next adjacent group. As the dark lines of hydrogen seemed to be common to all stellar spectra, the original plan called for labeling as Class A the stars with the most intense hydrogen lines, those with the next strongest hydrogen lines Class B, and so on down to Classes M and N, where the hydrogen lines are very weak. This scheme had to be modified for a number of reasons. Some of the classes, for example, C, D, H, had been derived from out-of-focus photographs and were spurious. Also the arrangement in order of decreasing hydrogen-line intensities produced discontinuities in the trends of other spectral lines. Class O, discovered later, was found to belong at the beginning of the sequence. As finally adopted, the classes follow the order O, B, A, F, G, K, and M. In addition a few stars classified as R, N, and S appear to represent side branches jutting off from the main sequence near class K. (If the reader finds it difficult to adjust his memory to this peculiar arrangement of letters, we venture to suggest that the sentence: "Oh, Be A Fine Girl, Kiss Me Right Now, Sweet!" has already proved its worth.)

The photographic plate shows such a wealth of detail that it has been necessary to divide each of the Harvard classes into subdivisions by affixing a number from 0 to 9 to each letter; thus the dark-line pattern of spectral Class A5 lies midway between those of A0 and F0. According to this system, the sun was classified as G0 in the Henry Draper catalogue (see Appendix A).

In Fig. 19 we have arranged a number of typical stellar spectra photo-

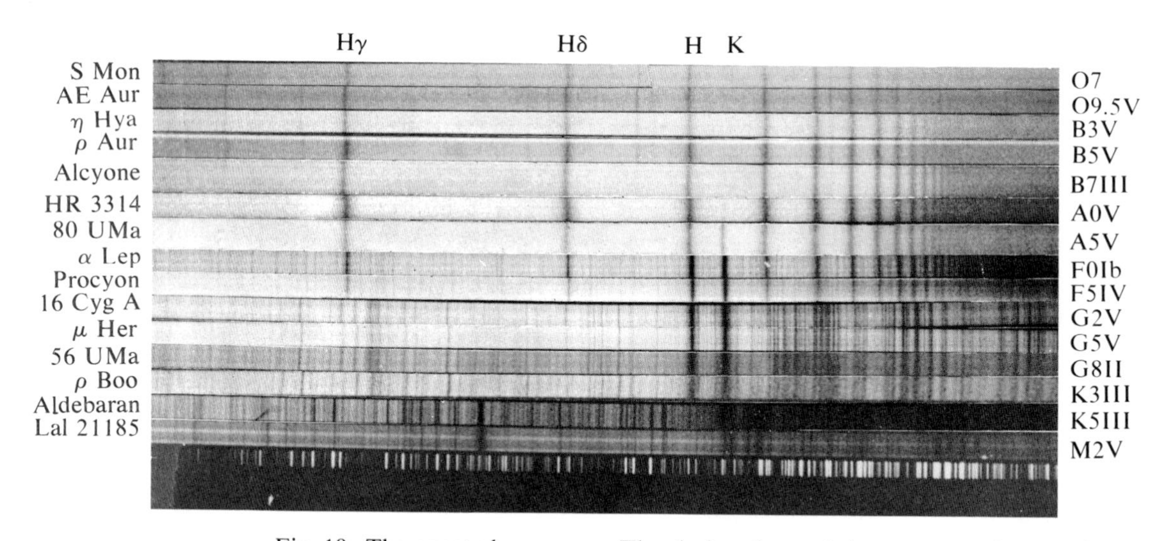

Fig. 19. The spectral sequence. The designations of the stars are given at the left, the spectral classes at the right. The Roman numerals denote luminosity classes (see Chapter 6). Notice the great strength of the metallic lines in the cool stars, while the hot stars are characterized by lines of hydrogen and helium. (From observations secured with the 24-inch reflector at Ojai Observing Station, University of California.)

graphed at the Ojai Observing Station to show the main features of the sequence. Owing to the difficulties in adjusting exposure times, the densities of the spectra vary. Thus, there is some spurious variation in the blackness of the lines, but the trends of excitation and complexity of the spectra are shown. Notice how the spectra grow in complexity from Class O to Class M. Beginning with Class O, the hydrogen lines grow steadily stronger, reach their peak of distinction at Class A0, and then sink into obscurity. Classes O and B bear the imprint of helium, which is absent from the spectra of later types. The lines of the metals like calcium, sodium, and iron are first noticeable in Class A and rapidly grow in strength and numbers through Classes F, G, and K. The broad bands of molecular compounds creep into the picture in Classes G and K, becoming the outstanding landmarks on the spectroscopic map in Classes M, R, N, and S. A very significant aspect of the spectral classification is that it also segregates the stars according to color. Furthermore, the colors along the sequence are arranged somewhat like those in a spectrum, the blue stars occurring at the beginning and the red stars at the end of the sequence. Thus the bright blue stars in the constellation of Orion are of Class B. Sirius, a whitish star, is of Class A0, while the southern beauty, Canopus, is of class F0. Capella, brightest star north of the celestial equator and yellow like the sun, is of Class G0; Arcturus, the bright orange star of spring and summer, is of Class K0; and Betelgeuse and Antares, red stars of Orion and the Scorpion, respectively, are Class M stars.

The actual determination of the spectral class of a star depends upon the relative intensities of certain lines. The helium lines (in the hotter stars), the hydrogen lines, the *K* line of ionized calcium (see Chapter 4), and the 4227-Å line of neutral calcium are among the lines used for this purpose. In the cool stars one employs the intensities of the titanium oxide bands in Class M, the zirconium oxide bands in Class S, and the carbon bands in Classes R and N (sometimes called Class C because they represent carbon stars).

These spectral differences are apparent even when the scale of the spectrum is very small, as it usually is with objective-prism plates or with plates secured with spectrographs intended for observations of very faint stars and distant galaxies. Spectral differences can also be recognized by accurate measurements of the color of a star, using combinations of filters and detectors to evaluate the star's brightness in three or more, preferably narrow, intervals of the spectrum. Classifications of spectra can also be made by use of a spectrum scanner, in which the photographic plate is replaced by a photocell that is moved relative to the spectrum.

We emphasize that this classification of stellar spectra was carried out solely on the basis of the appearance of the spectra themselves, without regard to physical causes that might be responsible. Many early workers believed these differences to be due to variations in chemical composition. Were the blackness of a spectral line dependent only on the abundance of the atom responsible for it, the stars could easily be arranged in order of steadily changing hydrogen abundance. It would be a most remarkable coincidence if stars arranged on this system also showed smoothly varying abundances of all other elements and if the hydrogen stars were always blue and the metallic stars red. We shall show in Chapter 4 that these variations are due, not to changes in chemical composition, but to changes in temperature and density. Such changes in chemical composition as actually exist are usually minor compared with effects of temperature or pressure. An exception is provided by certain cool stars in which carbon is more abundant than oxygen.

The Spectroscope as a Speedometer

The spectroscope reveals not only the compositions of stars, but also their speeds toward or away from the observer. To understand how the spectroscope can act as a speedometer, the reader should recall the high-pitched whistle that heralds the approach of a speeding train, and the sudden transition to a long-drawn-out wail that accompanies its passing and recession. The whistle emits sound waves of a definite frequency and wavelength, and the number of waves per second that strike the ear determines the pitch of the sound. When the train is in rapid motion toward the listener, the individual waves tend to crowd up on each other, and a greater number fall upon the ear every second. The increase in the number of vibrations per second is interpreted by the ear as a rise in pitch. Conversely, when the train is receding, the sound waves are drawn out and fewer of them per second strike the ear, which perceives that the pitch has fallen.

If light is propagated as a wave motion, a similar effect should operate, as was pointed out by Christian Doppler in 1842. Suppose that a source emits light of a certain frequency, which passes through the spectroscope and appears as a spectral line. The wavelength determines the position of the line; but when the light source is racing toward the observer, the light waves reach the spectrograph more frequently and the wavelength seems shorter. Consequently, the spectral line is shifted from its normal position toward the violet. And when the light source is receding, the line moves over toward the red. The magnitude of the shift, which is known as the Doppler shift, is related to the speed of the light source by the equation

$$\frac{\text{Change of wavelength}}{\text{Normal wavelength}} = \frac{\text{Speed of source}}{\text{Speed of light}}.$$

(It makes no difference whether the light source or the observer is in motion; the important thing is the rate at which the two are approaching or receding from one another.) Thus, for example, the speed of light is 186,000 miles per second and if the light source is receding at 18.6 miles per second the position of a line at 5000 Å is shifted by 0.5 Å, an amount easily detected.

To measure the speed of a star, the spectral lines of a laboratory source —for example, iron, titanium, or helium—are impressed on a photographic plate on either side of the stellar spectrum to serve as reference marks for measuring the positions of stellar lines. The astronomer then determines the displacements in angstrom units of the stellar lines with respect to the comparison lines. From these displacements he obtains the velocity of the star. The spectroscope thus gives the radial velocity, the rate of motion along the line of sight, just as the progressive displacement of a star's position on the celestial sphere measures its speed at right angles to the line of sight. The combination of the two completely defines the direction and speed of motion of the star relative to the earth. Figure 20 shows how the radial velocity of a star causes a shift in the positions of spectral lines. Radial velocities have received special attention at the Lick, Mount Wilson, Victoria, and Yerkes Observatories.

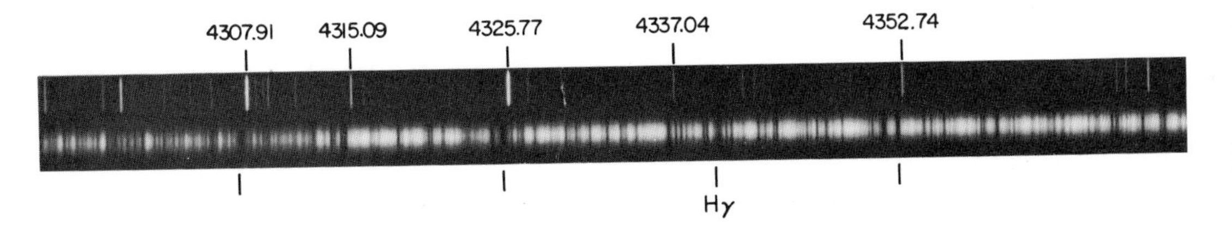

Fig. 20. A portion of the extremely complex spectrum of Arcturus, from which the radial velocity of the star can be determined. Lines of a comparison spectrum are shown at the top; corresponding lines in the stellar spectrum, several of which are indicated by arrows, are seen to be shifted toward the left, that is, toward shorter wavelengths, indicating that the distance between Arcturus and the earth is decreasing. In this cool star, the Hγ line of hydrogen at 4340 Å is no longer conspicuous. (Lick Observatory, University of California, 120-inch telescope with coudé spectrograph.)

The spectroscope as a speedometer has also had important application in studies of the orbital motions of double stars. The components of many double stars are so close together that they cannot be separated by direct observation. However, when the plane of the orbit is tilted even slightly in the direction of the line of sight, each star appears now approaching, now receding, as it whirls about its companion. If the two stars are almost equally bright, the spectrum will exhibit a periodic doubling of the lines, when one star is approaching and the other is receding. Usually, however, one star is so much brighter than the other that only a single spectrum is seen and the lines of this spectrum oscillate to and fro as the velocity of the star with respect to the observer changes. A star whose duplicity is recognized from its spectrum is known as a spectroscopic binary. Mizar, the star at the bend of the handle of the Big Dipper, was the first star of this class to be detected, by E. C. Pickering in 1889. Several hundred others have since been discovered. A catalogue by J. H. Moore and F. J. Neubauer of Lick Observatory gives the orbits of over 500 spectroscopic binaries, while Batten's catalogue lists 700 of these objects, many of them discovered very recently.

Polarized Light

Infrared, visible, and ultraviolet light, x-rays, gamma rays, and radio waves are all electromagnetic waves, that is, waves in combined electric and magnetic fields (Fig. 21). Imagine a small, free body carrying an electric charge placed in the path of such a wave: it would be accelerated up and down as the electric field pointed in first one direction and then the other. The oscillations of the fields take place perpendicular to the direction of propagation of the wave; we speak of such waves as transverse waves, as distinct from longitudinal waves, such as sound waves or compressional waves in a solid or fluid.

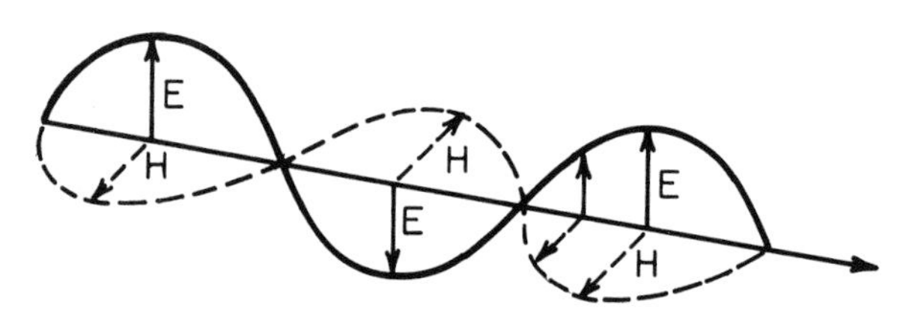

Fig. 21. An electromagnetic wave, consisting of a combination of an alternating electric field E and an alternating magnetic field H at right angles to each other.

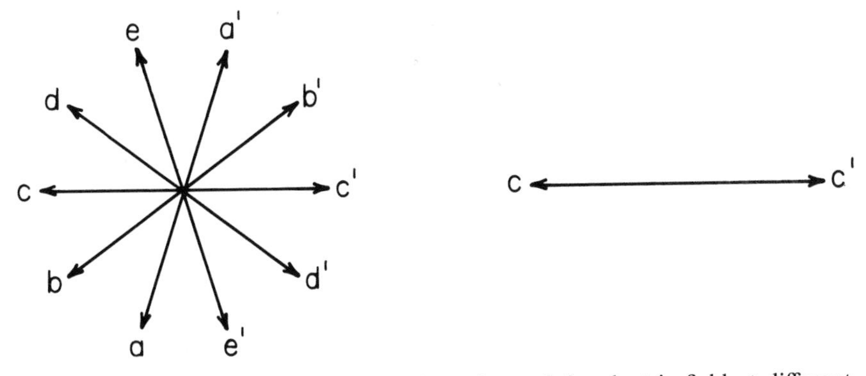

Fig. 22. Polarization of light: (*left*) orientations of the electric field at different instants; (*right*) all but one direction of vibration has been stopped by a polarizing filter.

An important property of transverse waves is that they can be polarized. Suppose we are looking along the direction in which a light wave is traveling (Fig. 22). At one instant the electric field will be along the direction aa', an instant later it may be along the direction cc', then along dd', changing randomly at a great rate. If a piece of Polaroid is placed in the beam, all directions except, say, cc' will be suppressed and the light ray is said to be plane polarized. By rotating the Polaroid, different directions of oscillation may be selected. If the light is initially polarized, the intensity of the transmitted light will be a maximum at some position of the Polaroid, zero at right angles thereto. If the light is partly polarized, there will be a variation of intensity as the Polaroid is rotated.

Another type of polarization is important. Suppose we looked along the direction of the beam and saw that the direction of the electric field rotated uniformly and with a frequency equal to the frequency of the light. The light would be said to be circularly polarized. Polaroid alone cannot distinguish such light from ordinary unpolarized light but with the aid of suitable auxiliary equipment the distinction can be made quickly. Polarization of radio waves, x-rays, and so on can be measured also.

3 Atoms and Molecules—Stellar Building Blocks

Atoms and Radiation

Where does light originate? When we press a switch at home, electrical energy flows through a wire and is somehow converted into light radiating from a tungsten filament. In some way, atoms, which are the tiny building blocks of all forms of matter, generate light of various colors or wavelengths when fed with fuel in the form of chemical or electrical energy. By what operation inside the atom is that light generated and why do different kinds of atoms radiate energy in different wavelengths?

Atoms are much too small to be seen; hence, experiments to find out their structure and behavior have to be conducted with large numbers of them. From the results of these experiments we may attempt to construct a hypothetical model of an atom that behaves like the true atom. Many such atomic models have been proposed in the past and have had varying degrees of success in reproducing the observed features of spectra. But all of them, at one time or another, have been contradicted by experiment. These failures have led to the conclusion that no purely mechanical model of the atom is entirely satisfactory; the laws of mechanics that govern the operations of large bodies break down when applied to ultramicroscopic particles. Entirely new laws of mechanics have had to be devised to cope with the behavior of atoms. These laws are embodied in the so-called wave mechanics, or quantum mechanics, which has thus far given a completely successful account of atomic behavior. The operation of these laws, although perfectly straightforward mathematically, is somewhat difficult to visualize. For this reason, even the scientist who makes his calculations according to the mathematical laws of quantum mechanics frequently thinks of the atom in terms of some simple mechanical model.

It may be worth while to recall here the differences between atoms and molecules. The chemists have shown that the many gases, liquids, and solids which make up the world are composed of the pure forms or combinations of fundamental substances, called elements, which may combine to form compounds. Thus water is composed of hydrogen (two parts by volume) and oxygen (one part by volume). The smallest particle of an element is an atom; the smallest particle of a compound is a molecule. The molecule of water consists of two hydrogen atoms bound to one oxygen atom, thus, HOH. We must distinguish between mixtures or alloys, such as brass, in which the atoms are loosely mixed with one another, and compounds, where the individual atoms that make up a molecule are tightly bound together.

What Atoms Are Made Of

Experiments in the laboratory have shown that the chief materials of atomic construction are three fundamental particles, which have been labeled electrons, neutrons, and protons. The electron, which carries a negative electric charge, is the lightest particle known in nature. It would take 311×10^{26} (that is, 311 followed by 26 zeros) of them to weigh 1 ounce. Expressed in grams, the mass of an electron is 9.11×10^{-28} (28.35 grams = 1 ounce). The neutron and the proton are equal in mass, weighing 1836 times as much as an electron, or 1.66×10^{-24} gram. A piece of dust 0.001 inch in diameter would still weigh 1000 million million times as much as a proton. The electric charge associated with atomic particles is conveniently expressed in terms of the charge of the electron, which is taken as -1. In the electrostatic system of units this charge is 4.803×10^{-10} esu; in the practical system of units it is 1.602×10^{-19} coulomb. The proton carries a positive electric charge, equal in numerical magnitude to that of the electron, or $+1$, whereas the neutron, as its name implies, is electrically neutral.

In every atom, protons and neutrons, often in nearly equal numbers, are tightly bound together to form a closely packed nucleus, which is surrounded by one or more outer electrons. A small amount of atomic matter occupies a relatively enormous volume of space, for the electrons are probably separated from the nucleus by distances of the order of thousands of times the diameter of the nucleus. The cement that binds this outer atomic structure together is the force of electrical attraction between the positive and negative charges. It is this attractive force that keeps atoms electrically neutral. Strip an atom of its electrons and the nucleus continually strives to capture others until the electrical balance is restored.

The number of protons and neutrons that constitute any nucleus, say that of an iron atom, may be learned from two observable quantities, namely, the mass of the atom and the number of outer electrons. Since the proton and the neutron weigh so much more than the electron, the total number of them in a nucleus determines the mass of the atom. Of this total, for a neutral atom, enough must be protons to equal the number of outer electrons and thereby provide electrical neutrality. The lightest of all elements is hydrogen, with a nucleus composed of a single proton, and with one outer electron; it has no neutron. The hydrogen atom weighs 1.673×10^{-24} gram, which is a bit more than the mass of one proton. A helium atom weighs approximately four times as much as hydrogen, and, with two outer electrons, must contain two protons and two neutrons within its compact nucleus (often called an alpha particle). Oxygen atoms are 16 times as massive as hydrogen atoms and have eight electrons; their nuclei consist of eight protons and eight neutrons.

The spectrum and the chemical properties of an atom depend essentially only on the number of its outer electrons. The differences in chemical properties between potassium, which has 19 outer electrons, and calcium, which has 20, are well known. Likewise the spectra emitted by calcium and potassium are entirely different. Disturbances of an atom's outer electrons by means of collisions with other atoms, or with a stream of electrons in an electric arc, produce the spectral lines we observe in a flame or an arc. To disturb the nucleus we must resort to far more drastic measures (see Chapter 8).

More than a hundred separate elements are known (Table 1). Of these, 88 appear on the earth as stable elements, radioactive elements such as thorium or uranium, or decay products of such elements. The remainder are unstable, but their nuclei have been created in the laboratory (see Chapter 8). Each atom has been given a number corresponding to the number of its electrons; thus, the atomic number of hydrogen is 1, that of helium is 2, of oxygen 8, and of uranium 92. The masses of atoms are usually expressed on a relative scale, which is based (by chemists) upon an adopted atomic weight of 16 for oxygen. Since oxygen contains 16 protons and neutrons, the mass of each of these particles must be unity. Why is it, then, if atoms are made up of integral numbers, 1, 2, 3, . . . , of the fundamental particles, that the atomic weights listed in Table 1 are not integers? Even the alpha particle weighs slightly less (by about 0.7 percent) than four protons. The reason (see Chapter 8) is that when helium is formed from hydrogen in the stars, part of the mass disappears as energy. But this mass deficiency (which is often expressed in energy units) is a relatively small fraction of the mass. How, then, can we explain the atomic weight of chlorine, which is 35.46, or of zinc, which is 65.38?

It so happens that two or more electrically neutral atoms may be of the same atomic number, and yet have different masses because they have different numbers of neutrons in the nucleus. Such atoms are said to be isotopes of the same element. The atomic weight of each isotope is nearly an integer, but, since each element may contain a mixture of stable isotopes, its average atomic weight need not necessarily be a whole number. Practically all elements have isotopes. Carbon, for example, has two stable ones, each containing six protons, but one with six neutrons and the other with seven neutrons; the atomic weights are 12.004 and 13.008 respectively. By far the most abundant carbon isotope is of atomic weight 12, hence the average atomic weight of ordinary carbon is 12.006. Carbon 13 (C^{13}) as it is called, is scarcely more than a trace of adulteration (1 percent) in the predominant carbon 12 (C^{12}). Since the spectra of atoms depend essentially on the numbers of their outer electrons, the spectra of different isotopes of the same element are nearly identical.

Some isotope nuclei and even nuclei of unique elements can be produced by bombardment by high-energy particles such as occur naturally in cosmic rays or are produced by accelerators (see Chapter 8). Often these nuclei are unstable, decaying to other nuclei in times ranging from a fraction of a second to many years. One of the best known of such isotopes is C^{14} (which contains six protons and eight neutrons). It decays to N^{14} with the emission of an electron from the nucleus. Cosmic-ray bombardment produces this C^{14} on the earth. Since it obeys the chemistry of ordinary carbon, C^{14} becomes involved in living material such as trees and bones, and steadily decays after the organism dies. W. F. Libby showed how the C^{14} content of old organic remains could be used to date artifacts of old civilizations and primitive man.

To explain how atoms radiate light, we shall rely on an atomic model that has served physicists for many years in the visualization of the be-

Table 1. The chemical elements.

Element	Symbol	Atomic number	Atomic weight	Element	Symbol	Atomic number	Atomic weight
Hydrogen	H	1	1.008	Technetium	Tc	43[a]	99
Helium	He	2	4.003	Ruthenium	Ru	44	101.07
Lithium	Li	3	6.94	Rhodium	Rh	45	102.91
Beryllium	Be	4	9.01	Palladium	Pd	46	106.4
Boron	B	5	10.81	Silver	Ag	47	107.87
Carbon	C	6	12.01	Cadmium	Cd	48	112.41
Nitrogen	N	7	14.01	Indium	In	49	114.82
Oxygen	O	8	16.00	Tin	Sn	50	118.70
Fluorine	F	9	19.00	Antimony	Sb	51	121.76
Neon	Ne	10	20.18	Tellurium	Te	52	127.61
Sodium	Na	11	22.99	Iodine	I	53	126.91
Magnesium	Mg	12	24.31	Xenon	Xe	54	131.30
Aluminum	Al	13	26.98	Caesium	Cs	55	132.91
Silicon	Si	14	28.09	Barium	Ba	56	137.35
Phosphorus	P	15	30.98	Lanthanum	La	57	138.92
Sulfur	S	16	32.06	Cerium	Ce	58	140.13
Chlorine	Cl	17	35.46	Praseodymium	Pr	59	140.92
Argon	A	18	39.95	Neodymium	Nd	60	144.25
Potassium	K	19	39.10	Promethium	Pm	61[a]	147
Calcium	Ca	20	40.08	Samarium	Sm	62	150.36
Scandium	Sc	21	44.96	Europium	Eu	63	151.96
Titanium	Ti	22	47.90	Gadolinium	Gd	64	157.25
Vanadium	V	23	50.95	Terbium	Tb	65	158.93
Chromium	Cr	24	52.00	Dysprosium	Dy	66	162.50
Manganese	Mn	25	54.94	Holmium	Ho	67	164.94
Iron	Fe	26	55.85	Erbium	Er	68	167.27
Cobalt	Co	27	58.94	Thulium	Tm	69	168.94
Nickel	Ni	28	58.69	Ytterbium	Yb	70	173.04
Copper	Cu	29	63.57	Lutecium	Lu	71	174.98
Zinc	Zn	30	65.38	Hafnium	Hf	72	178.50
Gallium	Ga	31	69.72	Tantalum	Ta	73	180.95
Germanium	Ge	32	72.60	Tungsten	W	74	183.86
Arsenic	As	33	74.92	Rhenium	Re	75	186.31
Selenium	Se	34	78.96	Osmium	Os	76	190.2
Bromine	Br	35	79.92	Iridium	Ir	77	192.2
Krypton	Kr	36	83.80	Platinum	Pt	78	195.10
Rubidium	Rb	37	85.48	Gold	Au	79	196.98
Strontium	Sr	38	87.63	Mercury	Hg	80	200.60
Yttrium	Y	39	88.91	Thallium	Tl	81	204.39
Zirconium	Zr	40	91.22	Lead	Pb	82	207.20
Niobium[c]	Nb	41	92.91	Bismuth	Bi	83	209.00
Molybdenum	Mo	42	95.95	Polonium	Po	84[b]	210

Table 1 (*continued*).

Element	Sym-bol	Atomic number	Atomic weight	Element	Sym-bol	Atomic number	Atomic weight
Astatine	At	85[a]	211	Americium	Am	95[a]	241
Radon	Rn	86[b]	222	Curium	Cm	96[a]	242
Francium	Fr	87[a]	223	Berkelium	Bk	97[a]	243
Radium	Ra	88[b]	226.05	Californium	Cf	98[a]	244
Actinum	Ac	89[b]	227	Einsteinium	Es	99[a]	
Thorium	Th	90[b]	232.12	Fermium	Fm	100[a]	
Protoactinium	Pa	91[b]	231	Mendelevium	Md	101[a]	
Uranium	U	92[b]	238.04	Nobelium	No	102[a]	
Neptunium	Np	93[a]	237	Lawrencium	Lw	103[a]	
Plutonium	Pu	94[a]	239				

[a] Artificial elements, not found on the earth, but created in the laboratory.
[b] Naturally occurring unstable elements.
[c] Niobium was formerly called Columbium, Cb.

havior of the electrons within an atom. In the representation that we shall describe, the electrons are pictured as revolving about the nucleus in much the same fashion as the planets revolve about the sun. But whereas the planets are prevented from escaping into space by the sun's gravitational attraction, the electrons are held within the atom by the force of electrical attraction between the positively charged nucleus and the negatively charged electron. Such a model closely reproduces the behavior of the hydrogen atom but must be modified to explain, even qualitatively, the behavior of more complex atoms.

When we speak of atomic behavior, we refer to the fact that each atom emits and absorbs light of certain wavelengths. Consider, for example, the hydrogen atom. The spectrum of this element in the discharge tube and in the stars is characterized by a precise regularity. The strongest line is the red line at 6563 Å, often labeled $H\alpha$, followed by the blue line $H\beta$ at 4861 Å, the violet line $H\gamma$ at 4340 Å, and a sequence of others $H\delta$, $H\epsilon$, . . . gradually drawing closer together until they merge near 3650 Å. Balmer, in 1885, showed that the wavelengths, λ, of this series of hydrogen lines could be represented accurately, by the simple formula

$$\frac{1}{\lambda} = R\left(\frac{1}{2^2} - \frac{1}{n^2}\right),$$

where R is a constant. The wavelengths of the successive members of the series, beginning with the red line, are computed from the formula by

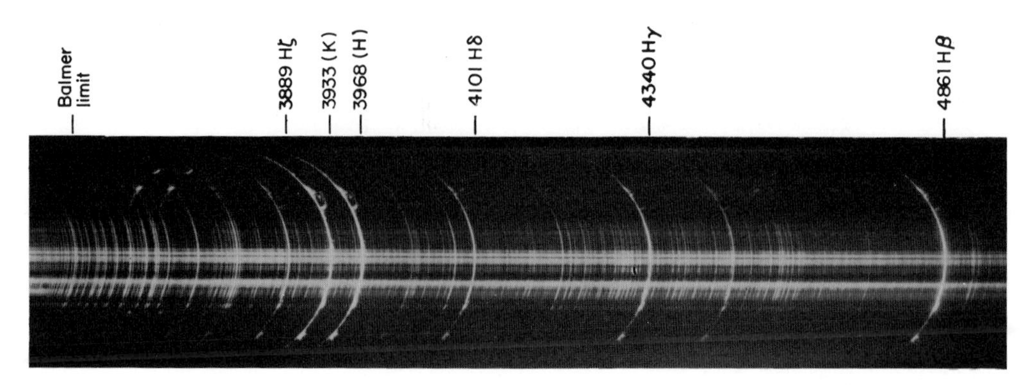

Fig. 23. The Balmer series of hydrogen in the solar chromosphere.

setting $n = 3, 4, 5, 6, \ldots$ Over 30 members of this series, known as the Balmer series, have been observed in the spectra of certain stars, of the sun's outer atmosphere, the chromosphere (Fig. 23), and of many gaseous nebulae (see Fig. 12).

Lyman found another series of hydrogen lines, in the far ultraviolet beginning at 1216 Å and ending at 912 Å, whose wavelengths could be represented by the formula $1/\lambda = R(1/1^2 - 1/n^2)$, with R is the same constant as before and $N = 2, 3, 4, 5, \ldots$ Paschen discovered a series of infrared lines which could be represented by $1/\lambda = R(1/3^2 - 1/n^2)$, with $n = 4, 5, 6, \ldots$, and Brackett discovered a far-infrared series that followed the formula $1/\lambda = R(1/4^2 - 1/n^2)$, with $n = 5, 6, 7, 8, \ldots$ Two additional series, still farther in the infrared, have been observed by Pfund and by C. J. Humphreys, respectively.

Bohr's Model of the Atom

In 1913, Niels Bohr successfully explained the various hydrogen series by suggesting an atomic model in which the electron travels in a circular orbit about the proton. In his picture of the hydrogen atom, the motion of the electron is subject to very specific traffic rules, for only a restricted set of orbits is allowed—those whose radii are proportional to the squares of integers from 1 to infinity, that is, to 1, 4, 9, 16, . . . (Fig. 24).

In each of these orbits the energy of motion, or kinetic energy, of the electron is just balanced by the force of attraction that is exerted by the nucleus and prevents the electron from escaping. If an electron traveling in a particular orbit is to be made to travel in an orbit more distant from the nucleus, it must be supplied with energy from some outside source,

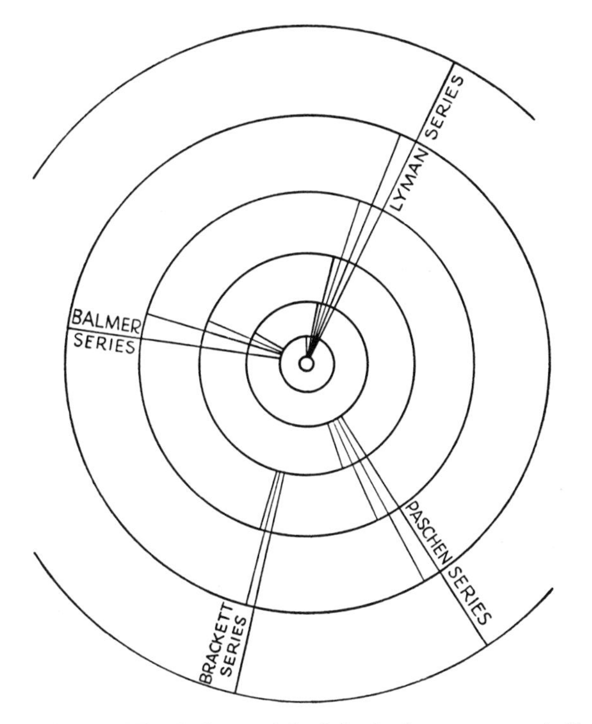

Fig. 24. The Bohr model of the hydrogen atom, indicating the first four spectral series. The radii of the successive orbits are proportional to the squares of consecutive integers, that is, to 1, 4, 9, 16, . . . ; the radius of the smallest orbit is 0.528×10^{-8} centimeter.

because work must be done to pull it away from the nucleus that attracts it. A jostling encounter with another atom or the seizure of a passing light pulse may suffice to do the trick. But atoms are fussy; the electron will not change orbits unless it takes up precisely the required amount of energy, no more and no less, to remove it to one of the other "allowed" orbits. Bohr showed that if the amount of energy required to pull the electron from the ground, or lowest, orbit entirely free of the nucleus is represented by W, the amount of energy required to pull the electron out of the second orbit is $W/4$, out of the third $W/9$, and so on. In other words, the energies required are proportional to W/n^2, where $n = 1, 2, 3, \ldots$ If, for convenience, we call the energy zero when the electron is completely removed from the atom and at rest, the energy when the electron is in the lowest orbit is $-W$ (minus because work must be done to remove it from this level). The energy when in the second orbit is $-W/4$, and so on. Hence we speak of the energies of the allowed orbits as being equal to $-W/n^2$, where $n = 1, 2, 3, \ldots$ The quantity W, which depends upon the charge and mass of the electron and other constants, may be computed from Bohr's theory. A positive value of the energy indicates that not only has the electron been removed from the atom but it is flying away in space with a velocity of its own. One important point is that, although the negative energies are restricted by the condition $E = W/n^2$, the positive energies are not restricted at all. This means, of course, that the electrons flying freely about in space are not constrained to move with special speeds but may travel about with random speeds and directions.

When an electron is removed from an atom, the atom is said to be *ionized*. The energy necessary to tear an electron from the orbit of least energy entirely away from the atom is called the ionization potential, which is measured in electron volts. The ionization potential of hydrogen is 13.60 electron volts, which means that if an electron is accelerated across a potential drop of 13.60 volts it will possess just enough energy, if it collides with a hydrogen electron to detach it completely from its orbit of least energy.

Bohr postulated further that an electron may switch from an orbit of higher energy to one of lower energy. Since the transfer involves a loss of energy, Bohr supposed that the atom simultaneously releases a pulse or quantum of light and that the frequency, and therefore the wavelength and color, of the emitted radiation must be related to the difference in energy between the two orbits, or

$$E_a - E_b = h\nu,$$

where E_a is the energy of the electron when it is in the larger orbit, E_b is its energy in the smaller orbit, h is a numerical constant, called Planck's constant, and ν is the frequency of the emitted radiation.

From these postulates Bohr was able to calculate the wavelength of the radiation resulting from any jump the electron might perform. We have seen that the relation between the frequency of light and its wavelength is

$$\text{Frequency} = \frac{\text{Velocity}}{\text{Wavelength}}, \text{ or } \nu = \frac{c}{\lambda},$$

and that by the Bohr theory the energy in the second orbit, for example, is $-W/2^2$ and in some higher orbit, say the fourth, the energy is $-W/4^2 = -W/16$. The reciprocal of the wavelength emitted by an atom when the electron "jumps" from the fourth to the second orbit should be given by

$$\frac{1}{\lambda} = \frac{h\nu}{hc} = \frac{W}{hc}\left(\frac{1}{2^2} - \frac{1}{4^2}\right) = R\left(\frac{1}{2^2} - \frac{1}{4^2}\right),$$

where we write R for the constant W/hc. When the value of W as calculated from the Bohr theory, along with those of h and c, is inserted, we obtain $\lambda = 4861$ Å, which is the wavelength of the blue hydrogen line! If we write down the formula for jumps or transitions from any orbit, say the nth, to the second, we obtain

$$\frac{1}{\lambda} = R\left(\frac{1}{2^2} - \frac{1}{n^2}\right),$$

where $n = 3, 4, 5, \ldots$, which is just the empirical formula for the wavelengths of the Balmer series, including the numerical value of R. Similarly, all electron jumps terminating in the lowest orbit produce a series of lines in the far ultraviolet, the Lyman series, whose wavelengths may be obtained from

$$\frac{1}{\lambda} = R\left(\frac{1}{1^2} - \frac{1}{n^2}\right),$$

where $n = 2, 3, 4, \ldots$, and the constant R is exactly the same as before. The transitions ending in the third orbit give the infrared Paschen series and those ending on the fourth level the far-infrared Brackett series.

We may conveniently represent the energies of the Bohr orbits by plotting them as horizontal lines, or energy levels, as shown in Fig. 25. Transitions between the various levels are indicated in the figure by vertical lines. In a neutral hydrogen atom, the electron spends the vast majority of its time in the lowest orbit. In this condition, of course, the atom cannot radiate. The electron may be driven into one of the outer orbits either by a collision with a rapidly moving atom or free electron, or by absorbing a quantum of light whose wavelength coincides with one of the lines of the Lyman series. When the electron is in one of the outer orbits, the atom is said to be *excited.*

Once an electron arrives in a higher orbit, say the fifth, it may decide to jump to any one of the four lower orbits. But the decision must be made rapidly, for the electron lingers in an excited state only about a hundred-millionth of a second. A return to the lowest orbit will be accompanied by the emission of the fourth line of the Lyman series in the invisible ultraviolet. If the electron chooses to stop at the second level, the third line of the Balmer series, at 4340 Å, will be emitted and we shall have a minute flash of violet light. Similarly, jumps from the fifth level to the third and fourth result in the second line of the Paschen series and the first member of the Brackett series, respectively, both of which are invisible infrared radiations.

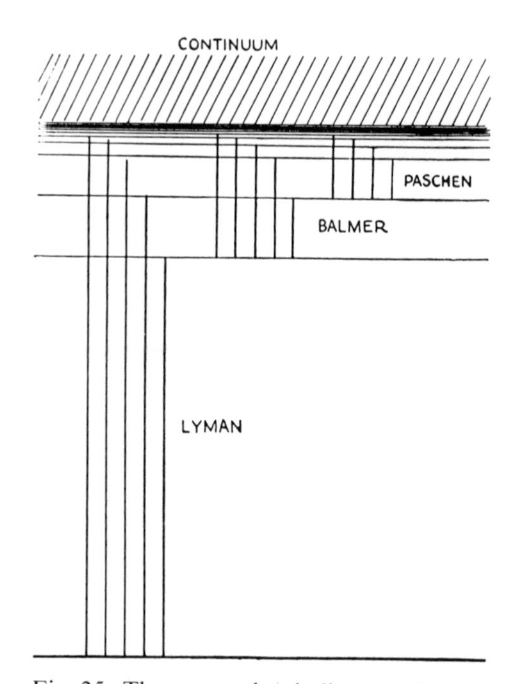

Fig. 25. The energy-level diagram for hydrogen. In this figure energies are plotted instead of orbits. Each level corresponds to an orbit in the Bohr model; the continuum above the uppermost level represents the energy that the electron may have when it has been completely detached from the proton forming the nucleus of the atom and is moving freely in space. Note that the greatest interlevel energy difference is that between the two lowest levels, corresponding to the two innermost Bohr orbits.

Figure 12 shows that the lines of the Balmer series crowd ever closer together toward the violet end of the spectrum until they terminate at the series limit, which is followed by a continuous spectrum. Figure 25 serves to explain the coalescing of the lines near the series limit. As we go to larger and larger orbits in the Bohr model, the difference in energy between proton and electron diminishes, until ultimately a minute amount of energy is sufficient to detach the electron completely. In our earlier discussion, we chose for the zero of energy the value that corresponds to the top of the series of horizontal lines in Fig. 25. The shaded region above represents a positive energy, the energy of the proton and electron after the election has been torn away. There are no longer any restrictions on the electron's speed; it may fly about in a carefree fashion, although excessively high velocities are improbable.

We have seen that to produce a transition between any two of the Bohr orbits, or, in terms of Fig. 25, between the corresponding two levels (below the shaded region), a discrete amount of energy must be emitted or absorbed. This explains why the hydrogen lines appear only at certain wavelengths and no others. But the electron may escape from the atom provided it absorbs any amount of energy above the minimum required for ionization. The excess energy is used up in imparting a velocity to the free electron. The upper portion of the shaded region in Fig. 25 thus represents free electrons with high velocities, the lower portion those with low velocities. Consequently, the ionization of hydrogen atoms will produce a continuous absorption spectrum. Conversely, the capture of free electrons by protons produces a continuous emission spectrum at the violet end of the limit of each series. Figure 32 shows the continuous absorption at the limit of the Balmer series in the star Canopus and Fig. 12 shows the continuous emission at the Balmer limit in a gaseous nebula.

Complex Atoms

We have discussed the spectrum of the simplest of all atoms, hydrogen. If we consider atoms with more than one electron, the problem becomes more involved. Each electron is free to travel in any number of allowed orbits, as before. But the energy of the atom depends upon the particular combination of orbits that are occupied by its electrons. The greater the number of electrons, the more numerous will be the possible combinations of orbits, and therefore the greater the number of spectral lines. The spectrum of iron (Fig. 26) is a good illustration of the intricacies of a complex atom. We find that a modified Bohr model is able to predict the exact number, but not the wavelengths, of the spectral lines that are observed for each atom.

The intricacy of an atom's spectrum, however, is not always in direct ratio to the number of its electrons. The reasons stem from the fact that there are limitations on the number of electrons that are allowed to move in orbits at the same distance from the nucleus. In the hydrogen atom, the electron normally moves in the smallest orbit. In helium, the two electrons revolve about the nucleus in orbits of the same size. Lithium has three electrons, two of which travel in identical orbits close to the nucleus, while the third moves in a larger, outer orbit. Beryllium has four electrons, two in the inner and two in the more distant orbit. As the number of electrons is increased through the elements boron, carbon, nitrogen, oxygen, fluorine, and neon, the additional electrons are all found in the second orbit. No atom ever has more than two electrons in the

Fig. 26. The spectrum of iron compared with that of a hot star. A small portion of the emission-line spectrum of iron is compared with that of the corresponding region of the hot star, τ Scorpii. Notice the complexity of the iron spectrum. None of these arc lines of iron appear in the star's spectrum, because the stellar temperature is about 28,000°K, while that of the arc is only about 4,000°K. In the stellar atmosphere the iron atoms have lost two or more electrons; hence there are no iron absorption lines corresponding to the emission lines in the arc, which are produced by neutral iron atoms. (Courtesy Mount Wilson Observatory.)

first orbit. Similarly, no atom can have more than eight electrons in the second orbit. We may think of each set of orbits as a *shell;* the electrons tend to arrange themselves in shells about the nucleus. The first shell is completed at helium with two electrons; the second contains eight, and is filled at neon with 2 + 8 electrons. If n denotes the number of a shell, $2n^2$ represents the number of electrons it may contain. It develops that the electrons in a closed shell are very tightly bound to the nucleus, and are excited to higher orbits only at the expense of a considerable amount of energy. When all but one of the electrons in an atom are in closed shells, it is the motion of this outside electron that is responsible for the spectrum. In this event the spectrum is roughly similar to that of hydrogen.

The sodium atom with its eleven electrons falls into this hydrogenlike category. The outstanding feature of the sodium spectrum is a pair of strong lines in the yellow, the famous *D* lines in Fraunhofer's map of the solar spectrum. If we regard the two lines as a unit, the *D* lines form the first component of a series, the higher members of which are shown in Fig. 13. Except for the doubling of the lines, this series closely resembles the Lyman series of hydrogen; the lines crowd closer and closer together and eventually approach a limit in the ultraviolet.

The doubling of each sodium line may be traced to the fact that the electron spins like a top at the same time that it revolves about the nucleus (Fig. 27). An electron, which is an electric charge, in motion is equivalent to a tiny electric current, which, as it flows in its closed circuit, generates a magnetic field. The revolution of the electron in its orbit about the nucleus generates one magnetic field, the spinning of the electron generates another. The energy of the atom depends upon the direction in which the electron is spinning. The energy is greater if, like the earth, the electron spins in the same sense as it revolves than if the spin is in the opposite sense. Both directions of spin occur in sodium

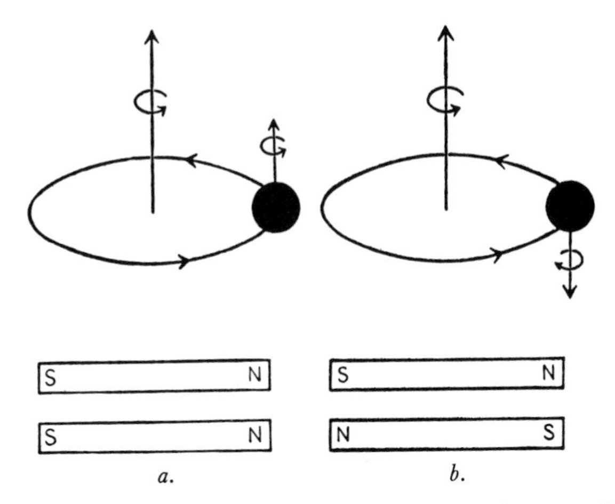

Fig. 27. The magnetic effects of spinning electrons. The curved arrows indicate the directions of revolution of the electron in its orbit and of its rotation about its axis. The straight arrows indicate angular motion in accordance with the right-hand rule: the arrow points in the direction of the thumb of the right hand when the fingers are curled in the direction of rotation.

atoms. As a simple, although not rigorous, analogy we may compare the behavior of the spinning and revolving electron with two bar magnets (Fig. 27). In case *a,* when the directions of spin and revolution are alike, the bar magnets are laid parallel, with their north poles side by side; in case *b* the magnets are antiparallel, with the north pole of one next to the south pole of the other. To shift the magnets from position *b* to position *a,* we must supply energy in order to overcome the mutual attractions of the opposite poles and the repulsions of like poles. The two positions therefore represent different energies. In the same fashion, each orbit in the sodium atom has two energies, corresponding to the fact that there are two directions in which the electron may spin. Consequently, each line appears doubled.

Now since the lines of sodium and similar atoms are doubled because of electron spin, and since the electron in the hydrogen atom is also spinning, we may ask why the hydrogen lines are not also split into components. The answer is that they do show a minute splitting, which has been detected in the physical laboratory, but which is too small to be observed in astronomical spectra. It will soon become evident, however, that the electron spin in hydrogen has an importance in astronomy which is quite out of proportion to the smallness of its effect on the visible spectrum.

There are two ways in which the spin of the electron may be deduced from observations of the spectrum. The first of these has already been described, namely, by the splitting of the lines, which comes about because, on the average, half of the atoms in a cloud of gas will at any instant have their electron spins oriented in one direction and half in the opposite direction. It can also happen that within a given atom an electron spinning in the same direction as it revolves about the nucleus may suddenly reverse the direction in which it spins. When this occurs, a quantum of radiation is released. Since the energy difference corresponding to the two orientations of the electron spin is very small, the wavelength of the quantum is very large, and may in fact be of the order of centimeters or meters, which is in the region of radio waves. When an electron is in an excited orbit like $n = 8$, the probability is tremendously greater that it will jump to a lower orbit than that it will reverse its spin. Hence it is much more likely to emit light waves than radio waves. In the hydrogen atom, the interaction between electron spin and orbital motion splits all of the excited energy levels ($n = 2, 3, 4, \ldots$) but does not affect the lowest energy level ar.d therefore is a very weak if not completely negligible source of radio waves.

A much more powerful mechanism for the emission of radio waves by hydrogen atoms results from the fact that the nucleus of the atom also spins about its axis. The spinning of the nucleus generates a tiny magnetic field which interacts with the field of the spinning electron and produces an additional slight splitting of the energy levels. The special significance of the nuclear-spin–electron-spin interaction is that it also splits the lowest energy level of hydrogen, unlike the orbital-motion–electron-spin interaction, which affects only the excited levels. The difference in energy between the two possible orientations of the electron spin in the first orbit of hydrogen is such that when the electron spin reverses its direction a quantum of radiation of wavelength 21.1 centimeters is emitted. Since hydrogen is by far the most abundant element in the universe and since in the gas clouds of interstellar space almost all of the hydrogen atoms are in the state of lowest energy, the 21-centimeter radiation is intense enough to be easily observed with radio telescopes, even though once any given atom is excited to the upper level it remains there on the average for 11 million years before jumping to the lower level. In Chapter 7, we shall describe how observations of this line have resulted in important discoveries bearing on the structure of the Galaxy and on the physical state of the interstellar gas clouds.

The Wave Atom

Since about 1925 the simple Bohr model of the atom has been replaced by a mathematical theory, which does not lend itself to pictorial visualization of the atom. The theory of the atom based on the laws of quantum mechanics shows that we cannot treat the electron as a point charge whose position in the atom at any instant may be strictly stated. Instead we may merely specify the likelihood or probability of finding the electron in any specified position. On this view the electron behaves for many purposes like a hazy cloud of electricity, as illustrated in Fig. 28, where the photographs are the sort we might expect to obtain of hydrogen if the electron could be seen as a point of light and were photographed with a time exposure. There are many points of correspondence between the Bohr model and the wave model of quantum mechanics, one of them being that the electron is most likely to be found at the same distance from the nucleus as the Bohr theory predicts. Nevertheless, the chance of finding the electron at some other distance from the nucleus may also be very good. The quantum-mechanical theory of the atom has enjoyed many successes, and has become well established.

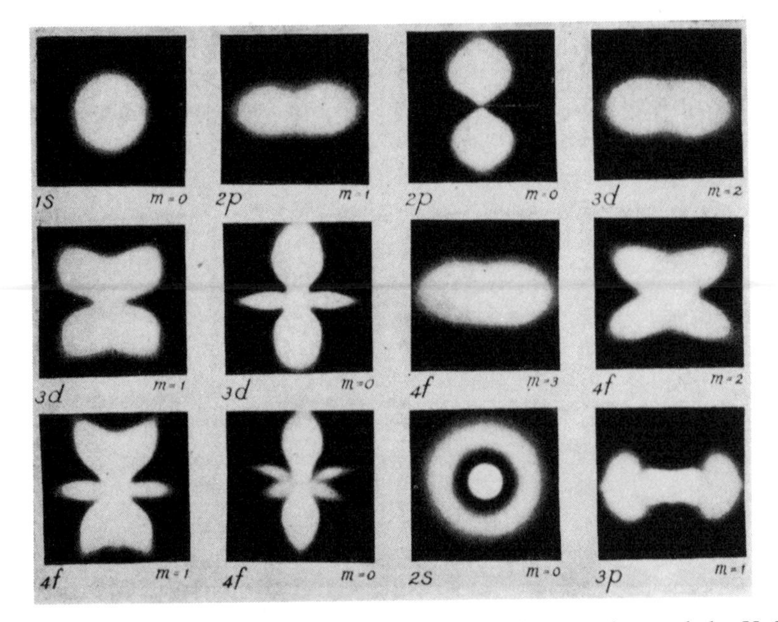

Fig. 28. The wave model of the atom. These photographs, made by H. E. White with the aid of a mechanical model, suggest the pictures that might be obtained if the electron could be seen as a point of light and were photographed with a time exposure.

In any event, whatever model is used, we may always think of the atom as possessing a number of discrete states or levels of energy. The transfer of an atom from one state to another, by the absorption or release of energy, gives rise to either an absorption line or an emission line, as the case may be.

Ionized Atoms

In Chapter 2 we made the statement that each of the atoms known in nature was distinguished by a unique and characteristic set of spectral lines. The statement is not strictly true, however, for by losing one of its electrons an atom effectively disguises its identity and radiates a completely new spectrum. E. C. Pickering, for example, examining the spectrum of ζ Puppis in 1896, found a series of unidentified lines, something like the Balmer series of hydrogen, at wavelengths of 3814, 3858, 3923, 4026, 4200, and 4542 Å, and concluded that they were "due to some element not yet found on other stars or on the earth."

The problem was clarified in 1913 by Bohr, who showed, in connection with his theory of the hydrogen atom, that the spectrum emitted by ionized helium atoms would closely resemble that of hydrogen, with the important difference that the ionized-helium lines in the visible portion of the spectrum correspond in origin to some of the infrared hydrogen lines. The energy of an electron in its orbit depends not only upon the number of the orbit but also upon the square of the nuclear charge. The charge of the helium nucleus is twice that of hydrogen. Consequently, each spectral series of hydrogen has its prototype in ionized helium, except that the wavelength of each helium line is one-fourth that of the corresponding hydrogen line. The lines that Pickering found correspond to the long-wavelength infrared Brackett series of hydrogen, consisting of electron jumps terminating in the fourth orbit. By an interesting coincidence, alternate members of the Pickering series fall within 2 Å of the Balmer lines and were therefore missed by Pickering. In 1922 H. H. Plaskett reported the discovery of these helium lines in the spectra of three class O stars. The prediction of Bohr, therefore, was brilliantly confirmed.

The similarity between the spectra of neutral hydrogen and ionized helium is one example of a general rule: the spectrum of an ionized atom is qualitatively similar to that of the neutral atom with the same number of electrons, but corresponding lines are displaced toward the ultraviolet. The analogue of the *H* and *K* lines of ionized calcium is a pair of lines of neutral potassium in the red region of the spectrum. Similarly, the close pair of ionized-magnesium lines observed in the "rocket" ultraviolet spectrum of the sun near 2800 Å correspond to the yellow sodium *D* lines.

When several stages of ionization are involved, Roman numerals I, II, III, . . . are used to denote the spectrum of the neutral atom, the first ionized stage, the second ionized stage, and so on. Thus, Fe I refers to neutral iron, Fe II to singly ionized iron, Fe III to doubly ionized iron, and so forth.

Molecules and Their Spectra

The world we live in is a world of molecules. The book you are reading, the hand that holds it, the chair you are sitting in, are all constructed of molecules. The hot stars, on the other hand, are worlds of atoms, ions, and electrons, where the complexity of coolness is replaced by the simplicity that accompanies high temperature. The link between our world and that of the stars is to be found in the cooler stars, where the pace is leisurely enough to allow atoms to unite in the fellowship of molecules. Even there, however, the atomic organizations are relatively simple. Carbon and nitrogen atoms join to form a fragmentary molecule, CN, which is called by chemists the cyanogen radical. Similarly, oxygen and hydrogen unite in the hydroxyl radical, OH, carbon and hydrogen in the radical CH, and so on. The most abundant conventional molecules include H_2 (hydrogen), CO (carbon monoxide), and H_2O (water). It is only under the comparatively frigid conditions on earth that atoms are permitted to give full reign to their organizing talents. The carbon atom on the earth, for example, is a master in the art of forming complex molecules. Some atoms of carbon, hydrogen, and oxygen cluster together in hexagons, others in long chains, like popcorn on a string. Carbon forms the base of all compounds found in living creatures; it is the atom that is mainly responsible for the complexities of the living world. It is in the cooler stars that simple molecules involving abundant elements may form.

Even the simplest molecule radiates a wonderfully intricate spectrum, whose appearance depends not only on the details of the molecule's structure but also on the local temperature. Like atoms, molecules can exist only in certain special energy states, and, like atoms, they emit light as they revert from a state of higher energy to one of lower energy. The energy states that are permitted a molecule, however, are vastly more numerous, and the relations between them more complicated, than those of any atom.

The two atoms of a diatomic molecule are bound together by strong attractive electrical forces to form a system resembling a tiny dumbbell with a slightly elastic connecting rod (Fig. 29). The molecule may rotate bodily in space about the axis *DC*, and the two atoms may vibrate toward and away from each other along the connecting line *AB*. In addition, the electrons within each atom may pursue a modification of any one of the many orbits normally permitted them. At any instant, the total energy of the molecule will depend not only upon the energies of the revolving and spinning atomic electrons, but also upon the distance between the two atoms and upon the speed of rotation of the molecule as a whole. Consequently, in place of each atomic line, corresponding to a particular electron jump, the spectrum of a molecule contains a system of bands, each consisting of a number of fine lines that converge slowly to a point known as the band head. The part of the spectrum in which the whole set or system of bands falls depends on the change of electronic energy in the molecule; the separations between the individual bands of a band system arise from changes of the vibrational states of the molecules, while the

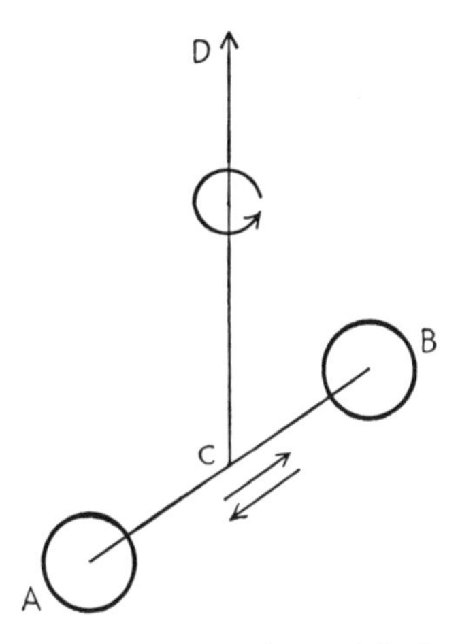

Fig. 29. Schematic model of a simple diatomic molecule. The arrows indicate that the molecule can rotate about the axis *DC* and vibrate back and forth along the line *AB*.

separation of the individual fine lines within each band is due to differences in rotational velocities.

In Fig. 30, the spectrum of cyanogen (CN) affords a beautiful example of the behavior of a typical diatomic molecule. All the bands in the figure and several others that are not seen in the photograph are analogous to one atomic line. We shall see later that molecules like cyanogen play an important role in the spectra of the cool stars.

Although the atomic spectra produced by two different isotopes of the same element differ so slightly that they rarely can be separated in stellar

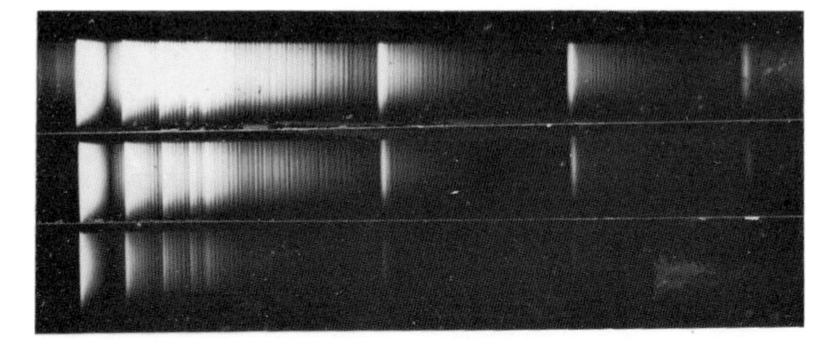

Fig. 30. The spectrum of cyanogen (CN). Note the individual lines that go to make up the bands. The overlapping of successive lines produces the phenomenon of band heads. There are three exposures, of different lengths, since the lines differ greatly in intensity. (Harvard Physics Laboratory.)

spectra, molecules involving different isotopes produce easily distinguishable spectra. The vibrational frequency of a molecule *AB* depends not only on the strength of the interatomic force but also on the masses of *A* and *B*. (A mechanical analogy would be a mass suspended by a spring and set in vertical motion; the heavier the mass the slower will be its rate of vibration.) Compare, for example, a C_2 molecule composed of two C^{12} atoms with one composed of a C^{12} and a C^{13} atom. The binding forces are almost exactly the same in the two molecules, since they depend only on the distribution of electrons in the outer cloud, but the masses involved are different. The $C^{12}C^{13}$ molecule will vibrate more slowly than the $C^{12}C^{12}$ molecule; hence the whole band system will be displaced.

With this brief description of the structure of atoms and molecules, we may proceed now to the story of how our knowledge of atoms and molecules, gained from the laboratory and from theory, can aid in unraveling the mysteries of stellar atmospheres and nebulae.

4 The Climate in a Stellar Atmosphere

In the preceding two chapters we saw how matter hidden away in the far corners of the universe is forced to reveal its chemical identity by means of its spectrum, and that the spectroscope can even measure the speeds of the stars and reveal their duplicities. But the story of its almost magical gifts of detection has barely begun. Indelibly recorded on every photograph of a stellar spectrum is a detailed account of the atmospheric conditions at the surface of a star. Strictly speaking, the spectrum tells us only which radiations the atoms are absorbing or emitting and how intensely. The atom, however, is a creature of climate; its ability to swallow up light depends upon the atmospheric conditions to which it is exposed. With present knowledge of atomic structure, the astronomer may now predict just what influence the stellar climate exerts on a particular atom, and thereby infer the stellar atmospheric conditions from the spectrum.

How Hot Are the Stars?

The most important attribute of stars, and indeed the one that makes it possible for us to see them at all, is high temperature. The stars are so hot that their material cannot possibly exist in solid or liquid form but must be entirely gaseous. We shall see that the effects of high temperature on the deportment of matter are often spectacular.

In physics and astronomy we employ what is called the absolute or Kelvin, temperature scale, which is reckoned from the lowest temperature that it is theoretically possible to attain. The absolute zero falls 273°C below the centigrade zero or freezing point of water. Thus temperatures are expressed on the Kelvin or absolute scale by adding 273° to the centigrade value; for example, the normal boiling point of water, 100°C, is 373°K.

A body at any temperature above the absolute zero always radiates energy. Although such emission is insignificant at low temperatures, it becomes very important for hot bodies, in accordance with Stefan's law:

$$\text{Rate of emission of energy} = \text{constant} \times (\text{Absolute temperature})^4,$$

that is,

$$E = \sigma T^4,$$

where σ is called the Stefan-Boltzmann constant, see Appendix C. For example, the average temperature of the earth is about 300°K or $1/20$ that of the sun, which therefore radiates 20^4 or 160,000 times more energy per unit surface area than the earth. We can measure the amount of energy received on the earth from a given star and if we also know the star's distance and its size, as is the case in certain eclipsing systems, we may calculate how much energy is leaving each square centimeter of the surface. This quantity in turn is related to the surface temperature by Stefan's law, and hence we have a method of finding the temperatures of stars whose brightnesses and angular sizes are known; see Appendix E.

For some stars of known apparent magnitude, it is possible to measure the angular diameter by means of the Michelson stellar interferometer or a new device called a photon-correlation interferometer. If the star's parallax is known, then in addition to the temperature we may also find the star's diameter, since

$$\text{Diameter in astronomical units} = (\text{Distance in parsecs}) \times (\text{Angular diameter in seconds of arc}).$$

Fortunately for our purposes, not only the amount, but also the quality, or color, of radiation is governed by the temperature. Everyone is familiar with the way in which the coil of an electric stove changes color as the current is increased. At first the coil glows a dull red, then turns a bright cherry color, and, if the current is imprudently increased still more, the color changes successively to orange, yellow, and white. This does not imply that only a single color is being emitted in each case, for we have already seen that an incandescent solid radiates light of all colors. But the proportions of the different colors are altered as the temperature increases.

We obtain a better insight into what happens by studying, with the spectroscope, light sources of different temperatures. With the aid of a suitable energy-measuring device, we could determine how much energy is contributed by each wavelength interval or color over a range of temperature from, say, 4000 to 20,000°K. Figure 31 illustrates the types of curves that would be obtained; actually, these curves are calculated by Planck's radiation law (see Appendix C), since stable terrestrial sources of radiation at accurately known temperatures in the range 4,000–

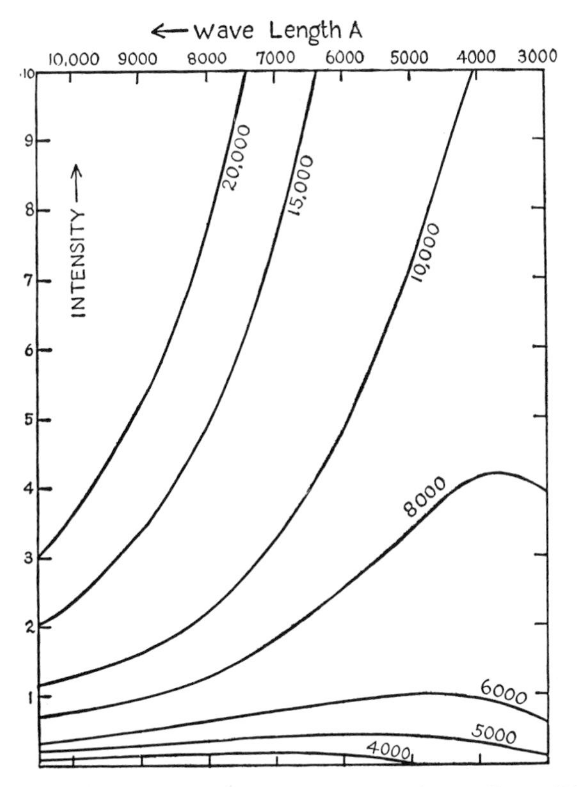

Fig. 31. The energy emitted by a perfect radiator (black body) at different temperatures. The curves indicate the relative energies radiated in different wavelengths for each temperature (°K). The range in wavelength, 3,000 to 10,000 Å, is that over which the energy of a star can be measured from the surface of the earth.

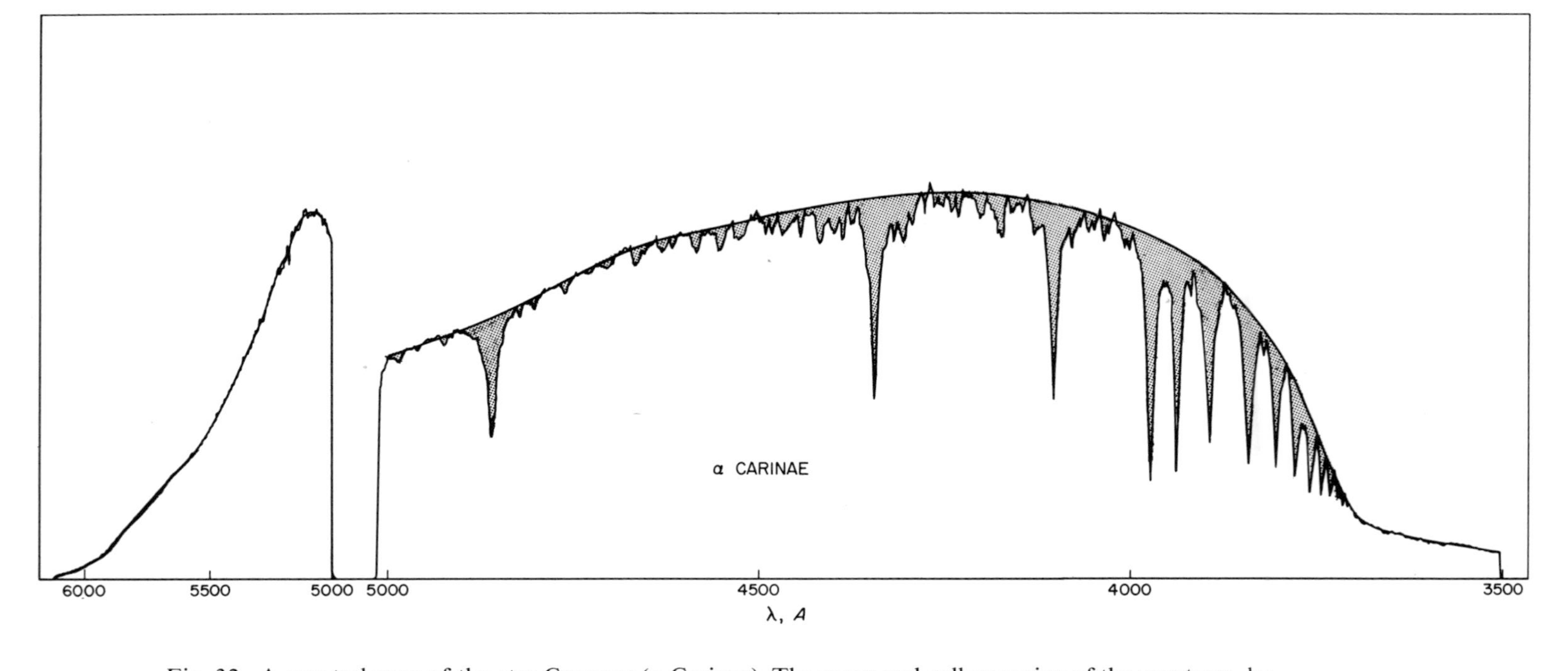

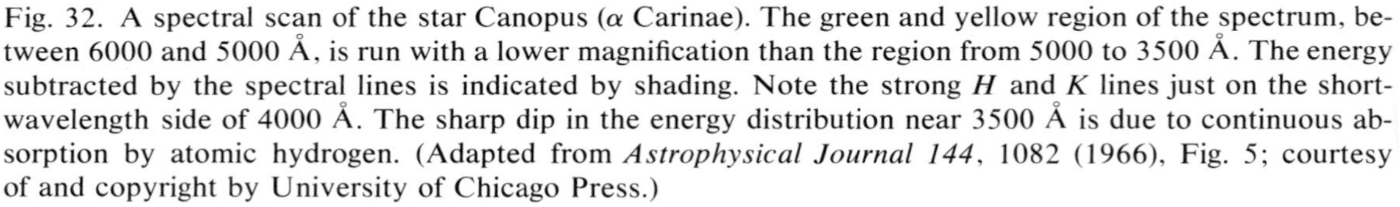
Fig. 32. A spectral scan of the star Canopus (α Carinae). The green and yellow region of the spectrum, between 6000 and 5000 Å, is run with a lower magnification than the region from 5000 to 3500 Å. The energy subtracted by the spectral lines is indicated by shading. Note the strong H and K lines just on the short-wavelength side of 4000 Å. The sharp dip in the energy distribution near 3500 Å is due to continuous absorption by atomic hydrogen. (Adapted from *Astrophysical Journal 144*, 1082 (1966), Fig. 5; courtesy of and copyright by University of Chicago Press.)

20,000°K are not easy to provide. Notice, that the shape of the energy curve changes with the temperature; the wavelength at maximum energy output decreases with increasing temperature, which means that the light as a whole appears bluer. For this reason the overheated coil appears to run the gamut of the spectrum as the temperature rises. Good absorbers of radiation are good emitters, and vice versa—a statement known as Kirchhoff's law. An ideal radiator, when cold, would appear as a perfectly black object. Hence, energy curves calculated by Planck's law are often called black-body curves. Experimentally, black-body or Planckian radiation may be obtained by uniformly heating an enclosure and allowing the radiation to escape through a small aperture.

Spectral-energy scans prove useful for evaluating the temperatures of the stars. Instead of using a spectrograph with a narrow slit and a photographic plate, the astronomer uses a slot broad enough to admit all the starlight and scans the spectrum with a photocell. With the same apparatus he also scans the spectrum of a standard lamp whose energy distribution has been established by comparing it with that of the radiation from an enclosure maintained at some known uniform temperature, for example, the melting point of gold. Spectral scans of the same star taken at different altitudes above the horizon permit a determination of the transparency of the earth's atmosphere, which is troublesome to evaluate, especially in the ultraviolet. These data (Fig. 32) enable the astronomer to compute the true energy distribution in a stellar spectrum from the observed energy distribution.

In Fig. 33 we compare the energy curve observed for the sun with a theoretical curve for a temperature of 5800°K. Although the shapes of the two curves are somewhat similar, the deviations are real and extremely significant in terms of the structure of the sun's outer layers, as will be made clear in the next section of this chapter. Quite generally, when proper allowance is made for distortions produced by the absorption lines which are important in cool stars, it is found that the stellar energy curves differ from those of an ideal radiator, that is, from black-body curves calculated by theory according to Planck's law (Fig. 31). The stars do not radiate as black or even gray bodies. There are two reasons for these deviations. One is that temperature increases with depth in the stars, so radiation from deeper layers corresponds to higher temperatures. The other is that the material in the star's atmosphere is not gray but may even be strongly colored; another way of saying this is that the emissivity depends strongly on wavelength. What then do we mean by the temperature of a star? By making the best fit of the energy curve to the theoretical black-body curve we could get some kind of color temperature but

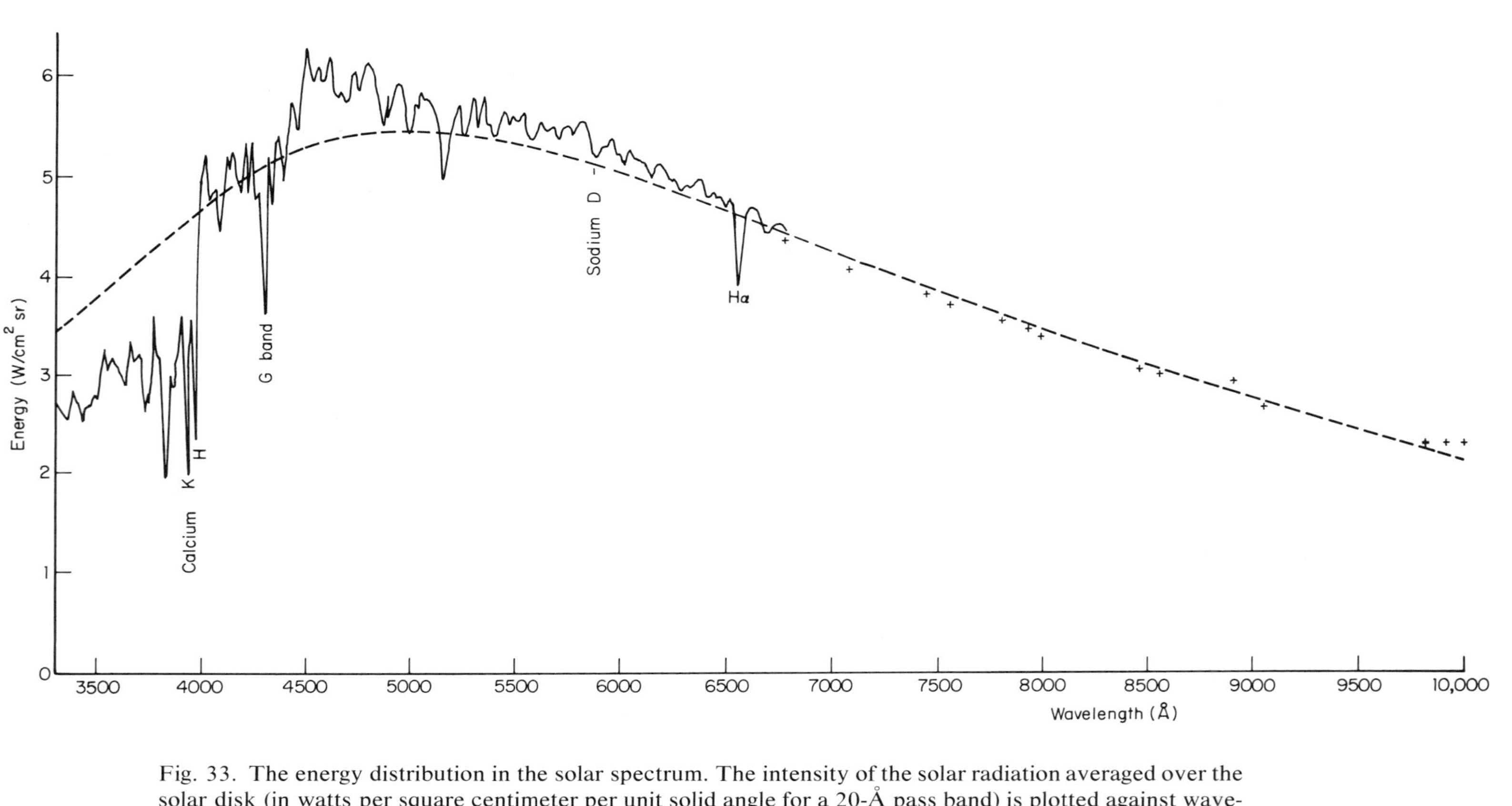

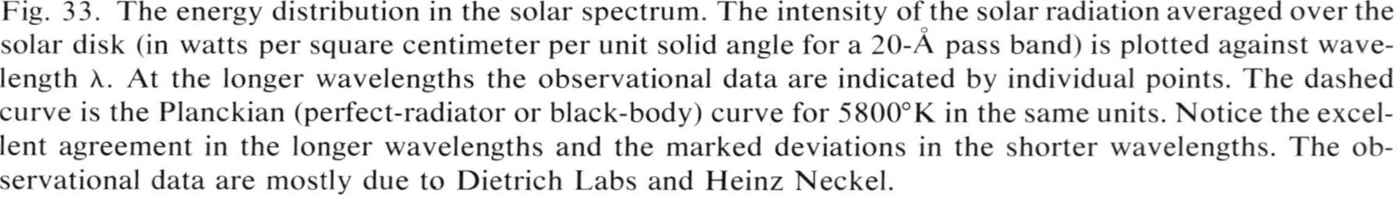
Fig. 33. The energy distribution in the solar spectrum. The intensity of the solar radiation averaged over the solar disk (in watts per square centimeter per unit solid angle for a 20-Å pass band) is plotted against wavelength λ. At the longer wavelengths the observational data are indicated by individual points. The dashed curve is the Planckian (perfect-radiator or black-body) curve for 5800°K in the same units. Notice the excellent agreement in the longer wavelengths and the marked deviations in the shorter wavelengths. The observational data are mostly due to Dietrich Labs and Heinz Neckel.

we would get different temperatures by making fits in different spectral regions. An alternative, but even less satisfactory, method of temperature determination consists in ascertaining the color in which the star radiates the most energy. The sun (temperature 5800°K) pours out the greatest amount of energy in the green region near 4800 Å. Altair, whose temperature is about 8100°K, has a maximum in the violet. Energy maxima of yet hotter stars fall in the "rocket and satellite" ultraviolet beyond 2900 Å. It is rather remarkable that the sun's energy curve can be represented as well as it can by a theoretical curve for 5800°K, inasmuch as the outgoing radiation comes from layers that vary in temperature between about 4400°K and 8000°K. The answer must be, of course, that the highest layers are too cool and rarefied to contribute very much, while the radiation from the very deep and hot regions is largely absorbed before it reaches the surface.

Another type of temperature can be inferred from the appearance of the line spectrum, as we shall describe shortly.

From many points of view the most satisfactory definition is the *effective temperature,* which is the temperature of a perfectly black sphere of the same size as the star that would have exactly the same total energy output. Effective temperatures can be measured directly for very few stars. We would have to measure the total energy received from the star above the earth's atmosphere and the star's angular diameter. As satellite observations improve, such a program will yield more and more accurate results. At present, the best method is to determine the energy received at the earth for stars whose angular diameters can be measured. Then from data for hot stars obtained from rockets fired above the earth's atmosphere, from measurements made with heat-radiation detectors for cool stars, and finally from theoretical predictions for very hot stars, we estimate the fraction of energy absorbed by the earth's atmosphere. These increments, which have to be applied to the observed luminosities in order to derive the true ones, are called bolometric corrections (see Appendix E). Fortunately, improvements in observation and theory are progressing very rapidly, so that accurate effective temperatures will be available from the hottest to the coolest stars. Why are effective temperatures so important? The reason is that the total luminosity L of a star is equal to its surface area $4\pi R^2$ times the energy emitted per unit area, that is,

$$L = 4\pi R^2 \sigma T_e^4,$$

where T_e is the effective temperature of the star and by Stefan's law the emissivity is σT_e^4.

In Chapters 8 and 9 we shall see that calculations of the life history of a star (commonly called its evolution) give for each stage of its development the star's radius and luminosity and hence its effective surface temperature. For most stars the radius cannot be observed directly, but the effective temperature can be inferred from the color or spectrum. From the apparent magnitude and distance, the absolute magnitude can be found. In order to get the actual luminosity we must find the bolometric corrections, which depend on the temperature of the star (Appendix E).

A Model Solar Atmosphere

Our discussion of stellar temperatures has proceeded on the implicit assumption that the intensity and spectral distribution of the radiation leaving the surface of a star can be represented by a single temperature. While this somewhat idealized view of stellar temperatures is most valuable for exploratory purposes, more detailed studies of stellar atmospheres must allow for the fact that the temperature increases with depth in the atmosphere.

Even an ordinary photograph of the sun in white light offers proof that the temperature of the sun increases inward. If the sun had a sharply defined radiating surface at a constant temperature, its brightness would be the same everywhere on the disk. Instead, the brightness decreases steeply toward the edge, or limb. As is explained in more detail in Chapter 5, the solar gases are highly opaque and therefore the radiation from the sun comes only from the outermost layers. If the top of the photosphere is defined as the point at which the gases become appreciably opaque, the opacity of the gases increases so rapidly with depth that no radiation is received from below a depth of about 400 kilometers. It is within this 400-kilometer layer that both the continuous spectrum and the absorption lines are formed. At the center of the disk the line of sight from the observer is radial and therefore extends deeply into the photosphere, but at the limb the radiation emerges tangentially from relatively high and cool levels in the photosphere, and hence the limb appears darker.

It is now clear why the observed energy curve for the sun cannot be fitted precisely to a theoretical curve with a single value of the temperature. The radiation that emerges from the surface is a composite of contributions from all depths in the photosphere. The extent to which radiation from each layer contributes to the total is determined by the absorptivity of the overlying layers. The problem is further complicated by the fact that the opacity of the solar gases varies with wavelength (see Chapter 5). If the temperature at each depth in the photosphere is known, together with the absorptivity of the gases at each wavelength, it is possible to predict from theory both the spectral-energy curve for the center of the sun's disk and the degree of darkening toward the limb at each wavelength. Conversely, if observations of the energy curve and of the limb darkening are available, the temperature gradient of the atmosphere and its absorbing properties may be derived. The first extensive measurements of solar-limb darkening were performed by C. G. Abbot, of the Smithsonian Institution. More recent and more accurate observations made by A. K. Pierce and others with a high-power spectrograph and photoelectric cells show that the degree of darkening diminishes toward longer wavelengths, chiefly owing to the fact that the contrast in brightness between radiators at two different temperatures becomes progressively smaller as the wavelength grows larger.

The analysis of the spectral-energy and limb-darkening observations has resulted in the calculation of numerous models of the photosphere, which are still in the process of improvement as both theory and observation continue to be refined. Table 2 gives a recent model of the solar atmosphere. Successive columns give the depth, the temperature T, the total gas pressure, the electron pressure, and the density. Notice the extremely low density throughout these visible layers.

Table 2. Model of the solar atmosphere.

Depth (km)	T (°K)	Total gas pressure, P_g (atm)	Electron pressure, P_ϵ (10^{-6} atm)	Density, ρ (10^{-6} gm/cm^3)
0	4730	0.0230	1.065	0.088
20	4770	.0281	1.32	.107
40	4830	.0338	1.60	.127
60	4910	.0404	1.98	.150
80	5040	.0488	2.64	.176
100	5150	.0560	3.60	.206
120	5310	.0705	5.40	.241
140	5490	.0845	8.0	.280
160	5710	.101	12.9	.320
180	5930	.117	21.5	.360
200	6230	.137	42	.398
220	6710	.158	111	.427
240	7310	.181	346	.450

The Relation Between the Temperature and the Spectrum of a Star

Now that we can answer the question, "How hot are the stars?" we should like to know whether there is any connection between stellar temperatures and the spectral classes that were described in Chapter 2. We recall that, purely on the basis of the appearance of the spectral lines, all stellar spectra could be arranged into one of the types O, B, A, F, G, K, M, R, N, S. From the fact that O and B stars are blue in color, A stars white, and G, K, and M stars yellow, orange, and red, respectively, we might suspect that the classes have been arranged in order of decreasing temperature. Stellar-temperature determinations by the methods we have described yield the results given in Table 3, which is based on the work of Hanbury Brown, Davis, Allen, and Rome, of Wolff, Kuhi, and Hayes, of Morton and Adams, of Popper, of Stebbins and Whitford, of Harold Johnson, of A. Code, of J. B. Oke, and of others.

With the knowledge that each spectral class corresponds to a different temperature, we suspect that the weakness of the hydrogen lines in the O and M stars, the former very hot, the latter cool, does not indicate a scarcity of that element; neither does the great number and intensity of iron lines in the spectrum of the sun necessarily point to an overabundance of iron. A more reasonable view is that the behavior of atoms, that is, their capacity for emitting and absorbing light, is regulated by the temperature. We shall see from what follows that temperature alone can produce the transformation from a rich M-type spectrum to a comparatively bleak spectrum of type B.

Temperature, Radiation, and Atoms

Suppose that in a large box, made of some hypothetical unmeltable substance, we placed an assortment of all kinds of elements—hydrogen, helium, oxygen, nitrogen, sodium, calcium, iron, chromium, lead, . . .— and that provision could be made for raising the temperature inside the box from the absolute zero to perhaps 50,000°K. What would happen to the elements as the enclosure grew hotter?

At the absolute zero all the matter is in the solid form; the individual atoms lie tightly packed, closely bound to one another in crystals or complicated molecular structures. The molecules are completely dormant, undisturbed by their neighbors or by any sort of radiant energy. As the temperature rises, the molecules begin to awaken from their leth-

Table 3. The temperatures of the stars.

Star	Spectral type	Temperature	Star	Spectral type	Temperature
θ' Orionis C	O6	40,000°K	θ Crateris	B9V	11,400
ξ Persei	O7	35,000	α Lyrae (Vega)	A0V	9,330[a]
λ Orionis A	O8	32,000	Sirius	A1V	9,900[a]
10 Lacertae	O9.5V	30,000	η Ophiuchi	A2V	9,300
τ Scorpii	B0V	28,000	Denebola	A3V	9,000
π Orionis	B1V	23,000	β Arietis	A5V	8,750
γ Pegasi	B2.5V	19,000	Altair	A7V	8,110[a]
η Ursae Majoris	B3V	17,500	γ Virginis	F0V	7,450
κ Hydrae	B5V	15,000	δ Bootis	F2V	7,100
Regulus	B7V	12,900[a]	Procyon	F5IV	6,470[a]

[a] The diameters of these stars are based on actual angular diameter measurements. All other temperatures are obtained by interpolation. (After John Davis)

Star	Spectral type	Temperature	Star	Spectral type	Temperature
β Virginis	F8V	6,120°K	ϵ Orionis	B0Ia	21,100°K[b]
Sun	G2V	5,780	β Orionis	B8Ia	11,200[b]
κ Ceti	G5V	5,570	α Carinae	F0Ib	7,510[b]
τ Ceti	G8V	5,340			
δ Draconis	K0V	5,150	π Cephei	G2III	5,300°K
ϵ Eridani	K2V	4,830	η Herculis	G5III	5,100
61 Cygni	K5V	4,370	η Draconis	G8III	4,840
	K7V	4,000	Pollux	K0III	4,670
	M0V	3,670	ψ Ursae Majoris	K2III	4,400
Lacaille 9352	M2V	3,400	γ Draconis	K5III	3,900
			Antares	M1Ib	3,200
			Betelgeuse	M2Ib	3,000

Note: The numerals Ia, Ib, III, and V are the designations of the Morgan-Keenan-Atlas of Stellar Spectra for supergiants, normal giants, and dwarf stars, respectively. See Chapter 6.

[b] Based on direct measurements of brightness temperatures (after Hanbury Brown, J. Davis, L. R. Allen, and J. Rome).

argy and to stir about sluggishly, occasionally jostling one another. Soon more volatile elements such as hydrogen, helium, oxygen, and nitrogen become first liquid and then gaseous, driven by the ever-increasing speeds of their molecules. As the temperature becomes yet greater, the elements liquefy and vaporize one by one. The pace becomes faster. Molecules dash madly about, colliding with one another and loading each other's electrons with energy, which is later lost in the form of radiation. Each molecule is assailed by flying particles and rapidly oscillating radiation waves. The molecules cannot long survive such brutal treatment. Eventually, one after another is torn apart into its constituent atoms. Some molecules, like the hydroxyl radical OH, are tied together more tightly than others, and may survive long after their contemporaries vanish from the scene. But they too are eventually disrupted, leaving only individual atoms with their electrons rapidly jumping back and forth between various excited levels, as each atom takes up energy from colliding electrons or ions and passes it on in the form of quanta of radiation. Some atoms, like hydrogen or helium, hold their electrons so tightly that only violent collisions or powerful pulses of energy are capable of raising the electron from its lowest orbit to a more distant one. Other atoms, like sodium, have only very loosely bound outer electrons, and much gentler encounters or weaker pulses of energy are sufficient to excite them.

As the gas grows still hotter, collisions become increasingly violent, and the supply of high-frequency radiation increases. The atomic electrons are now so heavily battered that one or more of them may be torn completely free of the parent nucleus, that is, the atom becomes ionized. In general, the metallic atoms sodium, iron, and so on, are much more easily ionized than are the light gases, hydrogen, helium, oxygen, and nitrogen. The relative amounts of energy required to ionize several of the more abundant elements are listed in Appendix G. Notice that helium is nearly twice as difficult to ionize as is hydrogen, which in turn is bound about twice as tightly as calcium. This means that calcium, hydrogen, and helium tend to lose electrons at successively higher temperatures. It should also be noted that atoms that are easily ionized are also more readily excited than those that are difficult to ionize.

Thus far, in describing the influence of stellar climate on the behavior of atoms, we have made no mention of the pressure or density. Once an atom becomes ionized, it acquires a positive charge and does its best to retrieve electrons so that the charge will be neutralized. Whether or not the ionized atom has a good chance of succeeding in its quest depends upon the number of electrons in the vicinity, or, in other words, upon the electron density. An atom, therefore, is more likely to radiate in the ionized condition when the density is low, and in the neutral form when the density is high.

The picture that we have drawn was first established on a quantitative basis by the Indian physicist Megh Nad Saha, in 1920. Saha not only showed that ionization would be favored by high temperature and low density, but was able to calculate exactly what fraction of atoms of a given kind would be ionized under specified conditions of temperature and pressure. His findings may be summarized by the formula

$$\frac{\text{Number of ionized atoms}}{\text{Number of neutral atoms}} = \frac{K}{\text{Number of electrons}},$$

where K depends on the kind of atom and the temperature. The degree of ionization of any atom thus depends directly on the temperature and is inversely proportional to the number of free electrons. The Saha type of formula may also be employed to calculate the degree of disruption of molecules into atoms if the temperature is known. The number of neutral atoms is replaced by the number of molecules, and the numbers of ions and electrons by the numbers of the two constituent atoms into which the molecules are broken up (see Chapter 5). A detailed discussion of the ionization formula and its application will be found in Appendix G.

The Meaning of the Spectral Sequence

We now ask: What are the consequences of this change in the structure of matter, from molecules to neutral atoms to ionized atoms, on the appearance of the spectrum at different temperatures? We have in a way already answered the question in Chapter 3. There we saw that the spectrum of a molecule, consisting of groups of closely spaced fine lines which blend together to form broad bands, is totally unlike that of an individual atom. Also the spectrum of an ionized atom is similar to that of a neutral atom with the same number of electrons, except that each ionized-atom line occurs much farther toward the ultraviolet than does the corresponding neutral-atom line. This fact was illustrated by the similarity between the spectrum of hydrogen and that of ionized helium.

With these facts in mind we may now venture an interpretation of the spectra produced in stellar atmospheres. For the time being, we shall assume that all stellar atmospheres are of the same density, and consider only the effects of temperature. At a temperature of 2500°, large numbers of atoms are still joined as molecules. Combinations such as titanium oxide (TiO), cyanogen (CN), and the hydrocarbon molecule (CH) imprint their intricate band patterns on the continuous spectrum. Of the elements that are present as individual atoms, those that are easily excited—the metals like calcium, sodium, and iron—are prominently featured. Surprisingly enough, in spite of the relatively large quantities of energy necessary to excite them, lines of atoms of hydrogen are also seen in the spectrum, and with considerable strength. The appearance of hydrogen must be due to its high abundance compared with other elements; in fact, as we shall see later, hydrogen accounts for about 92 percent of all the atoms in a star's outer envelope. Their great numbers compensate for the fact that at low temperatures only a small percentage are in a condition to absorb light.

As the temperature rises along the spectral sequence, more and more molecules become disrupted. At class K0, the bands of titanium oxide have already vanished. Some of the more stubborn molecules, such as CN, CH, and OH, persist as far as G0; they are easily recognized in the sun. Meanwhile, as increasing amounts of energy become available, the lines of hydrogen steadily strengthen. Even at low temperatures, some of the more loosely held electrons are broken off from their atoms, as is shown by the appearance of the strong *H* and *K* lines of ionized calcium in even the low-temperature M stars. This pair of lines is strongest near K0, but from that point on, the calcium atoms begin to lose a second electron (Fig. 34); the *H* and *K* lines weaken and fade away entirely at temperatures greater than 10,000°. They are still dominant, however, in Class G, as are the lines of neutral iron (Fig. 35), magnesium, and other metals, and the ever-growing hydrogen lines.

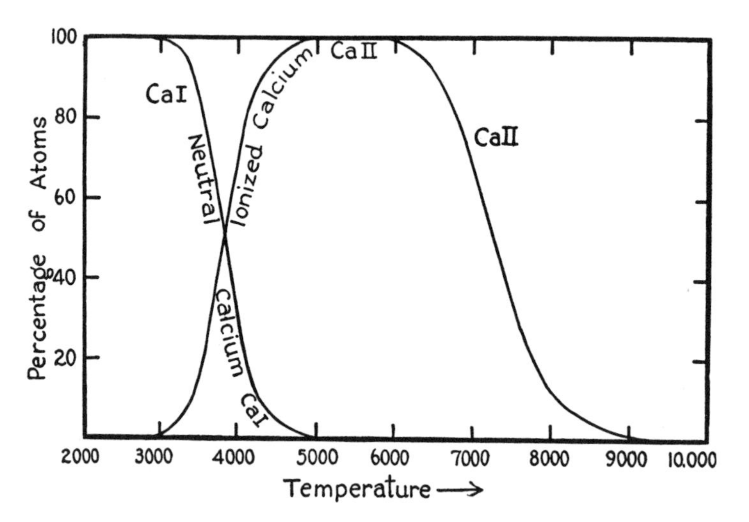

Fig. 34. The ionization of calcium. The curves show how the percentages of neutral and ionized calcium vary as the temperature rises. An electron pressure of 1/100,000 atmosphere has been assumed.

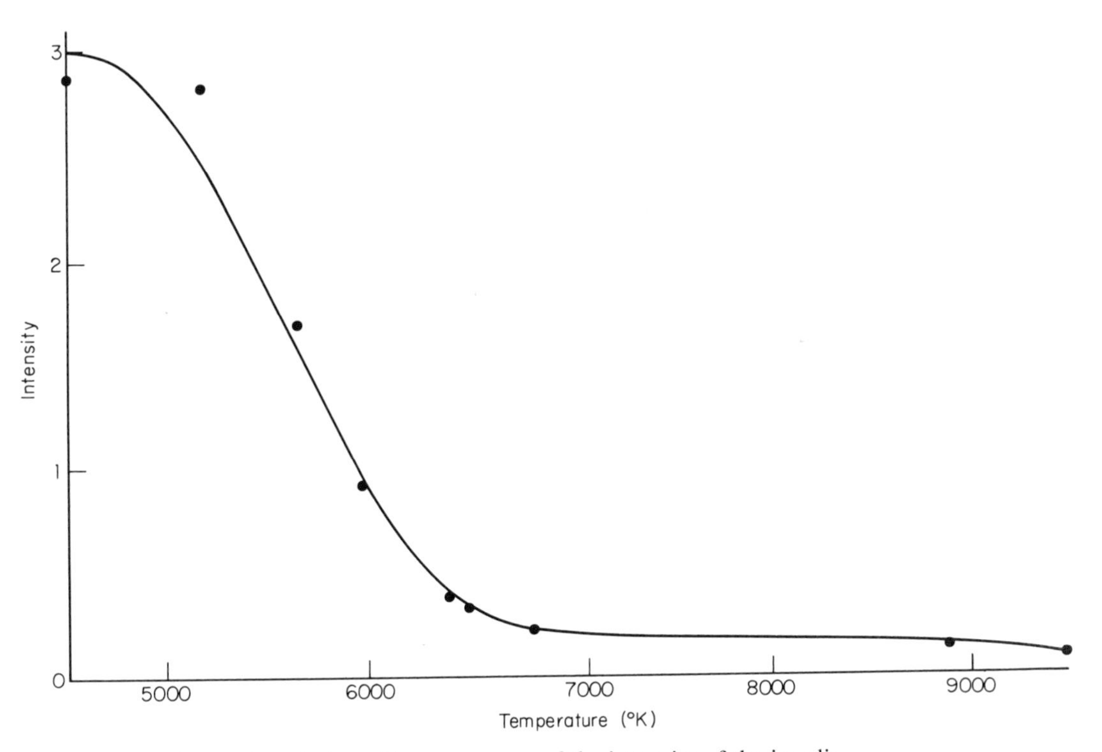

Fig. 35. The observed variation with temperature of the intensity of the iron line at 4383 Å.

In Class F, at a temperature of about 6500°, appreciable numbers of other metallic atoms part with their electrons; neutral iron and titanium weaken markedly; ionized iron and ionized titanium attain prominence until deprived of still a second electron, and then vanish as the temperature climbs beyond 10,000°. At Class A0, hydrogen attains its greatest glory, completely overshadowing all other atoms. But the inexorable march of temperature soon strips great numbers of hydrogen atoms of their single electrons, without which they are impervious to radiation, and their lines begin to fade. Here again, however, at the upper end of the temperature scale, hydrogen remains visible through sheer weight of numbers of atoms.

The very hot stars, in Classes B and O, range in surface temperature from about 15,000° to perhaps more than 50,000°. The advent of high temperature is signaled by the appearance, in Class B9, of neutral helium, the most difficult to excite of all the neutral atoms. The helium lines acquire their greatest intensity in Class B3, and then rapidly weaken as the atoms become more and more ionized. The spectra of the B stars also exhibit singly ionized oxygen and nitrogen.

In the very hottest stars of Class O, hydrogen is about as conspicuous as in Class M. Under the violent conditions prevailing, neutral helium disappears completely, giving way to its ionized form. Spectral lines of elements that are stripped of more than one electron mostly fall in the far-ultraviolet part of the spectrum. Since light of wavelengths shorter than 2900 Å is completely absorbed by ozone and other gases in the earth's atmosphere, these radiations can be observed only from rockets and satellites. However, some lines of O III (doubly ionized oxygen), N III (doubly ionized nitrogen), and Si IV (triply ionized silicon) fall in the customarily observed spectral regions.

The practical classification of stellar spectra begins with Class O5 rather than O0, in order to leave a place for still hotter stars that might be discovered later. Theoretically, at temperatures near 100,000°, all lines in the observable region of the spectrum should disappear, although the short-wavelength region from 100 to 2000 Å would be rich in lines of multiply ionized atoms. A star so hot that it shows no spectral lines would be placed in Class O0. Such stars are found in so-called planetary nebulae.

We have been making the tacit assumption in our discussion that all stars along the spectral sequence possess the same chemical compositions, for, obviously, lines of an element that is not present at all in an atmosphere will be absent from the spectrum. The intensities of an atom's spectral lines will depend upon its abundance as well as on the temperature and pressure, for large numbers of atoms will absorb more radiation than just a few. Theoretical calculations show our assumption to be correct for the vast majority of stars. Diversities in chemical composition play a secondary role in comparison with differences in temperature in shaping the appearance of stellar spectra. Suppose, for example, that we have calculated the chemical composition of the solar atmosphere (see Chapter 5). If we now take a hypothetical star and, with the aid of Saha's formula, predict its spectrum at different temperatures from 2,500° to 30,000°, we find that we can reproduce the observed features of the spectral sequence if, and only if, we adopt roughly the same mixture of elements as in the sun. This result was established about 1925 by the thorough studies of Mrs. Payne-Gaposchkin.

Since 1940, extensive studies have demonstrated that there exist significant variations in chemical composition between stars. These differences appear to be of two types: those that result from differences in the composition of the material from which the star was formed and those that result from element building within the stars themselves. We discuss these questions in later chapters.

We should comment finally that in our interpretation of the spectral sequence we have assumed, for simplicity, that all stellar atmospheres are of equal density. For a great many stars this assumption is correct, but it fails entirely for many others. We shall see just what effect the density has on the appearance of stellar spectra, and how we may turn this effect to good advantage in deducing the size and intrinsic luminosity of a star from its spectrum.

An additional illustration of the effect of temperature upon the appearance of a spectrum is provided by sunspots. Whereas the temperature of the bright surface (photosphere) of the sun is about 5800°K, that of a typical sunspot is close to 4500°K. In the sunspot spectrum, lines of neutral atoms are strengthened, those of ionized atoms are weakened, and the number and strength of molecular lines are enormously increased (Fig. 36). The strong magnetic fields present in sunspots produce additional effects on spectral lines (see Chapter 5), but the qualitative prediction of behavior on the basis of Saha's theory remains valid.

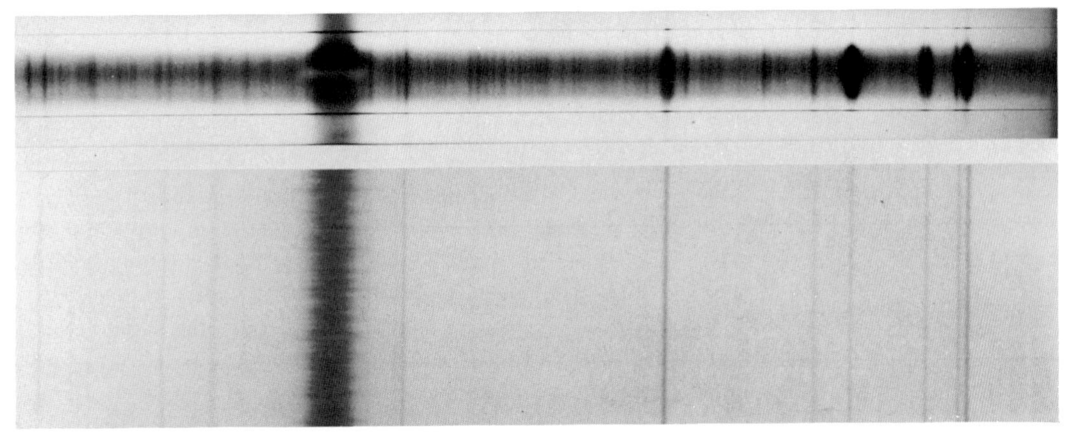

Fig. 36. A comparison of the spectrum of the solar photosphere (*below*) with that of a sunspot (*above*) in the neighborhood of the Hα line, 6563 Å. The irregularity in the lines of the solar spectrum is due to vertical motions of rising and falling columns of gas in the solar atmosphere. The lines at 6569, 6574, and 6575 Å are due to neutral iron (Fe I) and the line at 6572 Å is due to a combination of water vapor and neutral calcium. In the sunspot spectrum all the lines are strengthened and there is a great increase in the number of weak lines, many of which are due to fragmentary molecules. The bright streak in the center of the Hα line may be the result of electromagnetic flare activity. (Courtesy of Orren Mohler, McMath-Hulbert Observatory, University of Michigan.)

5 Analyzing the Stars

We have seen that, for stars of the same chemical composition, the appearance of each spectral line is regulated by the temperature and density of the star's atmosphere. We have found in this way that the main features of the spectral sequence are consistent with a series of stars of uniform chemical composition and varying temperature and density. Having thus made the preliminary exploration, we may now fix our attention on the detailed analyses of individual stellar atmospheres. Since the stars in each spectral class apparently share the same general physical characteristics, we have every reason to hope that studies of a small number of representative stars of each type will reveal the nature of the vast majority of stars in the Galaxy.

In making a detailed analysis of a stellar atmosphere, the problem facing us is to discover how the temperature, density, and chemical composition of the atmosphere may be deduced from the dark lines in the stellar spectrum. It must be obvious that the intensity, or blackness, of a spectral line is an index to the abundance of the element producing it. But in order to absorb, let us say, the first line of the Balmer series, a hydrogen atom must first be in the second energy level (Fig. 25). In short, the intensity of the line depends upon the temperature and density of the atmosphere as well as on the chemical composition. It will also depend on the temperature and density in another way through their influence on the fuzziness or broadening of spectral lines. Finally, it will depend on the process of formation of spectral lines.

The Widths of Spectral Lines

There are several reasons why spectral lines appear broad. In the first place, the sharpness of the lines is limited by the fact that the slit of the spectrograph is not infinitely narrow but has a very definite width. This "instrumental effect" can be overcome in studies of the solar spectrum by supplementing a grating spectrograph with special devices that enable one to attain spectra on what amounts to a very large scale. Then it is found that the lines in the spectrum of the sun, or indeed of any incandescent source, have a finite, measurable, intrinsic width. But even if we could observe the radiation from a single atom through an infinitely narrow spectroscope slit, the line would still appear to have a finite width. In other words, the atom does not radiate solely at a single wave-length, but may also radiate (or absorb) energy at adjoining wave-lengths. The line is said to possess a natural width, as shown in Fig. 37, in which

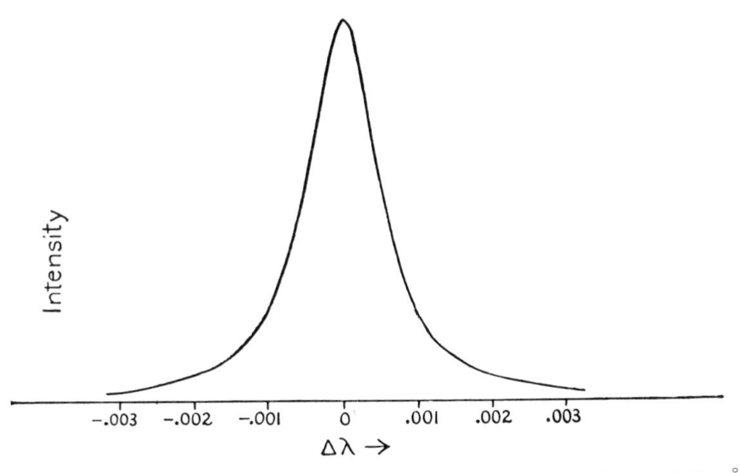

Fig. 37. The natural breadth of a spectral line (the iron line at 4383 Å). The intensity of emitted radiation is plotted against distance in wavelengths from the center of the line. The curve shows the shape that the emission line might have if we could observe the radiation from the atom at a temperature of 0°K, a condition we might approach experimentally by cooling the discharge tube with liquid helium. Since capacity to emit is proportional to the absorptivity, the curve also shows how the absorptivity varies in different parts of a spectral line.

intensity is plotted against distance (in wavelengths) from the line center. Notice that most of the radiation is at wavelengths close to the center of the line.

We may, if we like, visualize the atom as a tiny broadcasting station, and the spectroscope as a radio receiver. The station is usually assigned a specific broadcasting wavelength, but, owing to natural limitations on the broadcasting equipment, the wavelength of the signal is not perfectly sharp. There is one place on the dial where the reception is loudest, but the program may also be received, although less distinctly, at neighboring wavelengths on either side of the assigned wavelength.

Another important factor in line broadening is the Doppler effect. We recall from Chapter 2 that the wavelength of the light emitted or absorbed by a source that is in motion along the line of sight is displaced from the normal position by an amount proportional to the speed of approach or recession. The spectral lines of an approaching star are shifted toward the violet, those of a receding star toward the red end of the spectrum. The individual atoms in the atmosphere of a star are not at rest, but are flying about with different velocities (Fig. 38). Some atoms are approaching the observer at the instant they radiate; others are receding. Radiation emitted by approaching atoms will have a higher frequency than if the atoms were not moving, and radiation from receding atoms will have a lower frequency. The velocities should be entirely random as far as direc-

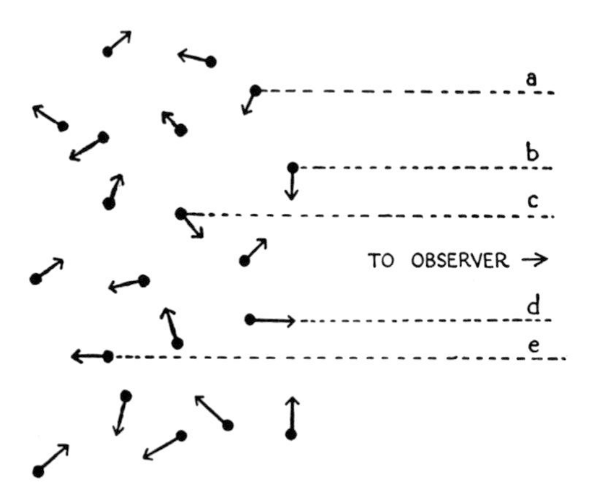

Fig. 38. Random motions of radiating atoms. Since atoms *a* and *e* are moving away from the observer, the wavelengths of their radiations are shifted toward the red end of the spectrum, whereas wavelengths from atoms *c* and *d* are shifted toward the violet end; the wavelength of the radiation from *b* is unchanged, since it is moving neither toward nor away from the observer. Compare Fig. 39.

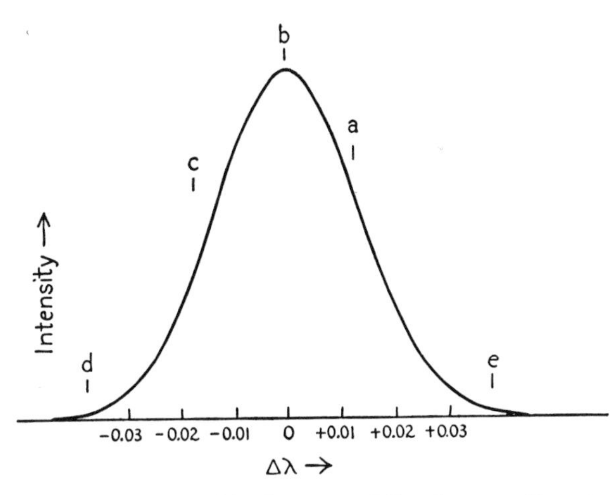

Fig. 39. The breadth of a spectral line for pure Doppler broadening (the iron line at 4383 Å). This is the profile we would observe if we examined the radiation from atoms emitting at a temperature of 5700°K (the temperature of the sun's atmosphere). The letters *a* to *e* indicate wavelengths emitted by corresponding atoms in Fig. 38. Compare this line profile with that shown in Fig. 37.

tions are concerned and, since an observed spectral line is the sum of the contributions from a great number of individual radiating atoms, the spectral line will appear widened (Fig. 39). The degree of blurring of a line depends on the velocities of the particles; thus hydrogen atoms move faster on the average than other atoms and hydrogen lines are widened more than those from heavier elements. At higher temperatures, also, the blurring is exaggerated because the atoms are moving more rapidly and the Doppler displacements are therefore larger. Even at laboratory temperatures, the physicist sometimes finds it necessary to cool his electric discharge tube with liquid air in order to narrow and thus separate spectral lines that are close together.

Electric and magnetic fields also widen the spectral lines emitted by atoms. We have seen that, when an atom undergoes a transition from one energy level to another, a single spectral line is normally emitted. If, however, the atom is placed near an electrically charged object or in a magnetic field, it is disturbed by the action of the electric or magnetic field. The energy of each atomic level may then be changed by a certain small amount, depending on the intensity of the disturbance. We say that each energy level is split into a number of sublevels. Each line is then divided into a number of components, the degree of separation depending upon the intensity of the field. The splitting of spectral lines in an electric

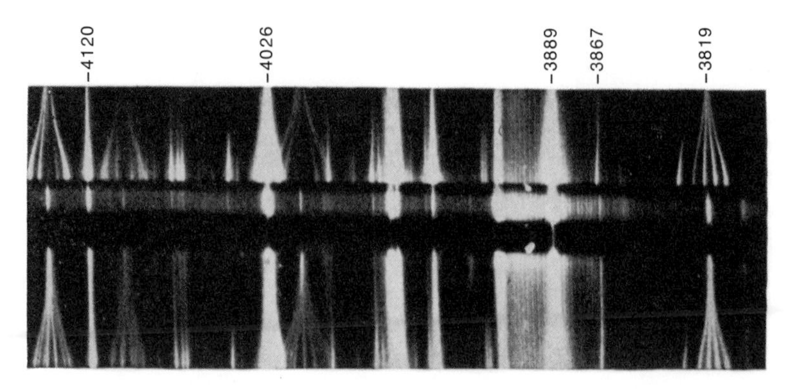

Fig. 40. The Stark effect in helium. Not only are the individual spectral lines separated into components, but the light from the components is polarized. The upper strip shows the lines polarized parallel to the electric field, and the lower strip those polarized perpendicular to the field. The field-free spectrum is shown in the center. (From a plate by J. S. Foster, courtesy of H. E. White.)

field is called the Stark effect (Fig. 40) and in a magnetic field, the Zeeman effect (Fig. 41). The separation of the components depends on the strength of the field. Furthermore, the components are polarized. In the Zeeman effect, for example, when the components are viewed at right angles to the magnetic field they are plane polarized; the inner components are polarized parallel to the magnetic field and the outer components perpendicular thereto. Viewed along the magnetic field the central components disappear and the outer components are circularly polarized. The important point is that the polarization (which can be measured by proper light-analyzing equipment) tells the direction of the magnetic field and the amount of the splitting tells the strength of the field.

Sunspots, which resemble great cyclones in the solar atmosphere, are always accompanied by powerful magnetic fields, from several hundred to 3000 gauss. (For comparison, the earth's field is less than 1 gauss.) Hence the Zeeman effect is a notable feature of the sunspot spectrum. The effect is particularly marked for the metallic lines. H. W. Babcock has detected the Zeeman effect in the spectra of a number of stars. Certain stars, particularly spectroscopic variables of Class A, show strong magnetic fields, of the order of several thousand gauss. Furthermore, the field intensity changes with the period of the variation in the light and the spectrum of the star. One star, HD 215441, has a field of 30,000 gauss, the strongest field known in nature.

The Stark effect is most pronounced in the lines of hydrogen and helium. Whenever the jumping electron is in a large orbit, and hence not

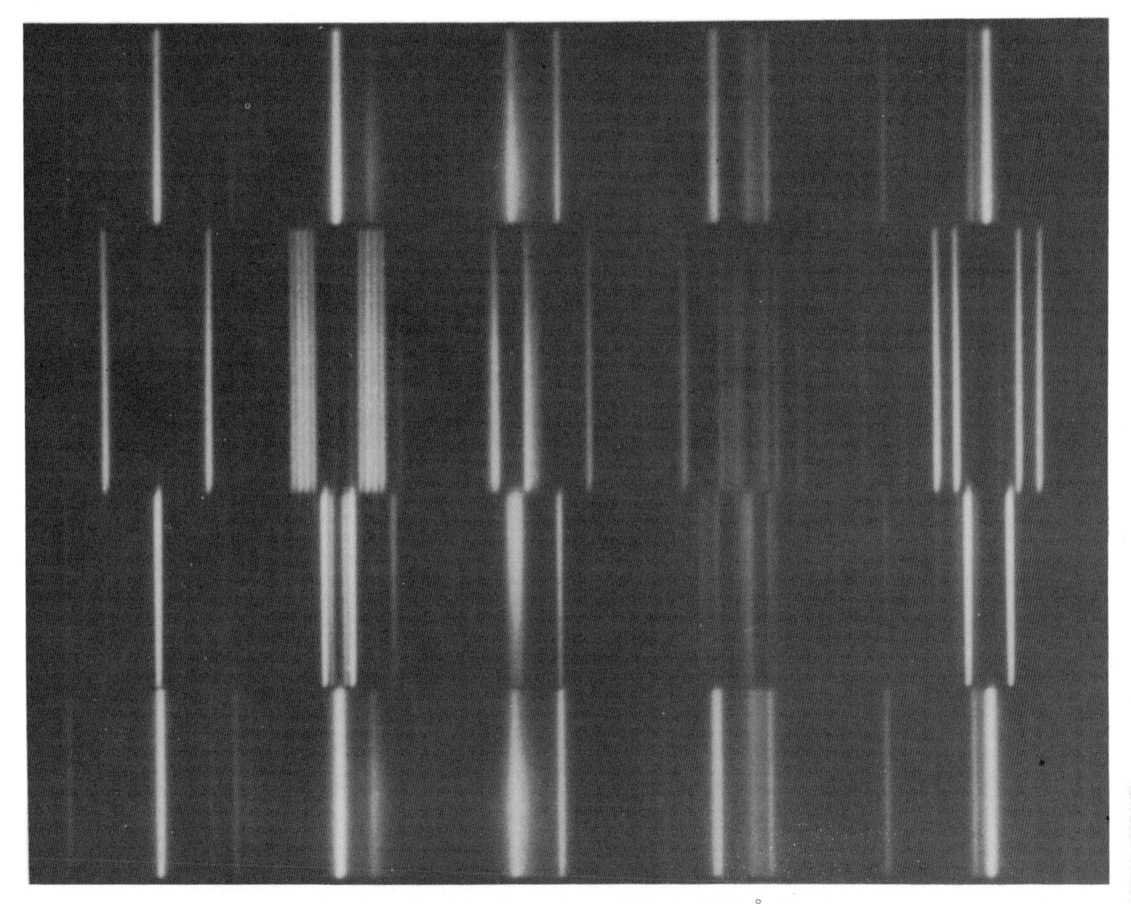

Fig. 41. The Zeeman effect in chromium, for lines from 4613 to 4626 Å. The top and bottom strips show the field-free spectrum. In a magnetic field of 31,700 gauss the lines are split and the components are polarized, as shown in the two middle strips; in the upper strip the components are polarized perpendicular to the magnetic field and in the lower one parallel to the field. Notice that the patterns differ from line to line, depending on the character of the energy levels involved. These Zeeman patterns are of great help in identifying the levels. (Courtesy Mount Wilson Observatory.)

firmly held by the attraction of the nucleus, it can be more easily dislocated by a passing charge, just as the outer moons of Jupiter are more seriously disturbed by the attraction of the sun than are the inner (Galilean) satellites. Thus the higher members of the Balmer series, Hδ (4101 Å), Hϵ (3970 Å), Hζ (3889 Å), Hη (3835 Å), Hθ (3797 Å), . . . , are more affected by Stark broadening than are the earlier members such as Hα (6563 Å) or Hβ (4861 Å).

The Stark effect observed in the hydrogen and helium lines in stellar spectra differs from that produced in the laboratory in one important respect. Electric fields produced by laboratory apparatus are constant over volumes billions of times larger than those occupied by the individual atoms. In a star's atmosphere, each atom is subjected to an individual field of its own, produced by the electrons and ions that happen to be dashing about nearby. At higher temperatures, the space surrounding each atom is filled with rapidly moving positively charged ions and negatively charged electrons whose velocities and positions are quite random. Each charged particle produces a field of different intensity at the radiating atom. At one instant the separate fields due to the ions and electrons may nearly cancel at the radiating atom; at the next instant a charged particle may make a close approach and the field may become very large. Consequently, the simple Stark splitting of a line such as is observed in the laboratory (Fig. 40) is not observed in the stars, since the fields acting on the radiating atoms there are not uniform but fluctuate rapidly in character and differ from atom to atom. Hence the superposition of the radiations from the different atoms of the same element will not coincide, but will overlap to produce a broad, fuzzy spectral line.

Figure 42, due to R. M. Petrie, shows how the appearance of the hydrogen lines depends on the gravity at the surface of a star. Very large (supergiant) stars have low surface gravities, relatively attenuated atmospheres, and rather narrow, weak hydrogen lines. Dwarf stars, which do not differ greatly in size and mass from the sun, have relatively broad, fuzzy hydrogen lines. The reason for this behavior is easy to guess. In the relatively dense atmospheres of a dwarf, radiating atoms and their disturbing charges are close together; consequently the momentary electric fields are larger and the lines become broadened. In the rarefied supergiant atmospheres, the density is usually so low as to render Stark broadening of minor importance. Hence the lines of hydrogen and helium, although broad and fuzzy in hot dwarf stars, are relatively sharp and narrow in the supergiants.

From the observed shapes, or profiles, of the stellar hydrogen and helium lines it should be possible to obtain information about the temperatures and densities in the strata where these lines are formed. In order to make progress in this direction it is necessary to simulate (under laboratory conditions) the temperatures and densities obtaining in stellar atmospheres. Fortunately, it is possible to study the broadening of these lines under controlled conditions. Furthermore, considerable advances have been made in our understanding of the theory of H and He line broadening. Helium shows particularly complex effects, some lines being

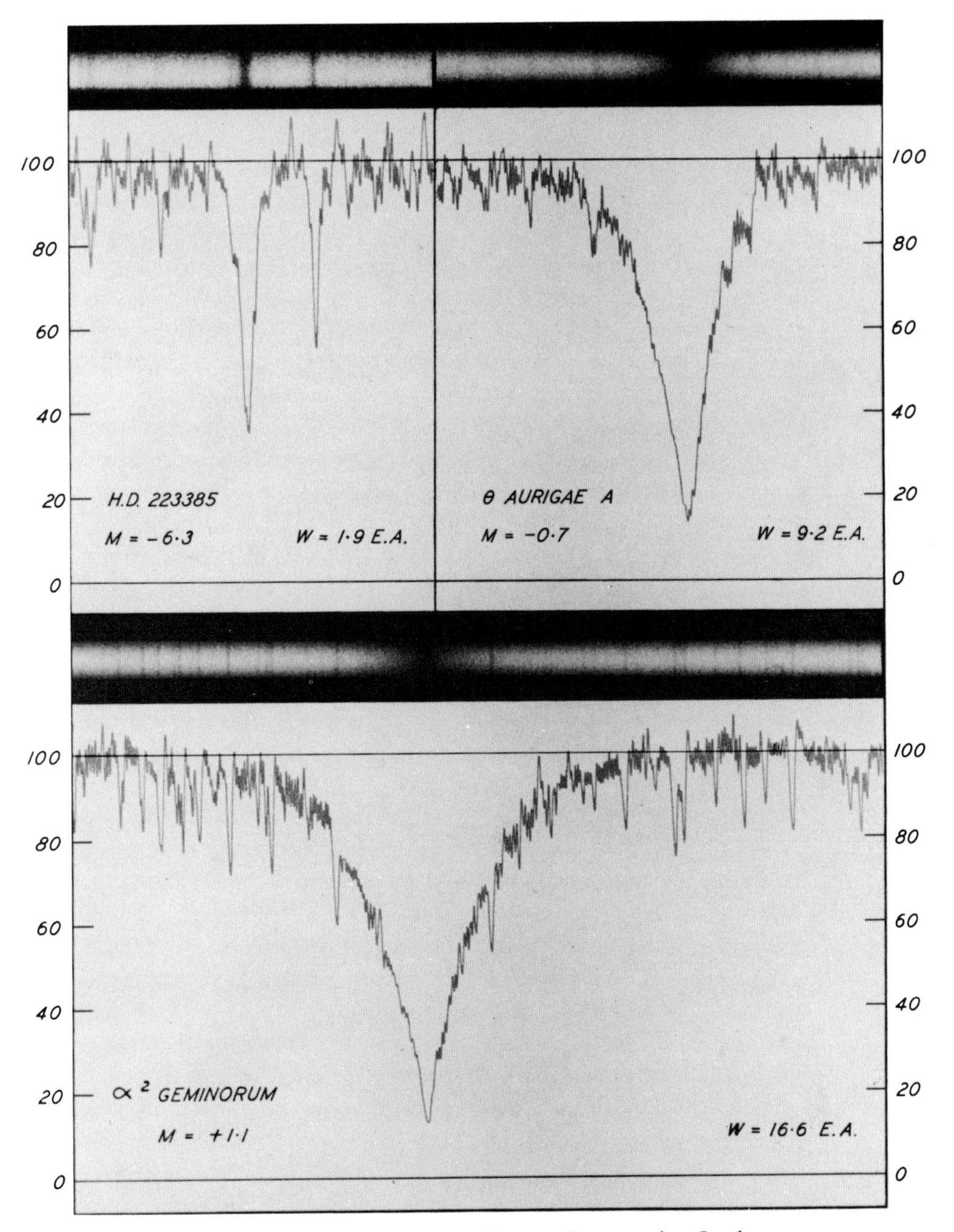

Fig. 42. Hydrogen lines in stars of high or of low surface gravity. In the very luminous star HD 223385, which is about 25,000 times as bright as the sun, the surface gravity and atmospheric density are very low. Hence the disturbing effect of charged particles on the radiating hydrogen atom is small and the line is extremely narrow. In θ Aurigae A, a star about 150 times as bright as the sun, the density is higher and the broadening effects produced by charged particles are more important. In Castor (α^2 Geminorum), which is 30 times as bright as the sun, the atmospheric density is high, so that the line is markedly broadened. (Courtesy Dominion Astrophysical Observatory.)

much more sensitive to electric fields than others. A small, second-order, or "quadratic," Stark effect occurs for lines of helium and heavier elements.

In one technique for studying Stark broadening, developed in the laboratory of Professor Lochte-Holtgreven in Kiel, Germany, an arc is struck down the axis of a hollow tube in which a rapidly whirling stream of water is flowing. The water near the axis of the tube is vaporized, decomposed, excited, and ionized. The spectral lines of hydrogen and of oxygen in several stages of ionization are observed. From measurements of these lines, interpreted by ionization theory, the temperature and density in the arc can be computed. The whirling-fluid arc tends to be unstable; in most modern work one uses a wall-stabilized arc.

Another technique involves the luminous-shock tube. In a long iron tube a gas such as hydrogen or helium at high pressure is separated from a mixture of a "noble" gas, such as argon, and hydrogen at low pressure by a thin membrane. When the membrane is pierced, a shock wave, traveling with a speed several times that of sound, rushes down the tube, strikes the far end, and is reflected back. The gas immediately behind the reflected shock wave is heated to incandescence and its spectrum may be observed. Pressure and temperature may be predicted accurately from the pressures, temperatures, and gaseous mixtures chosen for the experiment, and checked by separate measurements. In this way it is possible to ascertain the shapes of the hydrogen emission lines under different conditions of temperature and density.

In a relatively dense stellar atmosphere, such as that of the sun, radiating atoms may be bumped by passing neutral atoms, mostly H. Then the frequency of the emitted radiation is changed. Since these collisions occur randomly, the observed spectral line is broadened. In the sun, this collisional broadening is more important than natural broadening. In a given spectrum it is also more important for lines corresponding to electron jumps between the larger orbits.

In the early thirties, Struve and Elvey found that the spectral lines in many giant and supergiant stars were broadened by the Doppler effect in such a way as to indicate that the radiating gases were moving with velocities sometimes as high as 35 or 40 miles per second. Such velocities could not be attributed to the temperature of the gas, since these stars had temperatures of only 5,000 or 10,000°K, and the temperatures necessary to reproduce the line shapes would be in the millions of degrees. Struve and Elvey suggested that the atmospheres of these stars are not orderly quiescent envelopes but are subject to violent, large-scale, chaotic motions, which they characterized as turbulence. Independent evidence for large-scale mass motions of radiating gases is provided by the supergiant components of eclipsing binaries such as 31 Cygni. A modest degree of turbulence (amounting to at most only a few kilometers per second) appears to exist in the solar atmosphere, but the phenomenon is most developed in certain supergiants.

To summarize: exclusive of instrumental imperfections, stellar spectral lines are broadened by two classes of causes:

Intrinsic causes

(*a*) Natural width, which is due to the fact that an atom, like a radio station, cannot radiate at one sharp frequency, because the energy levels themselves have a certain width;

(*b*) Doppler effect, which is due to random motions of atoms in any heated vapor (see also *g*, *h*, and *i* below);

(*c*) Zeeman effect, which is the splitting of spectral lines by magnetic fields, as in a sunspot;

(*d*) Stark effect, which is the splitting of a spectral line by an electric field; in stellar atmospheres the lines are broadened because the fields acting on any radiating atom are momentary and random;

(*e*) Collisional broadening, which originates because radiating atoms may collide with neutral atoms and suffer a change in their radiated frequencies;

(*f*) Hyperfine structure; certain lines of various elements are observed to be split into a number of very close components as a consequence of a magnetic interaction between the spin of the nucleus and the total angular momentum of the electron. The phenomenon is analogous to the interaction of the magnetic field of the spinning electron with the field produced by its orbital motion (see Chapter 2) except that it is on a scale that is roughly a thousand times smaller.

Extrinsic causes

(*g*) Turbulence, or large-scale vertical motions of large masses of radiating and absorbing gases in a stellar atmosphere;

(*h*) Rotation of the star itself, which broadens all of the spectral lines; rotational speeds as high as 200–300 kilometers per second have been observed in A and B stars, whereas G and K dwarf stars, like the sun, appear to rotate slowly;

(*i*) Expansion of the stellar atmosphere itself. Some stars, such as those of the P Cygni type, certain "B emission objects," and "exploding stars," or novae, have expanding atmospheres or envelopes that produce broadened, unsymmetrical lines.

Before we turn to the interpretation of spectral lines in stellar atmospheres we must emphasize that the intensity of a dark line is a relative rather than an absolute quantity. A spectral line appears dark by contrast because the intensity at a given point in the spectrum is less than that at adjacent wavelengths. Thus, the intensity of the line is always measured relative to that of the bordering continuous spectrum, and hence the interpretation of dark-line intensities must be based on a prior understanding of the process by which the continuous spectrum is formed.

The Continuous Spectrum

As we have already explained in Chapter 4, there is no sharp dividing line between the main body of a star and its atmosphere. Looking down through successively deeper layers of the atmosphere, we arrive at a point where the gaseous material is completely opaque. This level is what we commonly refer to as the surface of a star. The thickness of the atmosphere thus depends upon the absorptivity of its material. In a dwarf star, where the gases are greatly compressed, we can penetrate through only a relatively shallow layer of material, and the depth of the atmosphere is thus small. In a giant star, however, the density is so low that we can see through a great depth of the atmosphere, which is said to be extended.

In our discussion of the chemical composition of the solar atmosphere (later in this chapter), we shall find that a surprisingly small amount of solar material, about 2 grams per square centimeter of the sun's surface, suffices to block the radiation from below the surface. The total amount of matter in the solar atmosphere, which is 10^{17} (100 thousand million million) tons, is huge only because of the sun's great size, for it represents only one part in 20 thousand million of the whole solar mass. The inference is that the gases in stellar atmospheres are exceedingly hazy. If the earth's atmosphere, with its relatively great density, were as opaque, one could hardly see as far as 50 feet.

The chief reason for the fogginess of the atmospheres of the hotter stars is that gases in the process of being ionized are highly opaque. We know, of course, that the atoms in a stellar atmosphere strongly screen radiations in the neighborhood of absorption lines, because an atom raised to higher energy levels absorbs energy corresponding to discrete wavelengths. But when an atom becomes ionized it may absorb energy of *any* frequency greater than the minimum amount necessary for ionization. Thus the ionization of hydrogen atoms whose electrons are in the second orbit produces a continuous absorption spectrum stretching to the violet of the Balmer series limit at 3650 Å, while the ejection of electrons from the third orbit screens off energy at wavelengths shorter than the limit of the Paschen series at 8210 Å in the infrared. It is clear that hydrogen atoms cannot contribute very much to the opacity of stellar gases unless a fair fraction of them become excited to the second and higher levels. Only those atoms excited at least to the second level can absorb radiation at wavelengths smaller than 3650 Å, and only those excited to at least the third level can absorb radiation in the part of the spectrum from 8210 to 3650 Å. At a temperature of 5700°, the theory of the excitation of atomic levels (see Appendix F) suggests that only four or five hydrogen atoms out of every 1000 million are in the second level. Thus, despite its great abundance, atomic hydrogen contributes but slightly to the opacity of the middle and high layers of the solar atmosphere. In the deeper layers, where the temperature is higher, absorption by neutral hydrogen is undoubtedly important. In the much hotter class A stars, many hydrogen atoms are excited to the second and higher levels, and the hydrogen becomes highly opaque. Figure 43 shows the strong absorption at the limit of the Balmer series in a class B star.

After it was established that the opacity of the atmospheres of the sun and of the cooler stars could not be accounted for by neutral hydrogen, it seemed logical to suppose that in these stars the ionization of metallic atoms was responsible. Alas, the abundances of the metals are too low to account for even a small fraction of the opacity of the solar atmosphere. The nature of the unknown source of opacity in the solar atmosphere was clarified in 1938 by R. Wildt, who pointed out that in stars as cool as or cooler than the sun a neutral hydrogen atom may acquire a second electron and thus become a negatively charged ion. Negative hydrogen ions are voracious absorbers of energy in the visible and infrared regions of the spectrum. The attachment of the second electron to hydrogen is exceedingly weak and only 0.75 electron volt of energy is required to remove it. Hence, in any given volume of the solar atmosphere, the proportion of negative hydrogen to neutral hydrogen is only about one part in 100 million. However, the hydrogen is so abundant that enough negative ions are present to render opaque the atmospheres of the sun and the cool stars. Precise calculations by Chandrasekhar and others have revealed that the absorptivity of the negative hydrogen ion varies with wavelength in a unique and interesting fashion, as shown in Fig. 44. Since the energy required to detach an electron from the negative hydrogen ion is so small, these ions can absorb radiation of wavelength shorter than

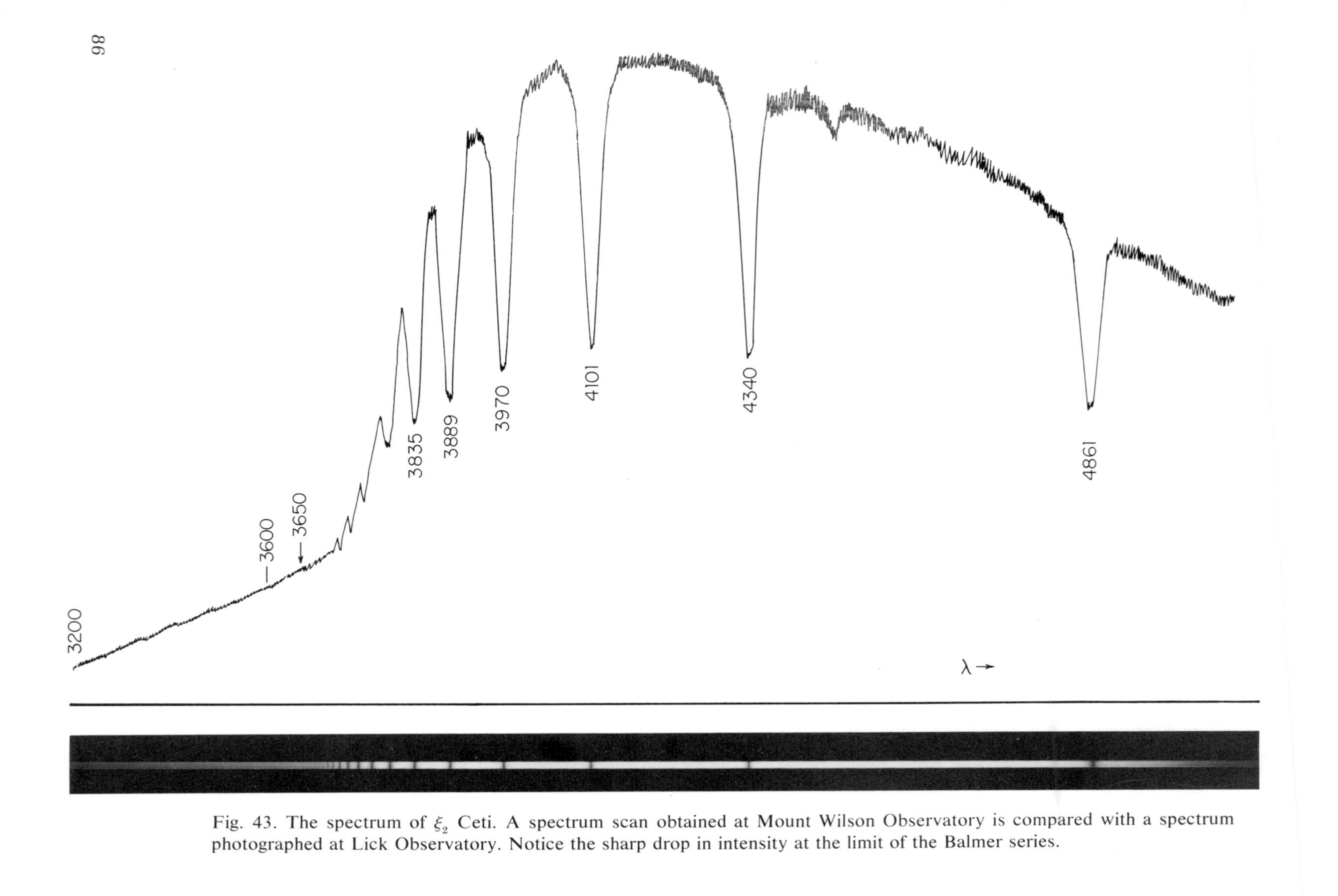

Fig. 43. The spectrum of ξ_2 Ceti. A spectrum scan obtained at Mount Wilson Observatory is compared with a spectrum photographed at Lick Observatory. Notice the sharp drop in intensity at the limit of the Balmer series.

about 16,000 Å in the near infrared. Hence, the detachment of electrons from H^- will produce an absorption continuum covering the near infrared and visible region. The curve labeled "bound-free" in Fig. 44 shows that the absorption by this process of electron detachment increases toward shorter wavelengths, with a maximum at about 8600 Å, and then declines toward still shorter wavelengths. Once the electron has been removed, it may be thought of as moving in a hyperbolic rather than a circular or elliptic orbit. Such free electrons may also absorb radiation and pass to a hyperbolic orbit of higher energy. This process gives rise to another continuous absorption spectrum, the intensity of which increases progressively toward longer wavelengths, as shown by the curve labeled "free–free" in Fig. 44. Finally, the heavy curve represents the combined absorption of negative hydrogen ions as a result of both processes. The observed continuous absorption in the solar atmosphere may also be

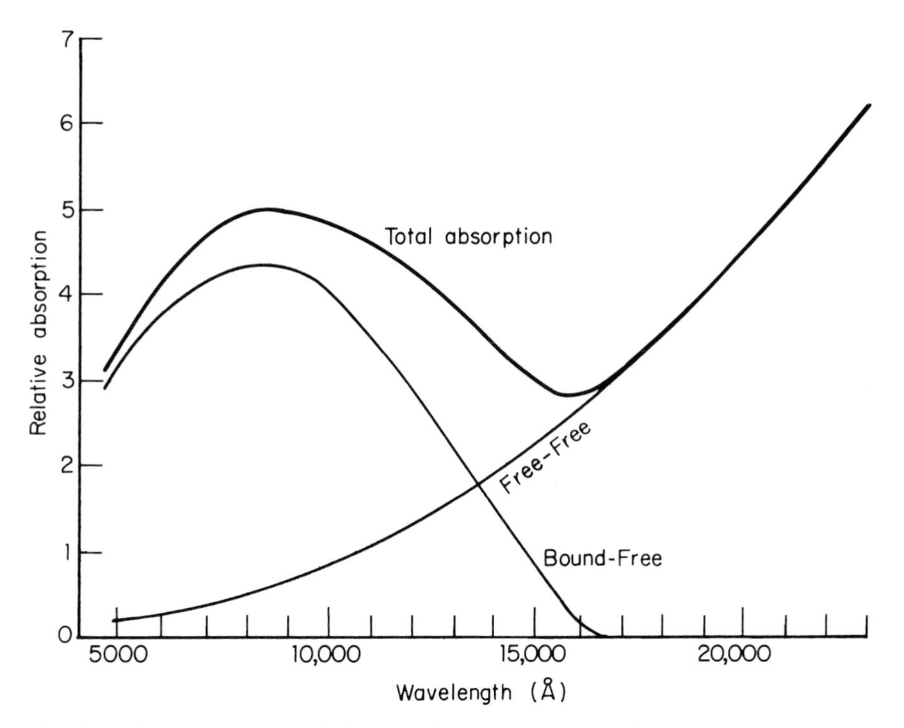

Fig. 44. The absorptivity of the negative hydrogen ion. Relative absorption is plotted against wavelength for the free-free, bound-free, and total absorption of the negative hydrogen ion for a temperature of 6300°K. Note the maximum near 8000 Å and the rise of free-free absorption in the infrared. (After S. Chandrasekhar.)

derived from observations of limb darkening, and the agreement of the observed curve with the theoretical curve in Fig. 44, though not exact, is sufficiently close to make it highly probable that the negative hydrogen ion is the principal source of opacity in the sun. In the ultraviolet, continuous absorption by silicon and metal atoms is important.

Probably also the negative hydrogen ion is the principal absorbing agent in the stars of spectral classes G and K. In the hotter stars, the ionization of neutral hydrogen atoms is undoubtedly the major source of opacity, whereas in the very cool stars the overlapping of closely spaced atomic lines and molecular bands may contribute significantly to the opacity, in addition to that due to negative hydrogen. Theory agrees with observation in predicting that the atmospheres of the hotter stars should be more opaque than those of the cooler stars. Thus the atmosphere of an A star is about 20 times as opaque as that of the sun.

In the early days of solar spectroscopy, it was supposed that the sun had a sharply defined radiating surface from which the continuous spectrum was emitted, whereas the absorption lines were produced as the radiation from the surface worked its way through a cooler atmosphere, or reversing layer. It is now clear that this picture was grossly oversimplified. In reality both continuous spectrum and absorption lines are formed in essentially the same regions of the atmosphere. The photosphere, in which (by definition) the continuous spectrum originates, is not a sharply bounded surface but a layer with a thickness of about 400 kilometers. In the giant stars the layer is much more distended, whereas in the dwarfs it is highly compressed. We have already seen that in the sun and in the cooler stars the continuous spectrum is formed as a result of absorption by negative hydrogen ions. The radiation from the lowest layers of the photosphere, which is at a temperature of about 8000°, is much more intense than the radiation from the higher levels, where the temperature is lower. The radiation from the deep levels is much more heavily absorbed than the high-level radiation. Hence it has to traverse a much greater thickness of absorbing negative hydrogen ions before it can escape into space. The result is that the radiation that finally does escape corresponds in color and quantity to an average temperature of about 5800°K.

How an Absorption Line Is Formed

The negative hydrogen ions are not the only hazards that must be faced by radiation working its way up to the surface. Stellar atmospheres contain, in addition to hydrogen, all of the chemical elements familiar to us on the earth. Each type of atom can absorb radiation in its own discrete set of wavelengths, and hence those quanta of radiation that are of the proper wavelengths are absorbed. It is true that every absorbed quantum is reradiated, but, whereas the beam of energy from the photosphere flows outward along the line of sight, the radiation emitted by atoms may be thrown off in any direction whatever, backward and sideways as well as forward. Hence the original outward beam is depleted in the absorbed wavelengths. Eventually the radiation leaves the star but only after being mutilated by its encounters with absorbing atoms. When properly interpreted, the marks that the spectrum bears yield a history of the passage of the radiation through the atmosphere—the types and numbers of atoms encountered, as well as the temperature of the gas. In the sun, most of the absorption in the so-called "wings" of a line occurs in the middle and upper regions of the photosphere, but the absorption at the centers of strong lines originates in the lower levels of the solar chromosphere.

The Number of Absorbing Atoms

The intensities of the dark lines in a stellar spectrum are usually measured on photographic plates or with the aid of suitable photoelectric devices. By the intensity of a dark line we mean the amount of energy that has been subtracted from the continuous spectrum at the position of the line. The unit of intensity is the *equivalent width,* expressed in angstrom units. Thus, an equivalent width of 1 Å signifies the removal of an amount of radiation equivalent to that contained in 1 Å of the neighboring continuous spectrum. The line intensities are governed chiefly by three properties of stellar atmospheres, namely, the chemical composition, the temperature, and the density. The problem that confronts the astrophysicist is to derive these properties from the observed intensities. The intensity of a dark line must depend upon the number of atoms per unit cross-sectional area along the line of sight that are absorbing at the wavelength under consideration. We call this quantity the *number of absorbing atoms* and proceed to clarify its physical meaning with the aid of Fig. 45. We suppose that energy is being absorbed by a hypothetical atom with four energy levels, *A, B, C,* and *D.* Let us assume that N_a atoms are in level *A* and N_b in level *B,* and that the atoms are being struck by radiation of the correct wavelengths to excite them to levels *C* and *D.* Assuming that the impinging radiation is equally intense in all four wavelengths, *A–C, A–D, B–C,* and *B–D,* how many atoms per second will absorb each of the four radiations? In the first place, the number of absorbing atoms

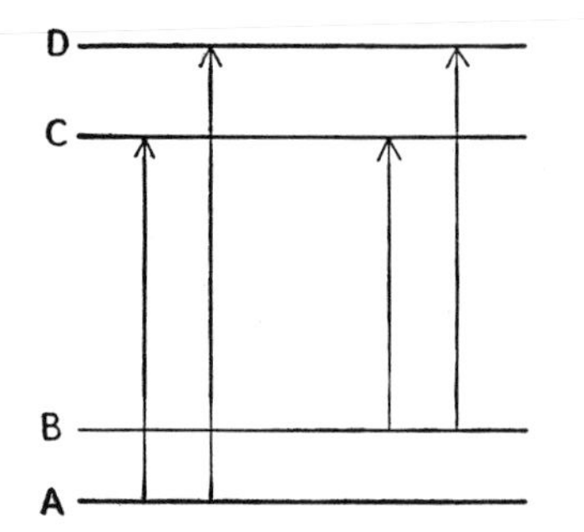

Fig. 45. Energy levels of a schematic atom.

will be proportional to the number of atoms, N_a or N_b, in the lower energy level. We must allow for the fact that certain transitions have a greater probability of occurring than others. An atom in level *A*, for example, will not usually have an equal preference for levels *C* and *D*. One or the other will be more inviting. This preference may be expressed by assigning each line a number, usually less than unity, which is known as the *f*-value for the line. The *f*-numbers depend only on the structure of the atom, and may be computed from theory or measured in the laboratory. They are defined in such a way that the number of absorbing atoms is proportional to the product of the number of atoms and the *f*-value, *Nf*. Thus the number of atoms absorbing the line $A-C$ is $N_a f_{ac}$. In hydrogen, for example, the *f*-values of successive members of the Lyman series, beginning with the first line, are 0.42, 0.079, 0.029, 0.014, 0.0078, and so on. Since all the lines originate from the same level, the first, we see that the number of absorbing atoms, *Nf*, diminishes rapidly along the series.

The Curve of Growth

The first step in the analysis of a stellar atmosphere is to evaluate the quantity *Nf* for each line from the observed intensities in the spectrum. To do so we must have a numerical relation between the intensity of the line and the number of absorbing atoms responsible for its production. At first sight it might appear that these quantities should be directly proportional to each other. Actually the relation is much more complicated, and depends on the mechanism responsible for broadening the lines. We have already mentioned that spectral lines are never perfectly sharp. Each line has associated with it an intrinsic natural width due to the fact that the energy levels themselves are so broad (they are zones rather than simple lines), and also a so-called Doppler width arising from the random motions of the absorbing atoms.

To illustrate our argument we shall revert momentarily to our simplified model of a stellar atmosphere and suppose that above the star's surface, which radiates a continuous spectrum, there exists a layer of perfectly motionless absorbing atoms. We suppose that the density of the gas is so low that collisional broadening is not important and only the natural width of the spectral lines is significant. Consider what happens to the radiation from the surface as it passes through a vertical column of the absorbing atmosphere. At the wavelengths corresponding to absorption lines, the radiation will be depleted by atoms of the atmosphere. If we now increase the length of the absorbing column, how will the blackness of the line increase?

The curves shown in Fig. 46 represent the shapes of absorption lines

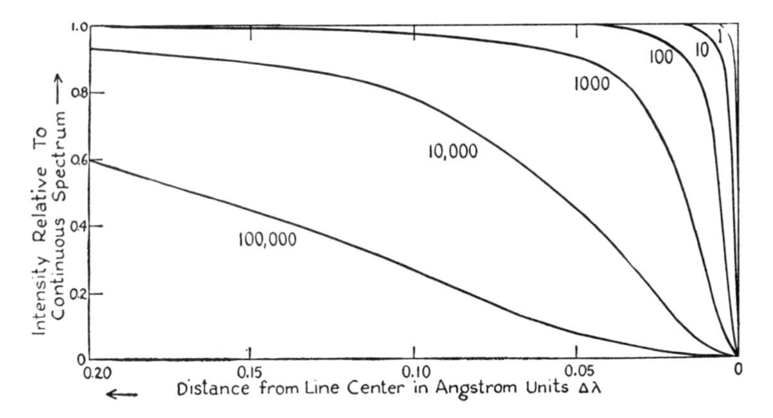

Fig. 46. Profiles of absorption lines showing natural width only. The curves show the change in profile as the relative number of absorbing atoms increases from 1 to 100,000. Since the profile is symmetric, we plot only half of it. Notice that as the number of atoms increases, the very strong "damping wings" come into prominence.

produced by successively greater relative numbers of absorbing atoms, from 1 to 100,000, as calculated from theory. Since each profile is symmetric, only half of it has been plotted. A large number of atoms act to produce a given spectral line; most of them absorb near the center, and progressively fewer at greater distances from the center.

Referring to Fig. 37 we see that, while the absorptivity of an atom is very high at the center of a line, it falls to about 2 percent of its maximum value at a distance of only 0.003 Å from the line center. Thus, even when relatively few atoms are present in the absorbing column, as illustrated by the curve labeled 100 in Fig. 46, the centers of the lines are completely black, but away from the center the intensity falls rapidly to zero. However, as the number of absorbing atoms increases, more and more radiation is removed away from the line center, where the absorptivity *per atom* may be $^1/_{100}$, $^1/_{1000}$, or $^1/_{10000}$ of its maximum value. Sheer weight of numbers of atoms overcomes the disadvantage of small absorptivity per atom and results in the removal of much radiation away from the center of the line, in the wings. When the number of absorbing atoms is so small that the line center is not yet completely black, the intensity, which is measured by the total area under each line profile, increases in direct proportion to *Nf*. As the number of absorbing atoms increases past the point at which the line center becomes black, the intensity increases at a slower rate, as the square root of *Nf*. To double the amount of energy absorbed, four times as many absorbing atoms are required. The resulting relation between the intensity and the number of absorbing atoms is shown in Fig. 47.

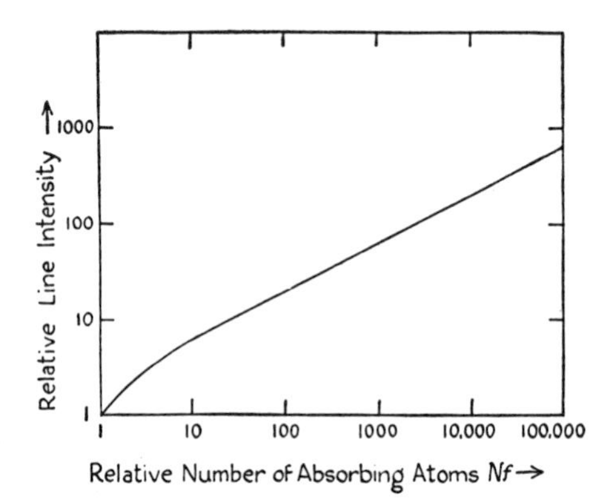

Fig. 47. The relation between relative line intensity and relative number of absorbing atoms for pure natural broadening. Except when the atoms are very few, the intensity is proportional to the square root of the relative number of absorbing atoms, *Nf*. The scales are logarithmic.

In most stellar atmospheres, the broadening of atomic energy levels by collisions between atoms is much greater than the so-called natural broadening. However, the collisional broadening of the lines of all elements other than H and He results in profiles of very nearly the same shape as those due to natural broadening, and hence the relation between intensity and numbers of absorbing atoms is also similar. The curve lies higher, however; that is, for a given number of atoms, the line is stronger

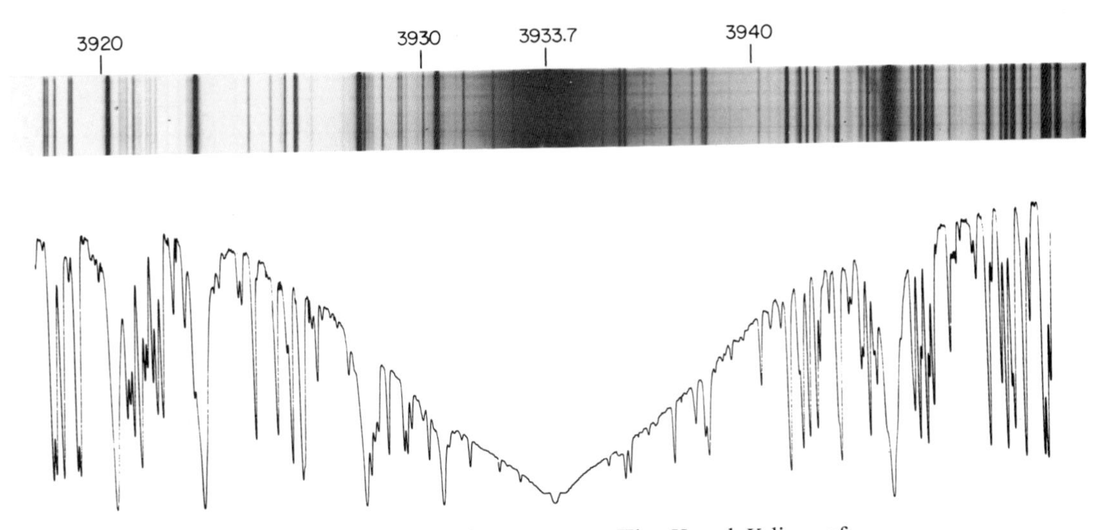

Fig. 48. The profile of the *K* line in the solar spectrum. The *H* and *K* lines of ionized calcium are the strongest lines recorded in the spectrum of the sun as observed from the surface of the earth. The wings are due to natural broadening and collisional broadening. The wavelength scale is given at the top of the figure.

than it would be for pure natural broadening. The *H* and *K* lines of ionized calcium, which are the strongest lines recorded in the spectrum of the sun, are very good examples of lines with very pronounced wings due to natural and collisional broadening. Figure 48 is a reproduction of a small section of a photograph of the solar spectrum in the vicinity of the *K* line. Above the photograph is an intensity tracing of the spectrum that reveals the enormous extent of the line wings. The irregularities in the profile arise from the presence of superposed weaker lines of other elements.

We now consider a column of atoms in rapid motion, so that broadening by the Doppler effect predominates (Fig. 49). We may also assume for the present illustration that collisions are absent and that the line has no natural width, each atom absorbing only at a wavelength determined by the speed of its motion along the line of sight. The shape of the resulting absorption line will therefore depend upon the relative numbers of atoms absorbing at each part of the line. The absorptivity is defined by a curve such as that in Fig. 39. The shape of the corresponding absorption line resembles that in Fig. 49. As in Fig. 46, the curves have been drawn for successively larger numbers of absorbing atoms. Notice that for large numbers of atoms the curves are bell-shaped, with flat tops and very steep sides. The physical meaning of the shapes of the curves is that, in a ran-

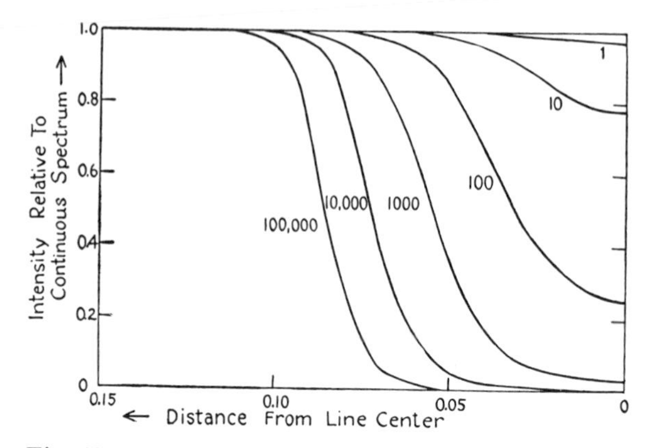

Fig. 49. Profiles of absorption lines with Doppler broadening and no natural width. As in Fig. 46, we show only half of each profile. In contrast to natural damping, the total absorption for large numbers of atoms increases very slowly as the number of atoms increases.

dom distribution of speeds, numerous atoms possess velocities near the zero value but only relatively few have excessively large velocities. Figure 49 shows that when the number of absorbing atoms is small the line is not very black, but broad. The energy absorbed is spread out over a wide range of wavelengths. This is because nearly as many atoms are absorbing slightly away from the center as at the exact center. Accordingly, when more atoms are added, a great deal of energy near the line center is still available for absorption, and the intensity of the line increases directly as the number of absorbing atoms. But the process does not continue indefinitely; the "growth" slows down. Eventually, as more atoms are added, the line becomes black at the center, and, since few atoms have high enough velocities to absorb very far away from the zero position, the line becomes "saturated." In other words, no matter how many additional atoms are added to the absorbing column, very little more energy can be extracted from the continuous background. The corresponding relation between intensity and number of absorbing atoms, Nf, is shown in Fig. 50. The shape of the curve evidently depends upon the temperature, for at high temperatures large numbers of atoms possess high speeds, more energy is available for absorption, and the line does not become saturated until a relatively high intensity is attained.

In reality, neither Doppler nor natural plus collisional damping operates independently. The two are combined, but in such a way that Doppler broadening prevails for small numbers of absorbing atoms and natural plus collisional broadening for large numbers of atoms. The resulting relation between intensity and number of absorbing atoms, which is known as the curve of growth, has the form shown in Fig. 51.

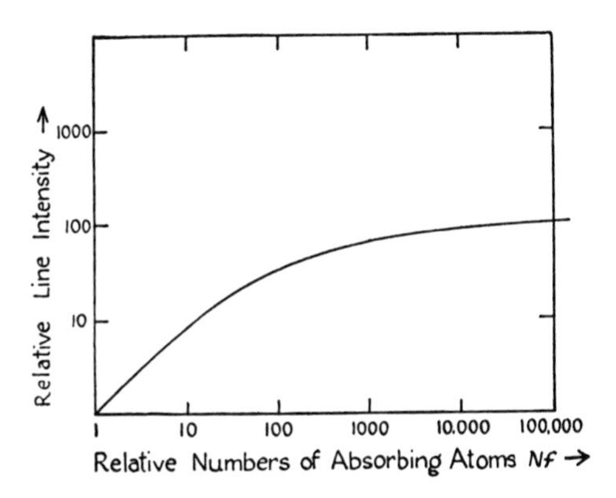

Fig. 50. The relation between relative line intensity and relative number of absorbing atoms for pure Doppler broadening. When the number of absorbing atoms Nf is small, the intensity is very nearly proportional to Nf, but when Nf is large, the intensity of the absorption line increases very slowly as more and more atoms are added. The scales are logarithmic.

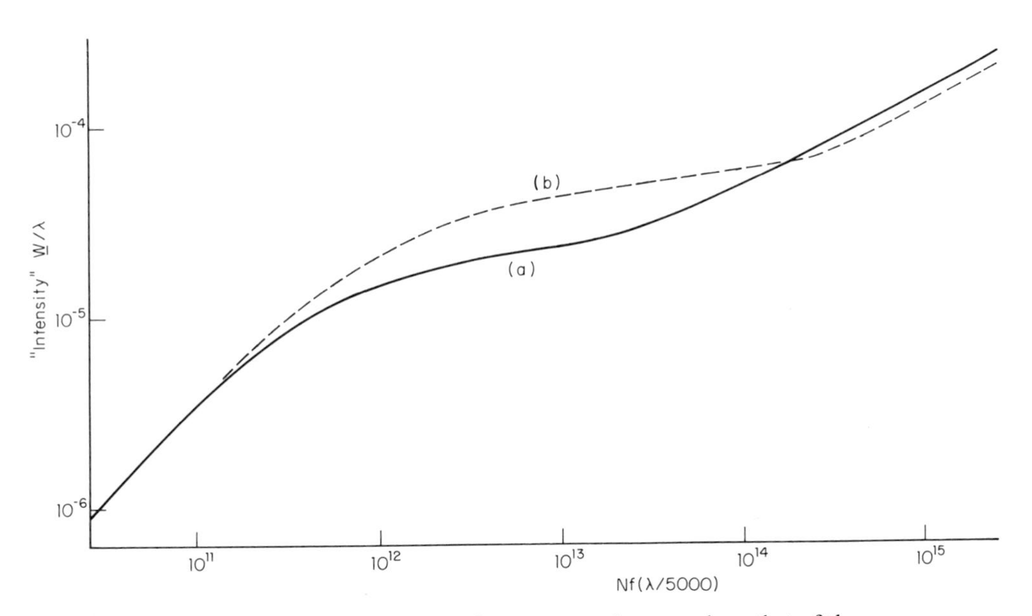

Fig. 51. Theoretical curves of growth: (*a*) for an atmosphere, such as that of the sun, with a high density and a low temperature; (*b*) for an atmosphere with a low density and a high temperature. The "intensity" W/λ, the ratio of the equivalent width of the line to the wavelength, is plotted against the quantity $Nf(\lambda/5000\ \text{Å})$, where N can be interpreted as the "number of atoms above the photosphere." Notice the pronounced effect of collisional broadening in (*a*), whereas in (*b*) it is negligible. Notice also the long interval over which the intensity of the line increases very slowly as the number of absorbing atoms is increased.

It will be noted that the curve of growth has three branches: (1) for very small values of Nf, the line center has not yet become completely black, and the intensity is directly proportional to Nf; (2) for intermediate values of Nf, the line center is black but absorption by the line wings has not yet become large and the intensity increases very slowly with Nf; (3) for very large values of Nf, the intensity is proportional to the square root of Nf. The relation between the various branches of the curve of growth is determined by the relative importance of Doppler versus natural and collisional broadening. The position of the left-hand branch of the curve is unaffected by the kind and magnitude of the line broadening. But the value of the intensity for which the curve begins to flatten out and the position of the right-hand branch of the curve are determined by the ratio of the combined natural and collisional broadening to the Doppler broadening. Since the Doppler broadening, which is governed by the random speeds of the atoms, depends on the temperature, and the frequency of collisions on the density or pressure, the ratio will be larger for an atmosphere of high density and low temperature than for one of low density and high temperature. These two extremes are illustrated by the curves marked (*a*) and (*b*) in Fig. 51. When the collisional broadening is very large, as in (*a*), the transition between Doppler and collisional broadening is rapid. When the collisional broadening is very small, as in (*b*), a very large number of absorbing atoms is needed to build up the line wings. Hence, over a considerable range in Nf, after the center of the line has become saturated and before the wings have begun to grow, a large increase in the number of absorbing atoms has very little effect on the line intensity. Actually, the lines are not totally black at the center as predicted by simplified theory. In a more realistic and physically correct model, predicted and observed central intensities agree.

The theory of the curve of growth has been verified by comparison with empirical curves derived both from laboratory experiments and from observations of stellar spectra. Two observed curves of growth for the sun are shown in Fig. 52.

Various refinements of the curve-of-growth theory must be introduced to allow for the presence of the other causes of line broadening: Stark effect, Zeeman effect, and turbulence. The Stark effect is of importance only for hydrogen and helium. The Zeeman effect exists only in magnetic fields, which in the sun and probably in most stars occur only in localized small areas of the surface. In recent years, however, H. W. Babcock has discovered that the intense magnetic fields must cover large areas of the surfaces of a relatively small number of so-called magnetic stars, of which

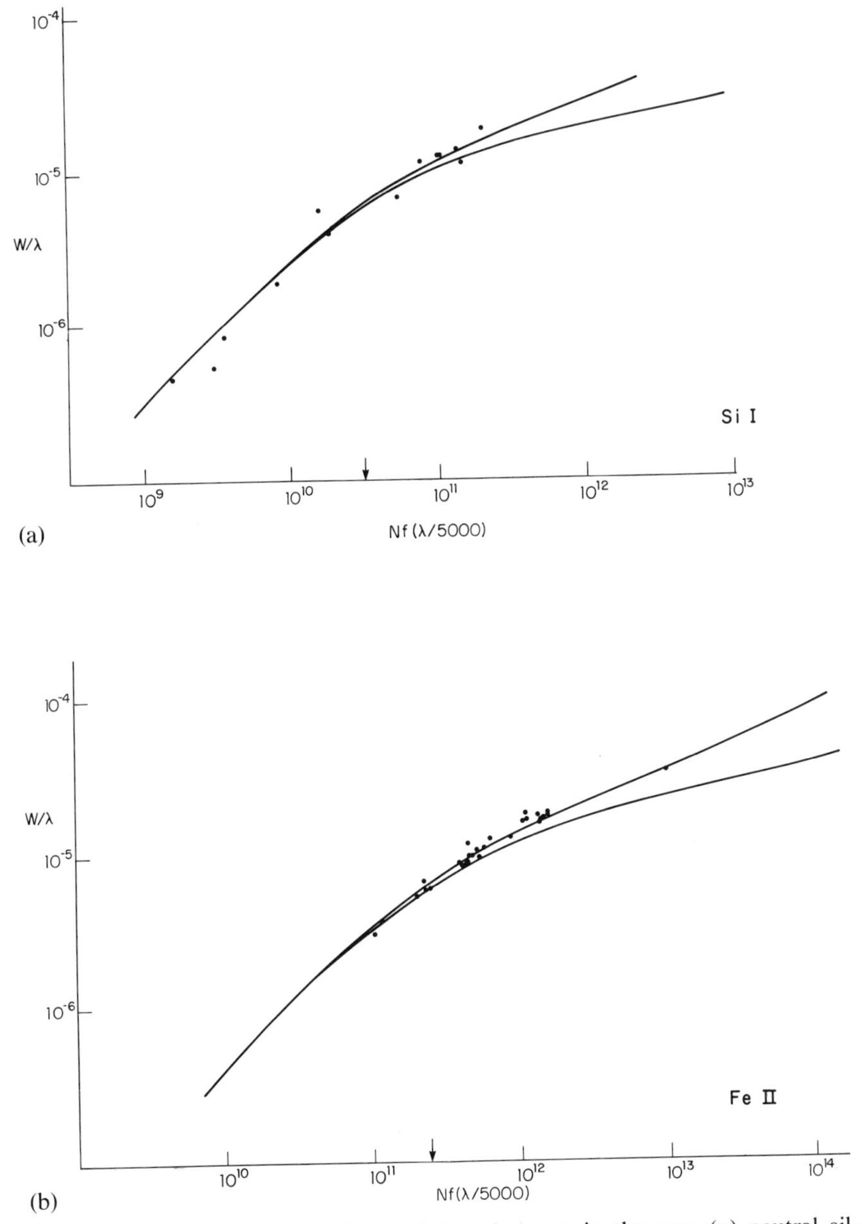

Fig. 52. Theoretical curves of growth for elements in the sun: (*a*) neutral silicon, Si I; (*b*) ionized iron, Fe II. As in Fig. 51, the ratio W/λ is plotted against $Nf(\lambda/5000$ Å). For each element, two branches of the curve of growth are shown, corresponding to two assumptions about the relative importance of collisional and natural line broadening.

HD 125248 and α_2 Canum Venaticorum are examples. The analysis of the spectra of the magnetic stars requires a major modification of the curve-of-growth theory. Turbulence affects the shape of an absorption line in the same fashion as Doppler broadening. In the calculation of the curve of growth it is necessary to use the mean of the vertical velocities of the large-scale masses of gas in addition to the mean of the velocities of the individual atoms. Stellar rotation tends to broaden the star's spectral lines, and makes the line profiles characteristically dish-shaped, but does not affect the shape of the curve of growth.

The theoretical curve of growth consists of a graph of W/λ, the equivalent width divided by the wavelength, plotted against Nf, the number of absorbing atoms. That is, it gives the relation between the spectral-line intensities and the number of absorbing atoms. In the simplified form used for preliminary analyses, it is assumed that one may employ an average temperature and pressure for the atmosphere. The temperature of any given star may be estimated from its color and the pressure from the broadening of the hydrogen lines, so that the theoretical curve may be computed. See Appendix G for an illustrative example.

Hence, with the aid of the curve of growth one may read off the value of Nf corresponding to the observed value of W/λ. Since the f-value is known from laboratory measurements or theoretical calculations, one may then obtain N, the number of atoms in the lower of two energy levels corresponding to the observed line. Occasionally, a given element, potassium for example, is represented in the ordinarily observable spectrum by lines arising from a single lower energy level. Many atoms, such as iron and titanium, display lines arising from a large number of different levels. Then one can calculate the temperature of the stellar atmosphere, for the relative numbers of atoms that are excited to the various energy levels of an atom are governed by the local temperature (see Appendix G). At low temperatures the higher energy levels are sparsely populated, while at high temperatures there may be an appreciable number of atoms in these levels. In either event, if we know the population of even a single level, we may calculate the population of all levels in that stage of ionization.

That is, we do not determine the total abundance of the element, but only the number of atoms that are neutral or ionized, depending on which lines are actually observed. For example, the overwhelming majority of sodium atoms in the solar atmosphere are ionized, but only the lines of the neutral element are observable. In Chapter 4 we saw that if the temperature and number density of free electrons were known we could solve for the ratio of neutral to ionized atoms by Saha's equation.

In the atmosphere of a star like the sun, whose temperature is known, we can estimate the electron density from the broadening of the hydrogen lines and from the relative numbers of ions and neutral atoms for metals such as calcium, iron, titanium, and barium, which show lines of both neutral and ionized atoms. Given the temperature and the ratio, say, N(ionized Fe atoms)/N(neutral Fe atoms) we compute the electron density, N_ϵ, from the ionization equation. Then, given the number of neutral sodium atoms we can compute N(ionized Na atoms)/N(neutral Na atoms) and thus get the total amount of sodium.

The method we have just described presupposes that absorption lines are formed in an atmosphere of constant pressure and density, whereas in reality both the pressure and the temperature of a stellar atmosphere increase inward. In order to allow for these effects, we must construct a model atmosphere, such as that described in Chapter 4, wherein pressure and temperature are specified at each point in the atmosphere. For the sun it is possible to use measurements of the energy distribution at the center of the solar disk and the limb darkening at different wavelengths, and from these data to construct a model of the solar atmosphere. For other stars, model construction is possible by theoretical means only. One simply requires that the atmosphere be in mechanical equilibrium, the pressure at each point sufficing to bear the weight of overlying layers, and that there be a constant flux of energy outward through each stratum. Such calculations have been carried out by Mihalas, Strom, Morton, and others. In the model-atmosphere method, one calculates essentially a separate curve of growth for each spectral line, including the effects of temperature and pressure variations. In this framework, the curve of growth relates the intensity of the line, not to the number of absorbing atoms, but directly to the abundance of the element.

Most analyses of stellar atmospheres have utilized total intensities of spectral lines (equivalent widths) rather than line profiles. Limitations in spectroscopic resolving power usually restrict line-profile investigations to very broad strong lines only, except for the sun, where the profiles of even weak lines can be measured. When it can be observed, the exact shape of a line profile gives much more information about the structure of a stellar atmosphere and elemental abundances than does the total intensity alone.

The procedure is to calculate the line shape, point by point, across the relevant region of the spectrum, taking detailed account of line-broadening effects, ionization, and excitation at each stratum through the atmosphere. The advantage is that one can allow for the effects of overlapping lines (blends), which cannot be handled by other methods. We illustrate

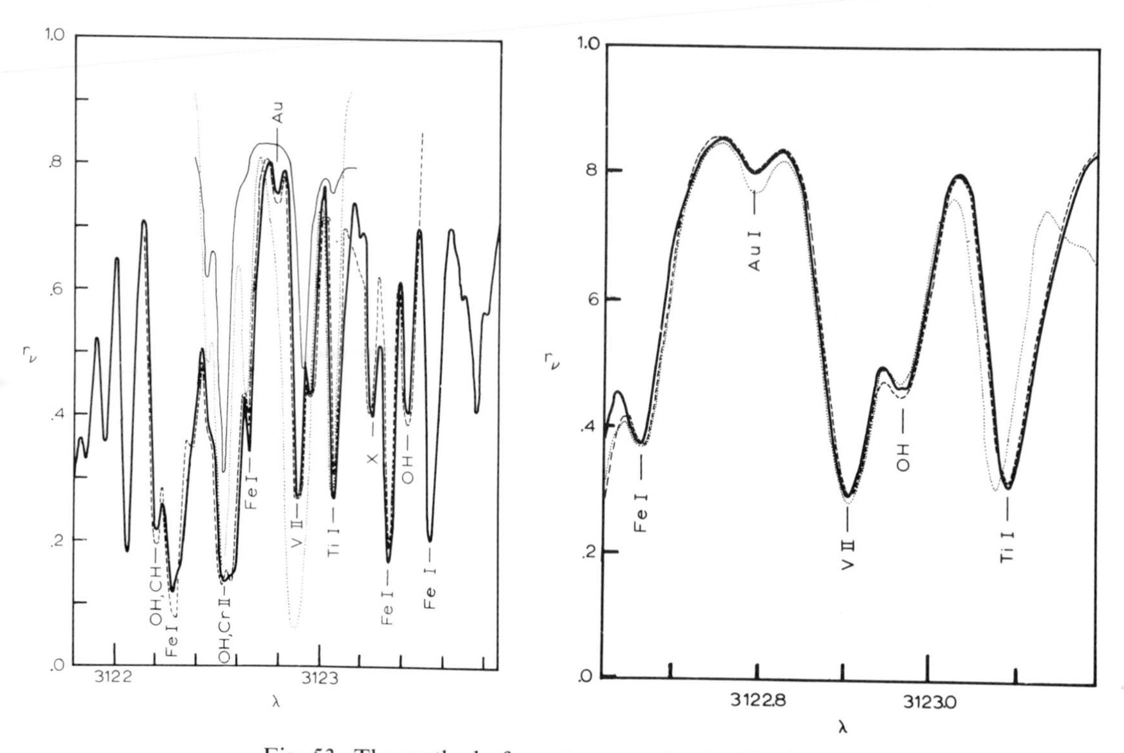

Fig. 53. The method of spectrum synthesis. The heavy solid line gives the observed solar spectrum as obtained from data secured at McMath-Hulbert Observatory and Mount Wilson Observatory. Dotted, dashed, and thin solid lines show successive approximations as the number of active atoms Nf and the influence of collisional broadening are adjusted for lines of different elements to reproduce the observed distribution of intensity in the spectrum. (*a*) Early stages of the approximation are shown by thin solid lines and dotted lines, while the later stages of approximation are denoted by dashed lines. (*b*) In the final stages of the calculation, the spectrum can be computed almost as accurately as it can be observed. In this instance the solar abundance of gold is found from the 3122.8-Å line to be 2 parts in a million million of hydrogen. (Courtesy of John Ross.)

the procedure in Fig. 53, due to John Ross, where a small section of the solar spectrum is reproduced by computing the overlapping profiles of lines of Fe I, Fe VII, Ti I, Cr II, OH, and Au in order to obtain the solar abundance of gold. One has to adjust the elemental abundances multiplied by the f-values for each of the lines and calculate the theoretical collisional broadening until this small section of the solar spectrum is reproduced. In this way the solar abundance of gold is found to be about 2×10^{-12} times that of hydrogen. Uncertainties arise from the effects of collisional broadening, the f-values, and possible unknown blends.

The Chemical Composition of the Stars

From the very beginning of stellar spectroscopy, it has been recognized that matter everywhere in the universe is essentially alike, and that the same chemical elements that make up the earth are also found in other planets, in the sun, in the stars, and in distant galaxies. Although the same chemical elements occur everywhere in nature, their relative proportions frequently differ, not only between planets and stars, but also often between types of stars. These abundance differences give important, though not yet fully understood, clues to the evolution of stars, the building of elements, and the history of the galaxy itself. We explore this problem more fully in Chapters 8, 9, and 12.

Let us describe briefly some of the principal results obtained from investigations of stellar chemical components. Of course, the sun has been most thoroughly studied. The pioneer investigation, carried out in 1929 by Henry Norris Russell of Princeton, before the invention of the curve of growth, clearly revealed the broad features of the sun's chemical composition. His work has been refined considerably for the sun and extended to various stars by numerous investigators in several countries. Let us review results obtained for the solar composition and then compare them with results derived for other stars.

Seventy chemical elements have been fairly certainly identified in the solar atmosphere, as shown in Table 4. Elements absent from the list should not be regarded as missing from the solar atmosphere. Some of them, such as mercury, have their strongest lines in the far ultraviolet, a spectral region that can be studied only by detectors flown above the earth's atmosphere. Elements added to the list since 1946 often have been recognized from their lines in the ultraviolet chromospheric and coronal spectra. Other missing elements, judging from their scarcity on the earth, may be present in the sun in such minute quantities that their lines cannot be found. An example in point is uranium, which is rare and all of whose numerous lines have small *f*-values.

Of great interest is the solar abundance of isotopes, particularly heavy hydrogen, or deuterium. Although there is some evidence that it may be produced in solar regions of great electromagnetic activity in the neighborhood of sunspots, there is no positive proof of its existence generally in the sun. The C^{13}/C^{12} ratio may be the same in the sun as on the earth, but this question needs careful further investigation.

Table 5 lists, for some of the elements in the solar atmosphere, the percentage abundance by number of atoms (that is, volume) and the total

Table 4. Elements present in the sun.

Hydrogen	Potassium	Niobium	Terbium
Helium	Calcium	Molybdenum	Dysprosium
Lithium	Scandium	Ruthenium	Erbium
Beryllium	Titanium	Rhodium	Thulium
Boron	Vanadium	Palladium	Ytterbium
Carbon	Chromium	Silver	Lutecium
Nitrogen	Manganese	Cadmium	Hafnium
Oxygen	Iron	Indium	Tantalum
Fluorine	Cobalt	Tin	Tungsten
Neon	Nickel	Antimony	Osmium
Sodium	Copper	Barium	Iridium
Magnesium	Zinc	Lanthanum	Platinum
Aluminum	Gallium	Cerium	Gold
Silicon	Germanium	Praseodymium	Lead
Phosphorus	Rubidium	Neodymium	Bismuth?
Sulfur	Strontium	Samarium	Thorium
Chlorine	Yttrium	Europium	
Argon	Zirconium	Gadolinium	

mass (in micrograms) that each element contributes to a column of the atmosphere 1 square centimeter in cross section extending vertically above the photosphere. Notice the great abundance of hydrogen and helium. Over 85 percent of the atoms in the solar atmosphere are hydrogen, and, since this high abundance persists in other stars, one can understand why the lines of hydrogen persist over such a huge range of temperatures in stellar atmospheres. We note, too, that although the *H* and *K* lines of ionized calcium are the strongest in the spectrum, several other elements are more abundant. These ionized-calcium lines are strong because they originate from the lowest energy level, in which most of the solar calcium atoms reside (calcium is mostly singly ionized in the solar atmosphere). The corresponding lines of abundant magnesium fall in the ultraviolet beyond the limit of transmission of the earth's atmosphere. The observable lines of carbon, nitrogen, and oxygen all arise from sparsely populated high energy levels. Helium is not represented at all in the dark-line solar spectrum. Its lines are observed in emission from the chromosphere.

Although it was once believed that most stars have nearly the same composition as the sun, the divergences being confined mostly to certain very cool stars, we now know that there are marked differences which are

Table 5. Abundances of some elements in the solar atmosphere.[a]

Element	Percentage volume abundance	Mass (10^{-6} gm/cm^2)
Hydrogen	92.5	1,090,000
Helium	7.3	347,000
Carbon	0.037	5,250
Nitrogen	0.0093	1,530
Oxygen	.068	12,500
Neon	.0074	1,750
Sodium	.00015	40
Magnesium	.0044	1,270
Aluminum	.00023	74
Silicon	.0029	1,000
Phosphorus	.000023	8.5
Sulfur	.00152	560
Potassium	.0000093	4.3
Calcium	.00022	100
Scandium	.00000013	0.07
Titanium	.0000059	3.3
Vanadium	.00000117	0.70
Chromium	.000037	23
Manganese	.000015	9.6
Iron	.00185	1,200
Cobalt	.0000039	2.7
Nickel	.000093	65
Copper	.0000008	0.59
Zinc	.0000008	0.63
Strontium	.000000074	.075
Yttrium	.00000002	.02
Zirconium	.000000052	.056
Molybdenum	.0000000062	.007
Silver	.0000000004	.005
Barium	.0000000073	.012
Lead	.0000000074	.018

[a] These results are based on investigations by D. L. Lambert, Brian Warner, B. J. O'Mara, E. A. Müller, Paul Mutschlecner, N. Grevesse, J. Ross, O. C. Mohler, and others, compiled in *Proceedings of the Astronomical Society of Australia 1*, 133 (1968). Revised Jan. 1971.

important for theories of element building and stellar evolution. The differences are of two types: those that are caused by differences in the chemical composition of the medium out of which the stars were formed and those that are produced by nuclear processes within the stars themselves. We discuss these matters in more detail in Chapters 8 and 9; here we describe some of the special problems encountered in stellar-composition studies.

The stars of high temperature offer certain intrinsic advantages for chemical-composition studies. The strongest lines are those of light elements; the opacity of the atmosphere is due to atomic hydrogen and helium and scattering of light by electrons. These stars are very young; their composition tells us that of the present interstellar medium out of which they were formed, recently. In principle the calculations are straightforward, but in practice verification is difficult because most of the stellar radiation is emitted in the far ultraviolet. Model atmospheres can differ enormously and yet predict nearly the same spectral features for observable regions. Furthermore, in stars of spectral class B0 and O the Saha equation and the Boltzmann equation (which relate the population of different energy levels to one another; see Appendix G) break down. Populations of atomic levels no longer depend simply on the local temperature, but are strongly influenced by details of the local radiation spectrum. Predictions of spectral line shapes and intensities, although still possible, become very difficult.

At the other extreme, spectra of cool stars are dominated by complex molecular bands, each of which consists of many fine components. Theoretically, each of these molecular fine components may be treated as a line produced by a certain number of absorbing molecules in much the same fashion as atomic lines. The problem is complicated by the fact that the lines overlap seriously as the number of absorbing molecules increases. Furthermore, except in spectrograms of very large scale, the lines are too close together to be separately resolved and analysis is difficult.

In analogy with the situation for atomic lines, the total blackness of a molecular band depends upon the abundance of the compound involved, and upon the temperature and density of the stellar atmosphere. In turn, the molecular abundance depends on the abundance of the constituent atoms. The problem is a classical chemical one, involving reactions between atoms and molecules. One assumes various mixtures of atoms and then predicts the numbers of molecules to be anticipated for each mixture.

Suppose that two atoms A and B, say titanium and oxygen, react to form the molecule AB, and that the molecule may also break down into its separate atoms. This reversible reaction may be expressed symbolically as

$$A + B \rightleftarrows AB.$$

These two inverse processes go on until the rates at which they occur just balance. We then say that the two reactions are in equilibrium. The resulting relative numbers of atoms and molecules depend on the concentration of atoms, the temperature, and the amount of energy required to dissociate the molecule. The process is similar to the ionization of atoms and may be handled by a formula similar to that derived by Saha for atoms, ions, and electrons (see Chapter 4):

$$\frac{(\text{Number of } A \text{ atoms}) \times (\text{Number of } B \text{ atoms})}{\text{Number of } AB \text{ molecules}} = K,$$

where K depends on the temperature and the kind of molecule. This equation is called the dissociation formula.

The first comprehensive study of molecules in cool stars was carried out by Henry Norris Russell. He first investigated an atmosphere somewhat similar to that of the sun, in which oxygen is more abundant than carbon. At temperatures well below that of the sun, molecular carbon C_2, the hydrocarbon CH, and cyanogen CN are very abundant. The most plentiful molecule of all is hydrogen, H_2, but its bands do not fall in the observable spectral region. At yet lower temperatures (about 3000°K), formation of the very stable carbon monoxide molecule CO, the familiar and lethal constituent of automobile exhaust fumes, steals the carbon away from other molecules. These theoretical studies seem to indicate that, at least as far as the oxygen/carbon ratio is concerned, the K and M stars have the same chemical composition as the sun.

Furthermore, Russell's theoretical calculations elucidated the greater strength of the cyanogen bands in giant stars of low surface gravity than in dwarf stars of high surface gravity. Among cool stars of the same temperature, the CN bands should be stronger in the giants than in the dwarfs. The propensity of the molecule to become dissociated at the lower density of the giant star's atmosphere is counteracted by the excessive haziness of the dwarf star's denser atmosphere. We observe to much greater depths, that is, look through much more material, in the giant stars than in the dwarfs. When we compare stars of the same spectral class, we find that the CN bands are much stronger in the giants, because dwarfs are hotter than giants of the same spectral class, and high temperature favors dissociation of the molecules.

It turns out that the compounds formed and the type of spectrum obtained are extremely sensitive to whether carbon is more or less abundant than oxygen. If carbon is more abundant in a cool star, practically all the oxygen will be tied up in carbon monoxide, there will be negligible amounts of TiO, and the excess carbon will permit the formation of large amounts of CH, CN, and C_2. The fact that the R and N or carbon stars show strong bands of C_2, CH, and CN indicates that they are objects in which carbon is more abundant than oxygen, an interpretation given many years ago by R. H. Curtiss. As the temperature is raised, the molecular bands disappear and the spectrum of a carbon star resembles that of a "normal" star of class K unless the composition is markedly abnormal.

Another remarkable class of stars are the cool so-called S stars, which show prominent bands of zirconium oxide rather than titanium oxide. These stars also show strengthened atomic lines of zirconium and prominent lines of neighboring elements in the periodic table: niobium, molybdenum, and ruthenium. Even more remarkable are the lines of technetium ($Z = 43$), an element that does not occur naturally on the earth but can be produced by bombarding neighboring elements with neutrons. Incredibly fantastic is the discovery by M. Aller and C. Cowley of the synthetic element promethium in the atmosphere of the A-type spectrum variable HR 465! This element has a half-life of 18 years; that is, after 18 years, of an original kilogram of Pm only 500 grams would remain. This and other elements must be produced by nuclear reactions occurring in the star's atmosphere itself.

Bizarre chemical compositions occur frequently in giant and supergiant stars, which are objects in advanced stages of their lives. They are not found in dwarf stars like the sun. Accordingly we defer further discussion of differences in stellar chemical composition until we have reviewed the problem of stellar evolution.

6 Dwarfs, Giants, and Supergiants

"One star differeth from another in glory."

The words of the Scripture apply not only to the apparent brightnesses of the stars as seen from the earth but to their true luminosities, sizes, and masses as well.

The total amount of radiation flowing from the surface of a star, that is, its intrinsic luminosity, is measured by its absolute magnitude, which is the apparent magnitude the star would have if it were at a distance of 10 parsecs, or 32.6 light years (see Chapter 1). Since the energy radiated per unit area depends only on the stellar surface temperature (which can be deduced from color and spectrum observations), two stars of the same size and temperature should have the same absolute magnitude. Or, if we compare two stars of exactly the same temperature, their surface areas will be proportional to their luminosities.

"The Whales and the Fishes"

All stars, however, are not uniform in diameter; they vary from diminutive objects one-fiftieth the diameter of the sun to mammoth stars a thousand times larger. In 1912 Einar Hertzsprung noted that among the hotter stars there did seem to exist a correlation between temperature and true brightness, in the sense that the hottest stars were also the brightest. Among cooler stars the situation was otherwise.

For example, the brighter component of Capella is visually about 132 times as luminous as the sun. It is also cooler, so that the power emitted per unit area is only 60 percent as great; hence the ratio of areas is 220 and Capella must have a diameter about 15 times that of the sun, or about 13 million miles. Hence it may be called a giant star. Arcturus is about 520 times as bright as the southern dwarf star ϵ Indi; since these two Class K stars have the same surface temperatures, the diameter of Arcturus must be about 23 times as great as that of ϵ Indi. The fainter component of the famous binary 61 Cygni B is 3700 times fainter than β Ursae Minoris (Kochab); hence the radii of these two K stars of the same temperature differ by a factor exceeding 60. The M giant β Pegasi is 450,000 times as bright as the faint M dwarf Lalande 21185 of the same temperature. The radii of these two stars differ by a factor of 670!

If one includes also extremely luminous stars such as Antares, Betelgeuse, μ Cephei, and VV Cephei, even greater disparities in dimensions are involved. Such stars are called supergiants. Even with the rather poor data available in 1912, Hertzsprung concluded that, as regards sizes, stars could differ from one another as the whales differ from the fishes.

The Hertzsprung-Russell Diagram

The great disparity in stellar sizes, temperatures, and brightnesses may be illustrated by a diagram of the sort known as the Hertzsprung-Russell (H-R) diagram, in which spectral types (or colors), which depend on temperature, are plotted against the absolute magnitudes, which measure the luminosity. Figure 54 shows the results obtained for the mixture of stars in our local region of the Galaxy. Notice that the stars are not scattered at random over the diagram but tend to be concentrated in certain well-defined zones or strips. The great majority of the stars fall in a narrow, continuous belt running diagonally downward across the diagram from very blue, hot, luminous objects at the upper left-hand corner to red, cool, intrinsically faint ones in the lower right-hand corner. This unbroken progression of stars is called the *main sequence* (sometimes the dwarf sequence). It includes brilliant, hot stars, such as ζ Puppis and 10 Lacertae, as well as nearer, more familiar stars such as Sirius, Vega, Procyon, α Centauri, the sun, 61 Cygni, and faint red dwarf stars such as Wolf 359 and Krueger 60.

Of special interest are the exceptionally luminous stars that lie above the main sequence and the unusually faint ones that fall far below it. The

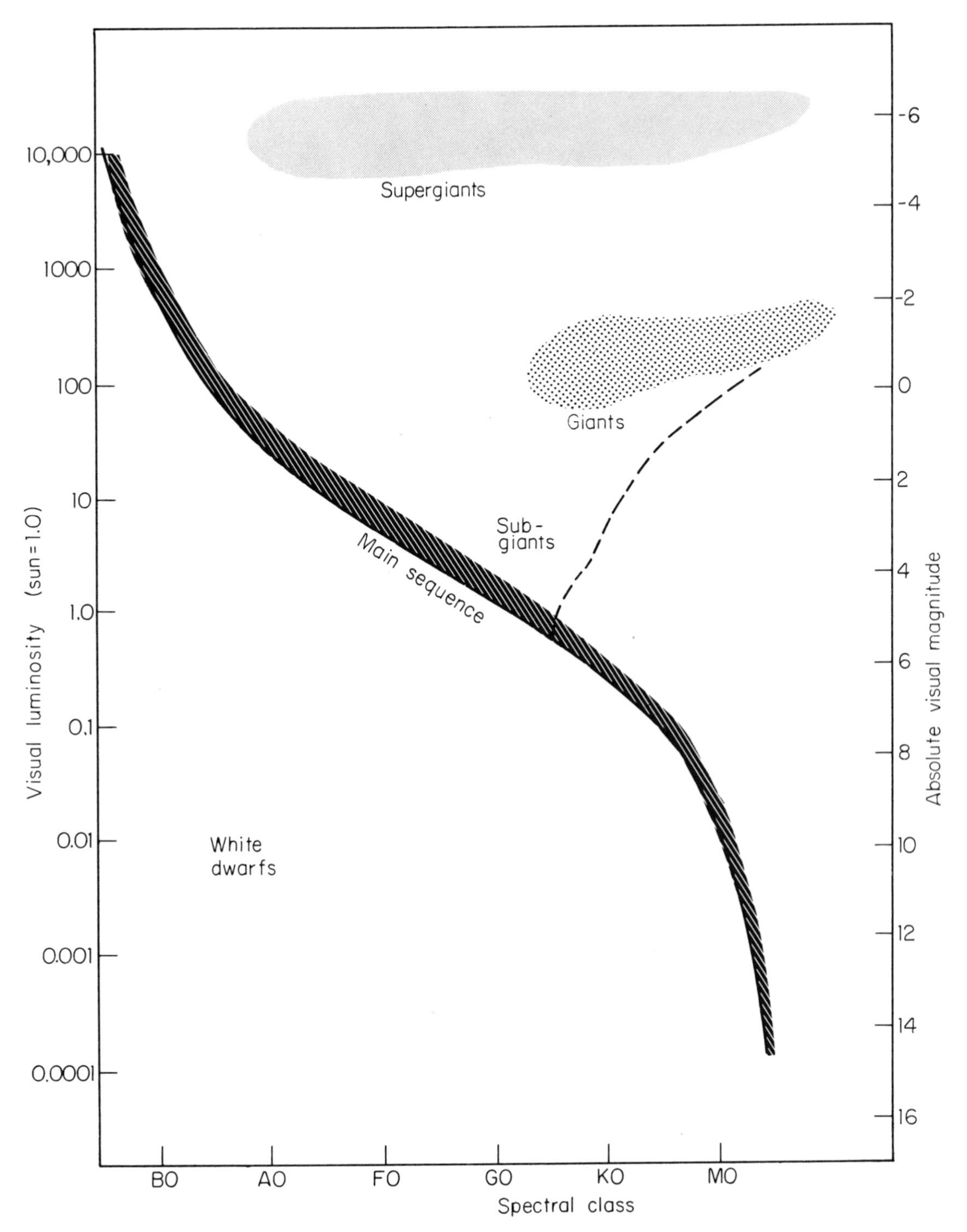

Fig. 54. The Hertzsprung-Russell diagram for stars in the neighborhood of the sun. The visual absolute magnitude (right-hand scale) or the visual luminosity (left-hand scale) is plotted against spectral class. Such a diagram was first plotted by H. N. Russell in 1913.

giant stars, such as Capella, Aldebaran, and Arcturus, have luminosities of the order of a hundred times that of the sun. They extend from class G to M and include representatives of spectral classes R, N, and S. There is a gap—the so-called Hertzsprung gap—between the giant stars and the main sequence. Scattered sparsely across the top of the diagram and including stars from about 300 times as bright as the sun to others 100,000 times as bright are the supergiants. Examples include the brightest Orion stars, Betelgeuse and Rigel, and also Deneb, Canopus, and β Centauri.

At the opposite extreme, far below the main sequence, are the dimly shining whitish stars, often of classes A and F, the so-called white dwarfs. Other groups have been identified. Between the giants and the F, G, and K main-sequence stars lie numerous stragglers which are called subgiants. Objects of this type are often included in eclipsing-binary systems.

Of great importance is the existence of certain regions of avoidance. For example, there is a large gap between the main sequence and the white dwarfs. The dashed line below the giant region and to the right of the subgiant region defines the edge of an excluded zone in which no stars are found.

One additional fact must be noted. As usually plotted, the H-R diagram presents a somewhat biased picture. Most of the stars visible to the unaided eye on a clear dark night are brighter than the sun, yet the overwhelming majority of the stars in a volume of, say, a million cubic light years in the neighborhood of the sun are fainter than the sun. Supergiants like Rigel are prominent at distances of 1000 light-years, but not a single class M dwarf star is visible to the unaided eye. Thus the main sequence is thinly populated at its brighter end but the number of stars increases steadily until the class M red dwarfs are reached. The vast majority of all stars belong to the main sequence. The next largest group is the white dwarfs. Then follow the subgiants and giants. The rarest of all stars are the lonely, luminous supergiants that adorn the vast domain of our galactic system.

One may construct H-R diagrams for different regions in the Galaxy or stellar clusters or aggregates. For example, the immediate neighborhood of the sun, galactic star clusters such as h and χ Persei and the Pleiades, globular star clusters such as M 92 and 47 Tucanae, the central bulge of the Galaxy, and the rich star fields of the Magellanic Clouds all yield H-R diagrams that differ in important ways from one another and provide clues to the histories of stars and stellar systems. The establishment of H-R diagrams and their interpretation occupies a central role in modern stellar astronomy. We must examine the question of stellar dimensions, temperatures, and masses more carefully and attempt to assess the role of chemical composition.

Stellar Masses and Dimensions

The scientist is constantly on the alert to find correlations between independently observed quantities, such as the masses, luminosities, and diameters of the stars. To illustrate the meaning of a correlation, suppose we were to make a study of the heights of boys of all ages ranging from several months to 20 years. We would find, on the average, that the tallest boys were also the oldest, although an occasional lad would be shorter, even though older, than another. We say that the two observed quantities, age and height, are correlated. Again, we know that the pressure of the earth's atmosphere diminishes in a regular fashion with increasing altitude. The two observed quantities, altitude and pressure, are said to be inversely correlated. The advantage of such correlations is that we need only to measure one quantity in order to get an idea of the other. Studies of double stars have yielded the important result that, for main-sequence stars, mass and luminosity are correlated in the sense that the heaviest stars are the brightest.

Table 6 gives the masses as compiled by Daniel Harris for a number of nearby binaries with good data. The luminosities are bolometric. In Fig. 55 mass is plotted against luminosity for these stars. Except for three objects, the points tend to define a curve. The white-dwarf companions of Sirius and o_2 Eridani B are much too faint for their masses. The subgiant ζ Herculis A is too bright for its mass. The mass-luminosity correlation works reasonably well for main-sequence stars and can be extended to brighter stars than Sirius by cautious use of eclipsing-binary data, but it fails for giants and subgiants and yields nonsense for white dwarfs. The luminosity rises rapidly with the mass along the main sequence.

For each of the stars listed in Table 6 we have given the spectral class, from which the temperature can be estimated. Then from the temperature we can calculate how much energy is radiated by each unit area of the surface. We may then calculate how large each star must be to provide its observed luminosity. Then from the mass and volume we find the average density, and from the mass and radius the surface gravity. Note the high densities and high surface gravities characteristic of the white-dwarf stars.

Table 7 gives the dimensions of certain giant and supergiant stars whose angular diameters in seconds of arc actually have been measured by Pease with an instrument called the *stellar interferometer*. This device, in-

Table 6. Mass and luminosity, with derived radius, density, and surface gravity for some dwarf stars. Mass, luminosity, radius, and density are expressed in terms of the sun as 1.0; luminosity is bolometric rather than visual. Surface gravity is given in terms of that at the surface of the earth.

Star[a]		Class[b]	Mass	Luminosity	Radius	Density	Surface gravity
η Cas	A	*G0V*	0.87	1.04	1.0	0.87	24
	B	*M0V*	.54	0.096	0.82	1.0	23
σ_2 Eri	B	*WD A*	.44	.0033	.022	43,000	25,000
	C	*M4V*	.195	.0050	.19	31	120
Sirius	A	*AIV*	2.14	27	2.04	0.25	14
		WD A	1.05	0.0027	0.020	130,000	73,000
Procyon	A	*F5V*	1.78	6.6	2.2	0.18	11
	B		0.65	0.0006			
α Cen	A	*G2V*	1.05	1.38	1.25	0.54	19
	B	*K0V*	0.89	0.70	0.94	1.1	28
ξ Boo	A	*G8V*	.83	.52	.85	1.35	32
		K4V	.725	.103	.42	9.9	115
ζ Herc	A	*G0IV*	1.12	4.9	2.2	0.11	7
		K0V	0.78	0.51	0.89	1.10	27
70 Oph	A	*K0V*	.89	.48	.88	1.30	32
	B	*K6V*	.66	.125	.72	1.75	34
Krueger 60	A	*M3V*	.256	.0077	.26	1.40	103
	B	*M4V*	.16	.0022	.13	82.0	270

[a] Components of a multiple system are denoted by A, B, C in order of brightness.
[b] WD denotes a white dwarf.

Table 7. Characteristics of some giant and supergiant stars whose diameters have been measured.

Star	Spectral class	Distance (light-years)	Diameter (Sun = 1)	Total luminosity[a] (Sun = 1)
α Bootis (Arcturus)	K1III	36	24	130
α Tauri (Aldebaran)	K5III	68	45	360
α Orionis (Betelgeuse)	M2I	600	650–900	85,000
α Scorpii (Antares)	M1I	420	560	28,000
β Pegasi (Scheat)	M2III	200	130	1,300
α Herculis (Ras Algethi)	M5II	500	480	1,600
o Ceti (Mira)	M6III	130	240	2,500

[a] The total or bolometric luminosity refers to the total power output of the star.

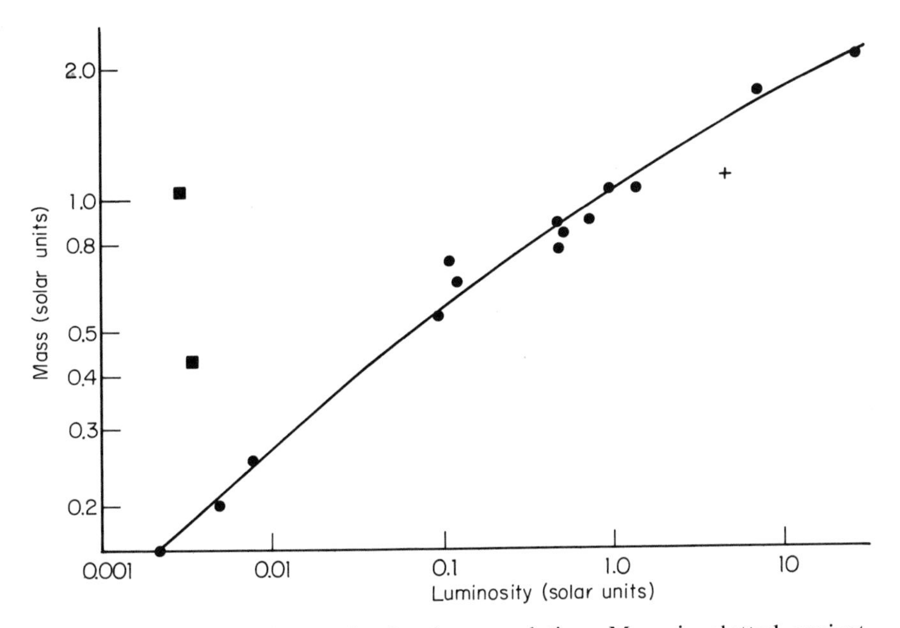

Fig. 55. The empirical mass-luminosity correlation. Mass is plotted against luminosity from data of visual binary stars. Note that the scales are logarithmic, and that the mass scale is 2.5 times the luminosity scale, because the luminosity depends so strongly on the mass.

vented by Michelson, is usable only for bright red giant or supergiant stars (Fig. 56). Fortunately, the "photon correlation" interferometer, due to Hanbury Brown and Twiss, provides complementary data on bright, hot stars such as Sirius or Vega.

If the distance of the star is known, its real diameter can be found at once from the angular diameter. This is the method used for getting the sun's diameter. The angular diameter, 1919.″26 = 0.009305 radian, is accurately measurable. The distance of the sun is also accurately known, 149,600,000 km (92,960,000 miles). Hence the diameter of the sun is 0.009305 × 149,600,000 km = 1,392,000 km (866,000 miles). Stellar angular diameters, however, range downward from 0.″056 and can be measured directly with only relatively small percentage accuracy.

Table 8 gives data for a number of eclipsing stars which have two spectra visible and for which the radii, separations, masses, and densities may be determined. Fortunately, in eclipsing-binary systems the derived dimensions are independent of the distance of the star; they depend only on the accuracy of the orbit.

Table 8. Dimensions and masses of eclipsing-binary systems. (Courtesy D. M. Popper.)

Star	Spectral class	Radius (Sun = 1)	Mass (Sun = 1)	Separation (10^6 km)	Separation (10^{-2} a.u.)	Period (days)
YY Geminorum	M1	0.60	0.58	3.7	1.75	0.81
(Castor C)	M1	.60	.58			
UV Leonis	G2	1.09	1.02	3.7	1.75	.60
	G2	1.05	0.95			
VZ Hydrae	F5	1.25	1.23	11.3	5.3	2.90
	F7	1.05	1.12			
WW Aurigae	A5	1.9	1.81	11.9	5.5	2.52
	A7	1.9	1.75			
RX Herculis	B9	2.4	2.75	10.6	4.9	1.78
	A1	2.0	2.33			
U Ophiuchi	B5	3.4	5.3	12.8	5.9	1.68
	B5	3.1	4.6			
Y Cygni	O9.5	5.9	17.4	28.7	13.3	3.00
	O9.5	5.9	17.2			

Table 9. Elements of some eclipsing binaries.[a]

Star[b]	Spectral class	Period	$R/R_\odot$	$M/M_\odot$	$\rho/\rho_\odot$	Separation ($R_\odot$=1)	$L/L_\odot$
β Persei (Algol)	B8V	$2^{\mathrm{d}}.87$	3.12	4.72	0.16	15.1	
	G4IV		3.68	0.95	.02		
ζ Aurigae	K3	$972^{\mathrm{d}}.16$	205	22.0	2.54×10^{-6}	1340	1400
	B8		3.5	10.2	0.24		220
31 Cygni	K3Ib	3780^{d}	174	18	3.41×10^{-6}	3100	5500
	B3V		4.7	9	0.087		1700
32 Cygni	K5	1148^{d}	195	21	2.82×10^{-6}	1360	2100
	B8		2.5	7.6	0.48		125
VV Cephei	M2Ia	$20^{\mathrm{y}}.4$	1620	80	1.87×10^{-8}	8130	
	B9		<88	40(?)			

[a] The radii, masses, densities, and luminosities are given in terms of the sun; the separation of the two stars of the binary is expressed in terms of the solar radius as unit. The radius of the sun is 695,300 kilometers (432,000 miles), the mass of the sun is 1.983×10^{33} grams (332,000 times the mass of the earth), and the density $\rho_\odot$ of the sun is 1.41 that of water.

[b] The data for Algol are from McLaughlin; for ζ Aurigae from O. C. Wilson (1960); for 31 Cygni from McKellar and Petrie (1958); for 32 Cygni from K. O. Wright (1952); for VV Cephei from B. F. Peery (1966).

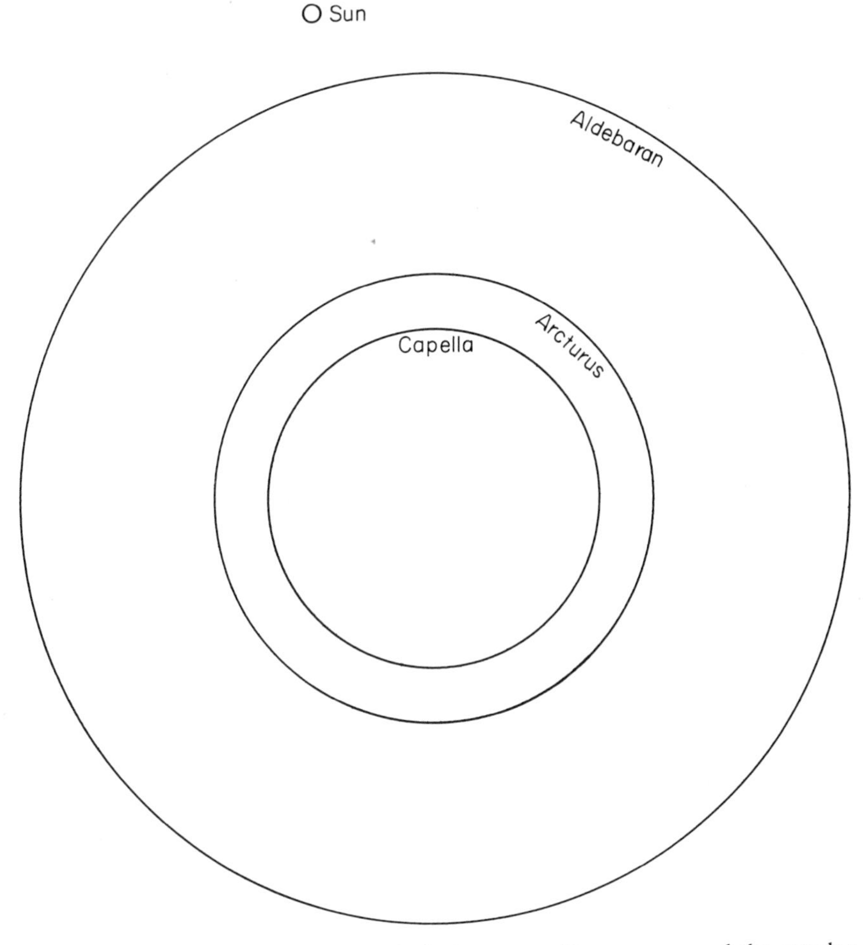

Fig. 56. Relative sizes of some typical giant stars. Pease measured the angular diameters of Aldebaran and Arcturus with the Michelson stellar interferometer. Since the distances of these stars are known reasonably well, their linear dimensions may be computed.

In Table 9 we summarize material for some interesting eclipsing binaries such as ζ Aurigae and VV Cephei. These unusual systems provide information not only about the dimensions of stars but also about the structure of stellar atmospheres.

Along the main sequence, the variation in radius is relatively small (Fig. 57). The stars range from about six solar diameters for the eclipsing system Y Cygni to about one-sixth that of the sun for Krueger 60B the fainter component of the binary Krueger 60.

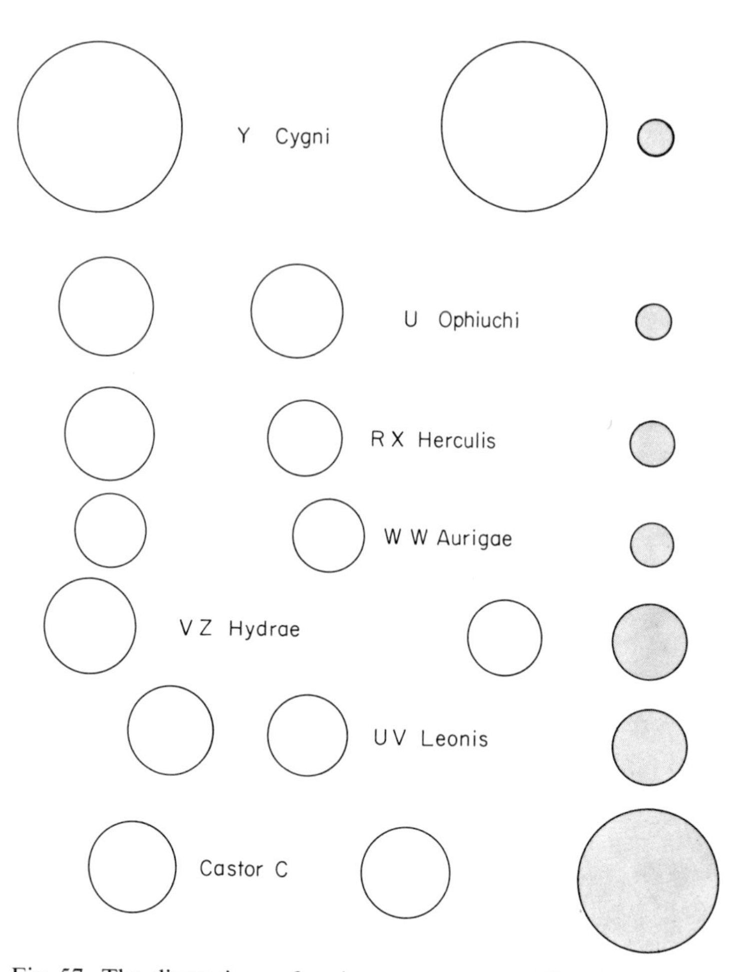

Fig. 57. The dimensions of main-sequence stars. The sizes and separations of typical eclipsing binaries are plotted to scale; in each case the sun is represented by the circle at the right of the diagram.

The supergiant Antares, however, is more than 500 times larger than the sun (Fig. 58). On the other hand, some of the white dwarfs are no bigger than the earth. Large stars tend also to be massive but the range in stellar masses does not begin to approach the range in sizes. Consequently, the stars display a startling variety of densities. The average density of the sun, for example, is somewhat greater than that of water—about equal to that of soft lignite coal—and most main-sequence stars have densities ranging from about 0.1 to about 3 times that of water. The radius of the supergiant Antares is about 560 times and its volume is

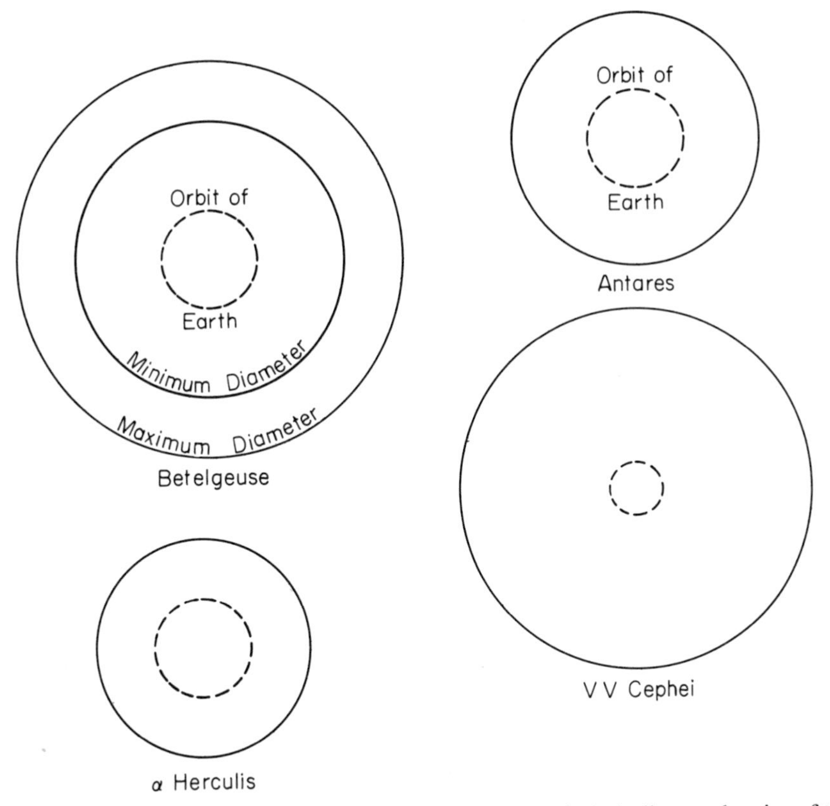

Fig. 58. Supergiant stars. In each case, the broken circle indicates the size of the earth's orbit, whose radius is 93,000,000 miles (149,600,000 kilometers). Betelgeuse varies irregularly in both light emission and radius.

560^3 or 175,000,000 times that of the sun. The mass of Antares probably does not exceed 50 solar masses; hence its average density must be less than one-millionth that of the sun. On the other hand, although the white dwarf o_2 Eridani B has only about 6 millionths the solar volume, its mass is about 44 percent that of the sun, which gives the star the amazing density of about 100,000 times that of water, or nearly 2 tons to the cubic inch! Another white dwarf known as "van Maanen's star," has about 3 ten-millionths the solar volume and a density that may be as high as 7 tons per cubic inch. We shall discuss these remarkable stars later in connection with the problem of stellar evolution.

A number of dwarf binaries are of particular interest. Many of these systems are built on about the same scale as the solar system. α Centauri, our nearest stellar neighbor, consists of two stars of nearly the same mass as the sun, plus a small distant companion, Proxima, 1 fifteen-thousandth as bright as the sun. Krueger 60B is one of the faintest stars whose masses are known, although several less massive "dark stars" (astrometric companions) are known from their gravitational effects on luminous stars. A planet would have to be about 4.5 million miles from Krueger 60B to receive as much light and heat as the earth gets from the sun. A habitable planet associated with van Biesbroeck's star (which is 1 millionth as bright as the sun) would have to be closer to it than half the distance of the moon from the earth. If, however, Arcturus (Fig. 56) were to replace the sun, we would be comfortable just outside the orbit of Saturn (886,-000,000 miles). With Betelgeuse as our central luminary, about nine times the distance of Neptune from the sun would be a good place for our abode. From this vantage point, Betelgeuse would appear about 30 times larger in our sky than does our present sun, but because of its low temperature its apparent total energy emission would equal that of the sun.

Relation Between Spectrum and Luminosity

Fortunately, the spectra of stars supply important clues to their luminosities, so that if a spectrum of sufficient scale is obtained one can make a good estimate of the star's intrinsic brightness. If one compares carefully the spectra of giants or supergiants and dwarfs of the same spectral type, one finds that the two groups of spectra are not exactly alike, as membership in the same spectral class would imply. Although the general features of the spectra may match very well, there will be certain easily recognizable differences. For example, the lines of ionized strontium are not prominent in the sun, but they are extremely intense in the spectrum of ζ Capricorni, a supergiant 6000 times as bright as the sun. Similarly (Fig. 59), a comparison of the spectrum of the supergiant Betelgeuse with that of the dwarf Lalande 21185, both of Class M2, shows that the line of neutral line at 4227 Å is enormously stronger in the latter. These

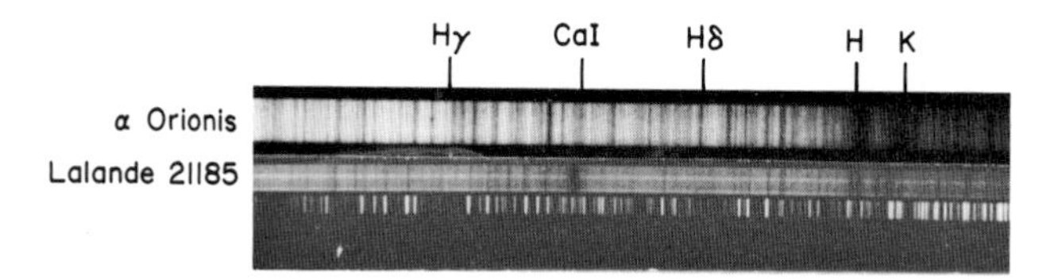

Fig. 59. Comparison of the spectra of the supergiant Betelgeuse (α Orionis) and the dwarf Lalande 21185. Notice that the lines in the supergiant (particularly the hydrogen lines) are generally stronger than in the dwarf, but that Ca I, 4227 Å, is very strong in the dwarf. (Ojai Observing Station, University of California.)

examples illustrate a general rule found by Adams and Kohlschütter, namely, that lines of certain ionized atoms tend to be strong in giant stars and weak in dwarfs, and that lines of certain neutral atoms behave in the opposite sense.

These luminosity or absolute-magnitude effects must be established empirically by a careful comparison of the spectra of stars that are known to differ in luminosity. The effects can be explained, qualitatively at least, in terms of the different physical conditions that prevail in the atmospheres of giant and dwarf stars. The chief difference is one of density. Tables 6, 7, and 8 show that the stellar material of the intrinsically faint dwarfs is relatively closely packed, and that the bright stars are very much more tenuous. The densities that we have given are, of course, average values for the whole body of each star, but the dense dwarfs may also be expected to have shallow, compressed atmospheres, and the tenuous giants rarefied and extended ones. In other words, the density of a star's atmosphere appears to be correlated with its size and luminosity.

We recall from Chapter 4 that the density has an important influence on the appearance of the spectrum. When the density is low, and free electrons are few and far between, atoms are more easily maintained in the ionized condition than when the density is high. Consider, for example, the behavior of an element such as calcium, which exists in both neutral and ionized forms in a great many stars. Neutral calcium atoms absorb a spectral line in the blue-violet region at 4227 Å; ionized calcium produces the well-known *H* and *K* lines in the far violet. Given two stars of the same temperature, one a tenuous giant and the other a dense dwarf, we would expect a greater percentage of calcium atoms to be ionized in the giant. Consequently, the line of neutral calcium should be weak in the giant and strong in the dwarf, whereas the converse should hold true for the lines of ionized calcium.

Now in practice we do not compare the spectra of two stars of the same temperature, but rather two stars of the same spectral appearance. The spectral type is judged from the intensities of the spectral lines and not from the intensity distribution in the continuous background. A given spectral type corresponds therefore to a certain degree of average ionization. High temperature as well as low density favors ionization; a dwarf star is hotter than a giant of the same spectral class, the higher temperature of the dwarf compensating for the lower density of the giant. If this compensation were identical for all elements, the spectroscopic discrimination between giants and dwarfs would be virtually impossible. Fortunately, this is not the case, and for some elements, such as strontium, the ionization is more sensitive to low density than to high temperature. Hence the ionized-strontium lines are strong in giants but weak in dwarfs of the same spectral class. Calcium is another example. The ionization of calcium is more sensitive to the low densities of the atmospheres of dwarfs. Hence the line of neutral calcium at 4227 Å is stronger in dwarfs than in giants of the same spectral class (see Fig. 59).

Ionization differences are only part of the story. The much higher densities prevalent in the atmospheres of dwarf stars produce profound effects of density broadening as discussed in Chapter 5. The 4227 Å calcium line in dwarfs is broadened by collisions. In main-sequence A and B stars the interatomic Stark effect markedly broadens Balmer lines such as Hγ and Hδ and causes the Balmer lines to be more washed out than in supergiant stars (see Fig. 42).

Atmospheres of giants and supergiants are often characterized by large-scale mass motions of the material, sometimes called "turbulence." These motions produce changes in the shapes of spectral lines that are easy to distinguish from the effects of density broadening but not always from the effects of stellar rotation.

Since the pioneer work of Adams and Kohlschütter, many investigators have studied effects of luminosity differences on spectral features. Atomic lines, molecular bands, and even features of the continuous spectrum have all been utilized. In modern work astronomers employ the Morgan-Keenan (M-K) system of spectral classification. The classical Henry Draper system assigned each star a spectral class with no indication of its luminosity. In the M-K system one adds a Roman numeral to indicate a luminosity. The designations extend from I*a* for the brightest supergiants to V for main-sequence stars. Thus Rigel (B8I*a*) is a very luminous supergiant, Betelgeuse (M2I*b*) and Antares (M1I*b*) are also supergiants, β Gruis (M3II) falls between supergiants and giants, Arcturus (K0III) is a giant; a GIV or KIV star would be a subgiant but a B5IV star (such as Achernar), an A1IV star (γ Geminorum), or an F5IV star (Procyon) are bright main-sequence stars. Vega (A0V), the sun (G2V), 61 Cygni AB (K5V and K8V) are dwarf stars.

Luminosity classes can be interpreted in terms of absolute magnitudes only when they are calibrated. Thus we must know the true distances of stars whose spectra we compare. The nearby dwarf stars offer no problems since their distances are easy to determine. Giant and supergiant stars can be handled when they are members of star clusters whose distances can be found.

The Colors of the Stars

With the advent of the photoelectric photometer it became possible to measure the color of a star more accurately than one can establish a spectral class. Furthermore, with a given telescope one can measure magnitudes and colors of much fainter stars than one can observe spectroscopically. Usually, the observer measures three colors with the aid of three different filters. The *U* filter transmits a broad band in the near ultraviolet, the *B* filter the blue region between about 3800 and 5000 Å, and the *V* filter in combination with the photocell gives a color sensitivity sufficiently similar to that of the human eye that photoelectric *V* magnitudes may be regarded as equivalent to visual magnitudes. We shall so treat them in subsequent discussions. With the brightness of a given star measured in all of the *U*, *B*, and *V* filter-photocell combinations, we may set up two different kinds of brightness differences or color indices, U – B and B – V.

By combining the advantages of photoelectric photometry with those of the photographic plate, accurate measurements of magnitudes and colors of large numbers of stars can be obtained efficiently. With a photoelectric photometer it is possible to observe only one star at a time. A photographic plate can record a whole cluster simultaneously. The astronomer measures photoelectrically the magnitudes and colors of a few selected stars over a range of color and brightness and uses these as standards with which to compare perhaps hundreds of other stars photographed on the plate.

Use of colors instead of spectral types has one severe drawback, however. The color (but not the spectral class) of a distant star may be reddened by particles in interstellar space, much as the setting sun is reddened by selective light scattering by molecules in the earth's atmosphere (see Chapter 7). A hot O or B star can appear as yellow as a K star. By measuring three (or preferably more) colors one can often correct for this effect since the brightness of a star in different regions of its spectrum will be affected differently by temperature than by coloring produced by interstellar smog.

Today, most H-R diagrams are constructed from photometric observations. From *U*, *B*, and *V* color measurements, one utilizes plots of $(B-V)$ vs. $(U-V)$ color indices to help assess effects of coloring and absorption by interstellar particles. Then one derives the corrected, or $(B-V)_0$, colors and plots them against *V* (also corrected for space absorption) for members of the cluster or stellar association. One then compares the plot of $(B-V)_0$ vs. V_0 with the standard H-R diagram, which gives $(B-V)_0$ vs. M_v. In this way one gets the *distance modulus* of the cluster, $V-M_v$, and hence its distance (see the following section).

Furthermore, for each value of the true color index $(B-V)_0$, the temperature may be established (see Appendix E). One may also assign for each value of $(B-V)_0$ and of M_v the corresponding spectrum and luminosity class. Hence a plot of $(B-V)_0$ vs. M_v can be converted to a spectrum-luminosity diagram. We can go further than that. For each luminosity class and temperature we can derive the bolometric correction needed to convert a visual absolute magnitude to a bolometric magnitude and hence obtain the true luminosity of the star (see Appendix F). Finally we can plot surface temperature against true luminosity, thus obtaining a relation that can be compared directly with the predictions of theory (see Figs. 60 and 61). As we shall see in Chapter 9, the theory of stellar evolution predicts for a star of given mass, rotation, and chemical composition the variation of its luminosity and radius with time. Since the luminosity depends on the surface area and temperature, a diagram giving luminosity and temperature can be converted to one giving luminosity and radius, or alternatively radius and luminosity can be converted to surface temperature and luminosity.

Let us summarize the steps needed to convert an observed color diagram of a star cluster to a meaningful array (since all members of the cluster are at the same distance from us, differences in apparent magnitude correspond to the same differences in absolute magnitude):

(1) Given the U, B, V color measurements for a cluster we compute the indices $(U-B)$ and $(B-V)$ for each star;

(2) Using the known effects of space absorption on colors and magnitudes, we convert to $(U-B)_0$, $(B-V)_0$, and V_0;

(3) We compare the cluster H-R diagram, namely, $(B-V)_0$ plotted against V_0 with the standard H-R diagram derived for similar kinds of stars to obtain the distance modulus, $V_0 - M_v$. The distance modulus, corrected for space absorption, is related to the distance by the equation $V_0 - M_v = 5 \log r - 5$, where r is the distance in parsecs.

The color of a star can be affected by factors other than temperature, luminosity, and space absorption. Not only the spectrum of a star but also its color can be affected by its chemical composition and to some extent by mass motions (turbulence) in its atmosphere. Now a color measurement, unlike a spectrogram, takes a bite of radiation over a big range in wavelength. If the star has numerous strong lines in this region, the energy received by the photocell will be diminished. If the lines are weak because of a low ratio of metal to hydrogen, the measured colors will be affected. Likewise turbulence effects, by modifying the amount of energy removed from the spectrum, can affect colors in the same way as would an increase in the metal-to-hydrogen ratio.

Although the *U*, *B*, *V* color system introduced by Harold Johnson has been of inestimable value in astronomy, many other photoelectric color combinations are extremely useful. We mention, for example, the six-color photometry of Stebbins and Whitford and more recently of Stebbins and Kron (which covers a wide range of wavelengths) and the system used by Strömgren. In the latter, a judicious selection of filters and photocells enables one to measure both narrow spectral intervals (a few tens of angstroms) and wide ones (100 Å or more), so chosen that by a suitable combination of the measurements one can obtain for each star its spectral class, luminosity class, and metal-to-hydrogen ratio, and the influence of space absorption.

Star Clusters and Associations

The power of the H-R diagram as a tool in astronomical research is exhibited in studies of aggregates of stars and particularly of star clusters.

There are two types of star clusters—the "open" or galactic clusters and the globular clusters. They differ rather fundamentally from one another, not only with respect to size, distribution in space, and numbers of stars but also with respect to the kinds of stars involved.

Galactic clusters are found mostly near the plane of the Milky Way. They may contain only a handful of stars, as in the Ursa Major cluster, or hundreds or even a couple of thousand stars, as in *h* and χ Persei. A few galactic clusters—the Pleiades, the Hyades, Praesepe, and Coma Berenices—are visible to the unaided eye, and many more can be seen with a field glass or small telescope. Some, like the cluster NGC 2244 in Monoceros or the cluster in Messier 8, have associated with them clouds of dust and gas. In others, there is no longer any trace of dust remaining. Their diameters range up to a few light-years. Sometimes the stars are so thinly spread that the cluster can scarcely be recognized against the background. When we consider the kinds of stars involved, or more particularly the individual H-R diagrams, we note a wide variety among open clusters.

From spectrum-magnitude plots constructed with apparent photographic magnitudes and spectra, Trumpler had discovered a great diversity among galactic clusters. Some, like the Pleiades, had only a main sequence starting around B3 and continuing on to fainter stars. Others, like the Hyades or Praesepe, had no B stars, but possessed a few giants; the main sequence began near A or F. Class B stars and giants seemed mutually exclusive in galactic clusters.

A great step forward in the study of star clusters came with the development of accurate techniques for the measurement of stellar colors and magnitudes. Conventional color-magnitude arrays are plotted with $B-V$ (corrected for space absorption) against V. In Fig. 60, however, we have plotted absolute visual magnitude against spectral type rather than color, giving the results in schematic form for seven galactic clusters; Fig. 61 shows absolute bolometric magnitude plotted against effective temperature. In NGC 2362 and h and χ Persei the main sequence extends to very luminous stars. The latter has also a few red supergiants. Notice that in the Pleiades the bright end of the main sequence veers to the right, that is, redward of the main sequence defined by h and χ Persei and other galactic clusters containing highly luminous stars. The main sequences of M 11, Praesepe, NGC 752, or the Hyades start with A or F stars; there is a gap between them and a handful of red giants. In NGC 188 the main sequence does not extend to bluer or brighter stars than G stars like the sun, but the "giant" branch is continuous with the main sequence.

We can define a sort of limiting main sequence that corresponds to that of NGC 2362. The younger the cluster, the more closely do the main-sequence stars adhere to this limiting main sequence, which is referred to as the "zero-age main sequence."

Before we elaborate further on these observations, let us discuss briefly the globular clusters. Whereas there are probably thousands of star groups in the Galaxy that can qualify as open clusters, the number of known globular clusters is only about 100. Most of them are concentrated toward the central bulge of the Galaxy but they pay no attention to the galactic plane. A typical globular cluster will have a diameter of 40 or 50 light-years and may contain upward of 100,000 stars.

It is in stellar content, however, that the globular clusters differ most strikingly from the galactic clusters.

In 1944 Baade called attention to the existence of two stellar population types. Population I is associated with spiral arms and particularly with clouds of gas and dust grains such as are found in the region of Orion. This population contains bright supergiants and main-sequence O and B stars. Population I stars are prominent also in the Large Magellanic Cloud, in the spiral arms of the Andromeda spiral M 31, and throughout most of the Triangulum spiral M 33. Population II contains no blue main-

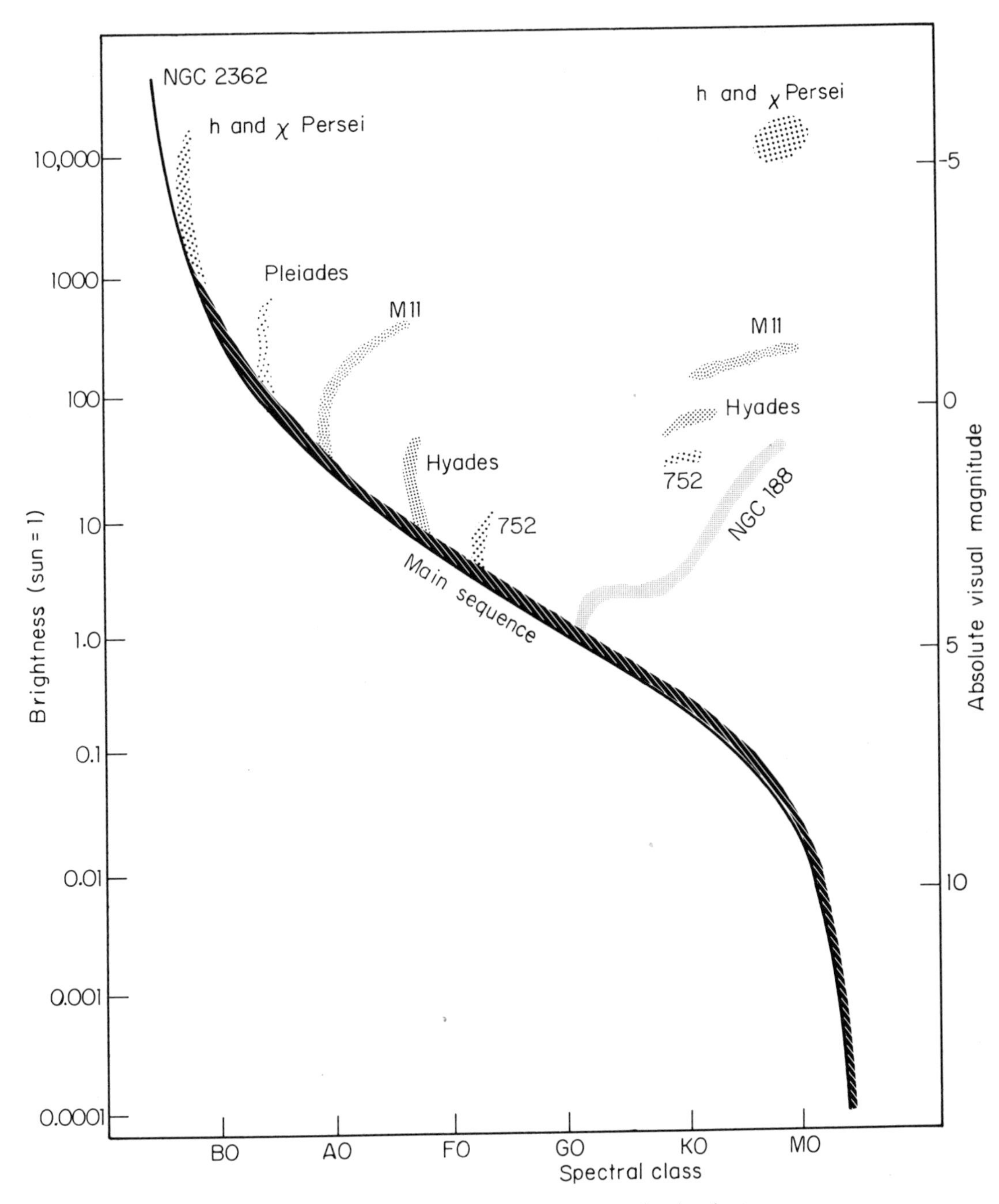

Fig. 60. Schematic Hertzsprung-Russell diagrams for seven galactic clusters; absolute visual magnitude is plotted against spectral class. The heavy line that marks the lower boundary of the main sequence defines what is called the zero-age main sequence, ZAMS.

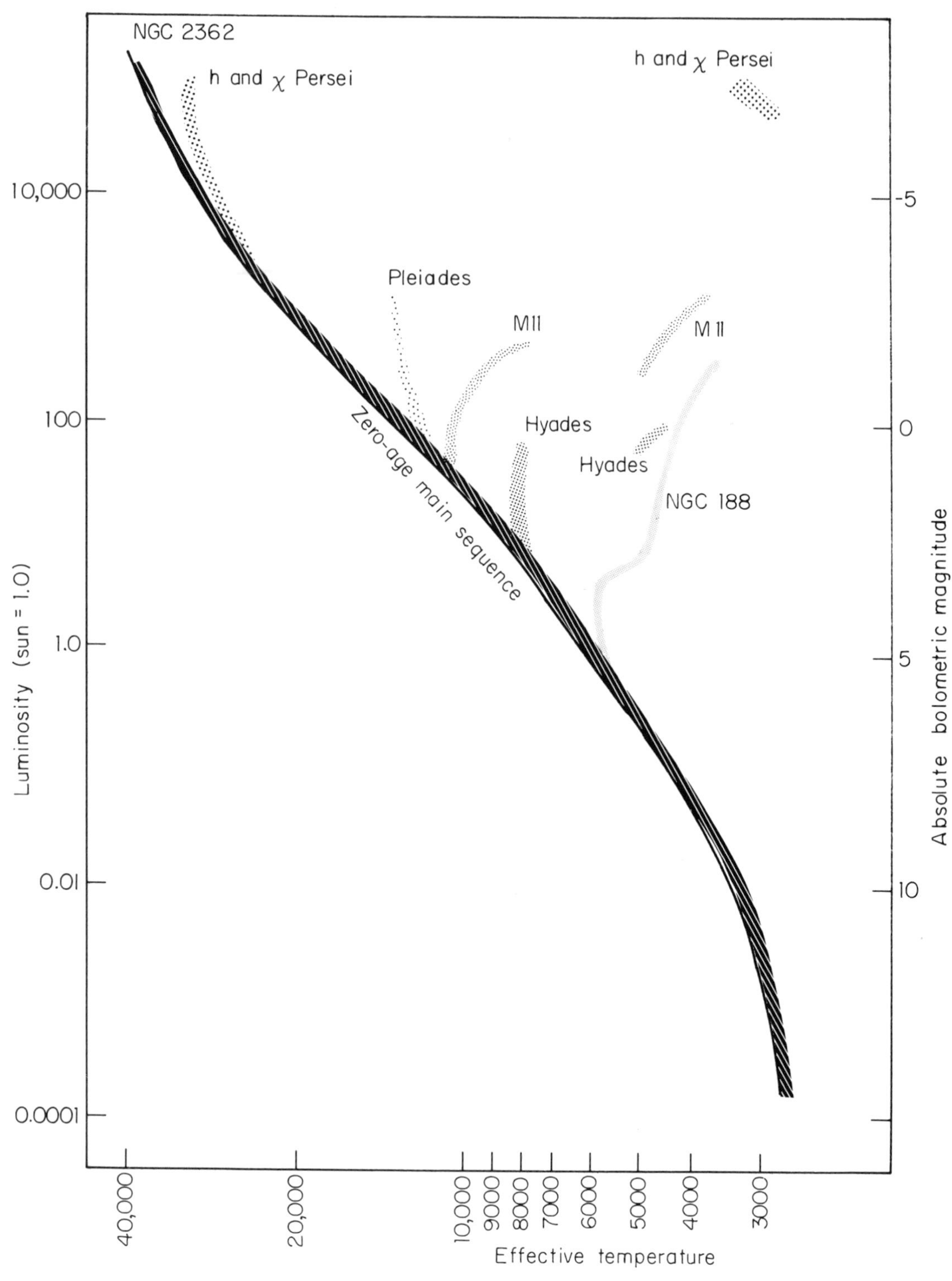

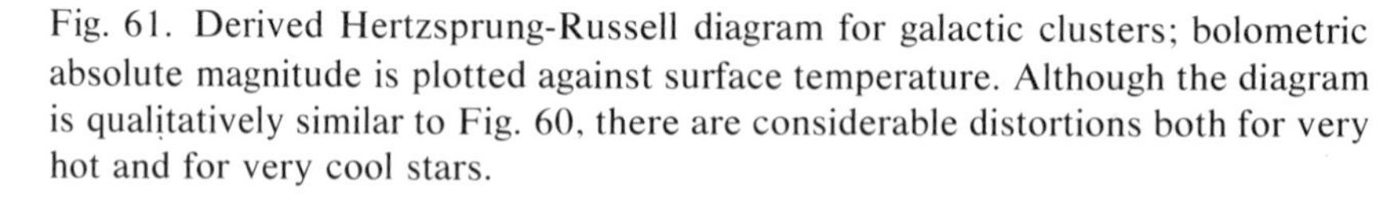
Fig. 61. Derived Hertzsprung-Russell diagram for galactic clusters; bolometric absolute magnitude is plotted against surface temperature. Although the diagram is qualitatively similar to Fig. 60, there are considerable distortions both for very hot and for very cool stars.

sequence stars. The main sequence breaks off in spectral class F and is joined to a giant branch which reaches an absolute magnitude as bright as −2. Population II stars are characteristic of the central bulge of our Galaxy, the thin halo of stars enveloping our galactic system, the elliptical galaxies such as the companions to Andromeda, and the giant galaxy M 87 (NGC 4486) in Virgo, which greatly exceeds our own stellar system in size. Population II stars are associated with very little dust and gas, and the relatively few high-temperature stars that exist in this population do not tend to fall on the main sequence.

Data on stellar motions are sometimes useful for separating the two populations. Our Milky Way system is in rotation; the sun is traveling about the center in an orbit that is probably not far from circular with a velocity of about 220 kilometers per second, completing a revolution in a period exceeding 200,000,000 years. Population I objects moving in similar nearly circular orbits will appear to have relatively low velocities as observed from the earth. Consider now stars moving in greatly elongated orbits (thin ellipses) about the galactic center. As they cross the orbit of the sun their motion will be almost entirely inward or outward, so that with respect to the sun they will appear as high-velocity objects. These stars belong to Population II; they are associated with distant regions of our Galaxy. Often they travel to large distances above the galactic plane or to the central bulge. All high-velocity stars belong to Population II, but not all members of Population II necessarily have high velocities.

More particularly, the globular-cluster stars are the prototypes of Baade's Population II, whereas the stars found in open clusters represent the prototypes of Population I.

Color-magnitude diagrams for some typical globular clusters are shown in Figs. 62 and 63. There are virtually no main-sequence stars brighter than $M_v = 3.7$, that is, more than about one magnitude brighter than the sun. The giant sequence is joined to the main sequence by an almost vertical bridge and there is sometimes a narrow branch of white and blue stars about absolute magnitude 0. When color-magnitude arrays of globular clusters are compared, significant differences are noted. For example, in M 92 the vertical and giant branches are shifted to the blue as compared with M13, NGC 4712, and 47 Tucanae. These systematic displacements of the giant and subgiant sequences seem to be associated with the metal-to-hydrogen ratio, as we shall now explain.

The spectra of the brighter stars in many galactic clusters can be investigated in considerable detail, but individual stars in globular clusters

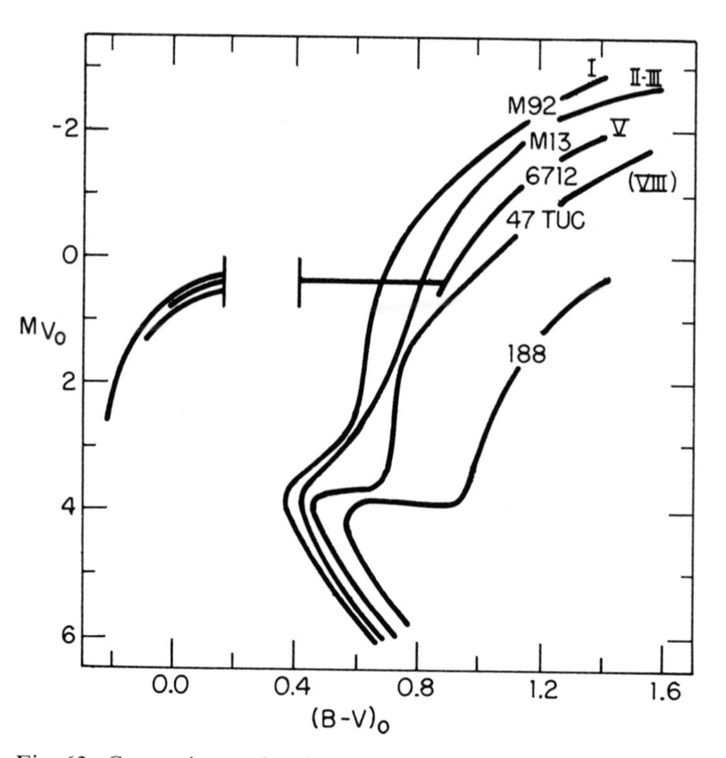

Fig. 62. Comparison of color-magnitude diagrams for globular clusters; the colors and the magnitudes are corrected for space absorption and for the distances of the clusters. The color-magnitude curve for the galactic cluster NGC 188 as determined by Sandage is shown for comparison. The curves for M 13 (H. C. Arp and H. L. Johnson, A. Sandage), NGC 6712 (A. Sandage and Lewis Smith), and 47 Tucanae (Wildy, Tifft) show the effects of differing metal contents. The Roman numerals refer to Morgan's spectral groups. The gap at M_v = +0.5 $(B - V)_0$ = 0.2 to 0.4 indicates the position of the RR Lyrae variable stars. (Courtesy Allan Sandage and Lewis L. Smith, *Astrophysical Journal 144* (1966), 890; copyright University of Chicago Press.)

can be studied only with difficulty. Following the pioneer work of D. M. Popper on ω Centauri, examination of various clusters with the 200-inch telescope showed that the spectra of stars of a given intrinsic brightness in different globular clusters differed not only from those of stars near the sun but also from one cluster to another. The differences were in the sense that the metal-to-hydrogen ratio often was smaller in globular-cluster stars than in the sun. This metal deficiency was more marked in some clusters than in others, ranging from a virtually "normal" (that is, solar) composition to a metal-to-hydrogen ratio a hundred times lower than the solar value.

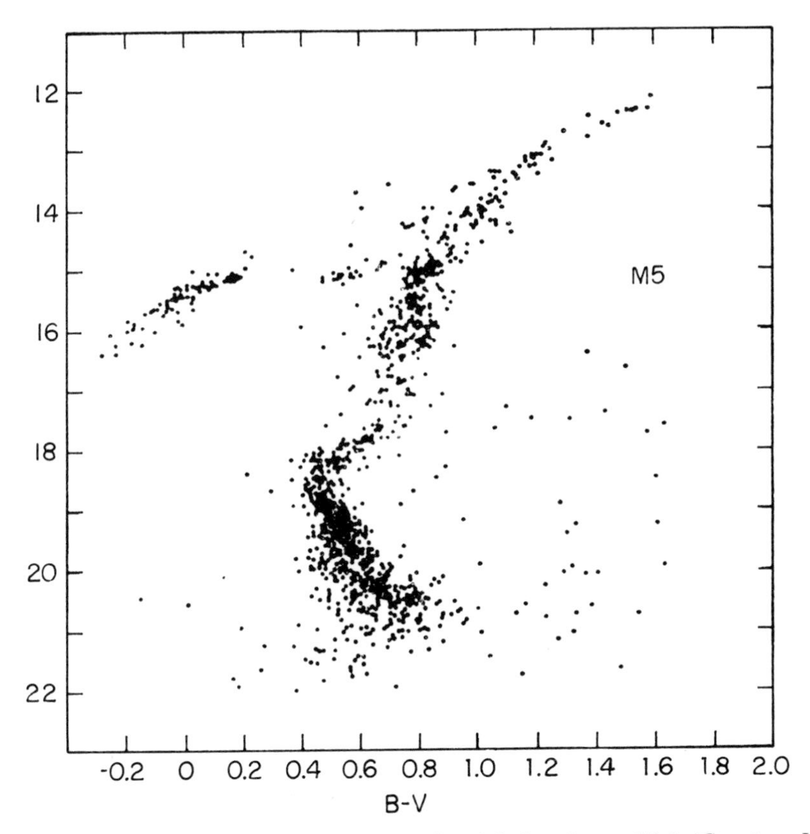

Fig. 63. Color-magnitude diagram for the globular cluster M 5. (Courtesy H. C. Arp, *Astrophysical Journal 135* (1962), 311; copyright University of Chicago Press.)

Stars of similar metal-deficient composition, the so-called subdwarfs, have been recognized in the neighborhood of the sun. When the metal-to-hydrogen ratio is a hundred times lower than in the sun (as in the star HD 140283), a star of solar temperature (5800°K) may show such weak metallic and strong hydrogen lines as to invite classification as an A star (T = 8000–9000°K). Considered as an A star it falls well below the main sequence; properly considered as a G star it will fall near the main sequence. Hence the misnomer "subdwarf"; such objects should be called "metal-deficient dwarfs."

Although detailed abundance studies require spectrograms secured with high dispersion, it is fortunate that metal-deficient stars can be recognized from their colors. The spectrum of a yellow star like the sun is terribly crowded with metallic lines below 4000 Å. Hence it is much fainter when it is observed with a U rather than with a B filter. A metal-deficient star has fewer and weaker lines in this region, with the result that the difference between magnitudes measured with U and with B filters is smaller. That is, metal-deficient stars of about the sun's temperature are brighter in the near ultraviolet than are stars of the same B (blue) magnitude that have normal metal abundances. Also the $B-V$ color indices are affected.

Thus, differences in the color-magnitude arrays for clusters such as 47 Tucanae and M 92 arise largely from differences in the metal-to-hydrogen ratio. These differences enter in a subtle and complicated form. Not only does the relation between the $U-B$ and $B-V$ color indices and the effective temperature depend on the metal-to-hydrogen ratio but the actual internal structure of the star may be likewise affected (see Chapter 8).

Effects of metal deficiencies can also be recognized from a comparison of estimated spectral types with measurements of energy distributions, even for star clusters. Since the composite light from the cluster consists of the superposed effects of many individual stars, neither its spectrum nor its energy distribution will correspond exactly to that of any star. Yet, if the cluster's composition is "normal," the energy distribution will agree approximately with the spectral class. If the cluster is metal deficient, the metallic spectrum will be too weak and the observer will classify it as too "early" (that is, corresponding to too high a temperature), but the energy distribution will correspond to its true temperature. Thus 47 Tucanae, which has a nearly normal composition, has an energy distribution corresponding to its spectral class. In metal-deficient ω Centauri, the published spectrum is too early.

Interpretation of Cluster Color-Magnitude Diagrams

We have discussed in some detail the interpretation of disparities between the color-magnitude diagrams for different globular clusters in terms of differences in chemical composition. Can any interpretation be placed on the shapes of the color-magnitude arrays themselves? Indeed it can! These curves can all be explained in terms of stellar evolution (see Chapter 9).

The concepts can be grasped more easily by looking at the H-R diagram for galactic clusters from the following point of view. Star clusters represent groups of stars of differing masses that were formed at about the same time. Suppose that the masses range from 10 solar masses to 0.1 solar mass, and that all the stars initially were shining on the main sequence. Then their luminosities would be correlated with their masses, but the more massive the star the more rapidly it would liberate energy (recall Fig. 55). Each gram of the most massive stars in our group would liberate energy about a hundred times as fast as a gram of the sun.

Now let us assume that the total amount of energy that can be squeezed out of each gram is the same for all matter everywhere. The more massive stars will exhaust their fuel more quickly, leave the main sequence, and eventually disappear. In this interpretation the cluster NGC 2362 is the youngest and in h and χ Persei the brightest stars are beginning to depart from the main sequence. As we proceed to the Pleiades, M 11, the Hyades and Praesepe, NGC 752, and finally NGC 188, the main sequence is "rolled back." We interpret these as increasingly older clusters in which the brightest, most profligate stars have used up their energy resources and disappeared, at least from the main sequence. We might speculate further and guess that giants and supergiants were stars that had evolved from the main sequence and that white dwarfs represented the final stage of evolution.

One important implication must be mentioned. Although we have not assigned ages to any of these clusters, it is quite clear that NGC 188, which contains no main-sequence stars bluer than the sun, must be much older than, say, the Pleiades, which in turn is older than h and χ Persei. Therefore, star formation must have been occurring in a more or less continuous fashion; presumably it is going on right now. A significant clue is that all these clusters belong to Population I—which is associated with dust and gas of the interstellar medium. To this topic we now turn our attention.

7 Gaseous Nebulae and the Interstellar Medium

That stars are continuously being formed is indicated by the existence of objects such as Canopus, Rigel, ζ Puppis, and γ Velorum. These luminous stars are pouring out energy at such a rate that they must have lifetimes measured in mere millions of years; some of them are younger than the race of man! Such bright blue and supergiant stars are found also in external galaxies such as the Magellanic Clouds and the Triangulum and Andromeda spirals. It is significant that all these stellar systems contain great clouds of dust and gas. Another clue to the possible locale of star formation is groups of highly luminous stars, known as stellar associations. If the velocities of these stars are carefully measured, it is found that all of them appear to have emerged from a small volume of space a few million or hundreds of thousands of years ago.

A glance at any ordinary photograph of the Milky Way shows that space is not empty. The bright patches of nebulosity, the great rift in Cygnus, and the inkiness of the Coal Sack all testify to the reality of the great clouds of occulting matter that fill up "empty" space. The reader may judge for himself by inspecting the famous nebula around η Carinae, shown in Fig. 64. He will see that the dark lanes are not holes through which we look into empty space but rather clouds of some material, prob-

Fig. 64. The η Carinae nebula, photographed at the Mount Stromlo Observatory.

ably a mixture of fine dust and gas, that obscures the light of background stars.

Before discussing the nature and quantity of the material in space, we find it interesting to note that an estimate of the total mass of the interstellar cloud may be obtained from a study of the motions of stars in the solar neighborhood. According to the rotation theory of the Milky Way, the stars revolve in its plane about a galactic center located in the direction of Sagittarius. If all the mass of the Galaxy were confined to a small region near the center, the motions of stars distant from the center might be restricted purely to orbits in the galactic plane, in analogy with the revolution of planets about the sun. The sun, however, is 25,000 or 30,000 light-years from the center of the Galaxy, and the motions of the stars in its neighborhood are governed not only by the distant and massive galactic nucleus, but also by the total amount of matter in our immediate vicinity. One effect of this surrounding material is to set stars oscillating to and fro in a direction perpendicular to the plane of the Milky Way. Oort's studies of the speed of this oscillatory motion indicated that the total density of matter in the solar vicinity is about 6×10^{-24} gram per cubic centimeter. The known stars contribute about half of this figure, which leaves a density of about 3×10^{-24} gram per cubic centimeter to be accounted for by dust and gas in our immediate celestial neighborhood.

An important characteristic of the interstellar medium is its extreme patchiness. Not only are the dust and gas concentrated in the spiral arms, but in the arms themselves the material tends to collect in features ranging from globules a few times the size of the solar system to vast clouds 100 light-years or more across. The density within these clouds may be 10 or 100 times greater than the density in intervening space.

Throughout most of its volume, far away from bright hot stars, the gas is cold—150°C below the melting point of ice or even colder. In the neighborhood of one or more bright hot stars the material becomes heated and largely ionized, to produce such features as the Orion or η Carinae nebulae or the 30 Doradus nebula in the Large Magellanic Cloud. What can we learn about the physical state and chemical composition of the interstellar medium from a study of these nebulae? From careful measurements of the spectrum of a nebula we can obtain information on the density, chemical composition, temperature, and motions of the gas. The problem is analogous to that described earlier in connection with the interpretation of the spectra of stars, except that here one deals with a bright-line spectrum, an extremely rarefied gas, and extremely low levels of radiation intensity.

Diffuse nebulae such as that in Orion, η Carinae, the Trifid, M 8, or 30 Doradus are complicated irregular structures in which gas and dust are intimately associated. Hence, in order to understand how the astronomer analyzes the spectra of gaseous nebulae, it is better to turn to less complicated examples in which occur fewer intricacies of the type that arise when a hot star appears near a chance agglomeration of dust and gas. A seemingly simpler type is the planetary nebulae, so called because they often show small disks similar in appearance to Uranus and Neptune. The name has been retained although these objects have nothing in common with the planets of our solar system.

Many planetaries are symmetrical and all appear to contain very hot stars near their centers. These stars are in advanced stages of their evolution and seem to be evolving into white dwarfs after having ejected their atmospheric layer into space. Thus they differ fundamentally from the bright blue stars in Orion, which are young objects, only recently formed from the interstellar medium.

The hot central stars or "nuclei" of planetary nebulae emit great quantities of ultraviolet radiation, which after absorption by gases in the nebulous shell produces the observed light emission. The physical processes involved have been widely studied in efforts to analyze and interpret the somewhat bizarre spectra of planetary and other gaseous nebulae. Planetaries are rare; intensive searches extending to great distances in the Galaxy have turned up only several hundred (as compared with hundreds of millions of stars). They are found in greater numbers toward the center of our Galaxy, where the star density is, of course, greater. Planetaries are by no means uniform in size, but a typical bright one has a diameter exceeding 10,000 times the distance of the earth from the sun; yet its total mass is probably not much more than one-fifth that of the sun.

Recent work has tended to substantiate the conclusion, reached some decades ago by Bowen and Wyse at the Lick Observatory, that the chemical compositions of many planetary nebulae do not differ markedly from those of ordinary stars like the sun, although some, such as the planetary in the globular cluster M 15 (which was studied by O'Dell, Peimbert, and Kinman), may share deficiencies of oxygen, neon, and possibly other elements as well. Hydrogen and helium are always the most abundant gases. Other permanent gases, such as oxygen or nitrogen, are normally rather abundant and metals are present in small amounts. Neon (a rare gas terrestrially) is often abundant although perhaps not as much as it is in diffuse gaseous nebulae like that in Orion.

Some idea of the nature and tenuity of a planetary nebula may be gained from the following illustration. Imagine an ordinary drinking glass filled with hydrogen gas at room temperature and atmospheric pressure. Add half a thimbleful of ordinary air and a few dust particles to provide some metallic atoms and other impurities. Now seal the glass and allow it to expand until it is as tall as Mount Everest and some 2 miles across. The vastly expanded contents of the glass would then be fairly comparable in density and composition to the gas of a planetary nebula. Diffuse galactic nebulae are often 10 or 100 times more rarefied than a typical planetary nebula. It is only because gaseous nebulae are often so extremely large that their light is perceptible to the astronomer.

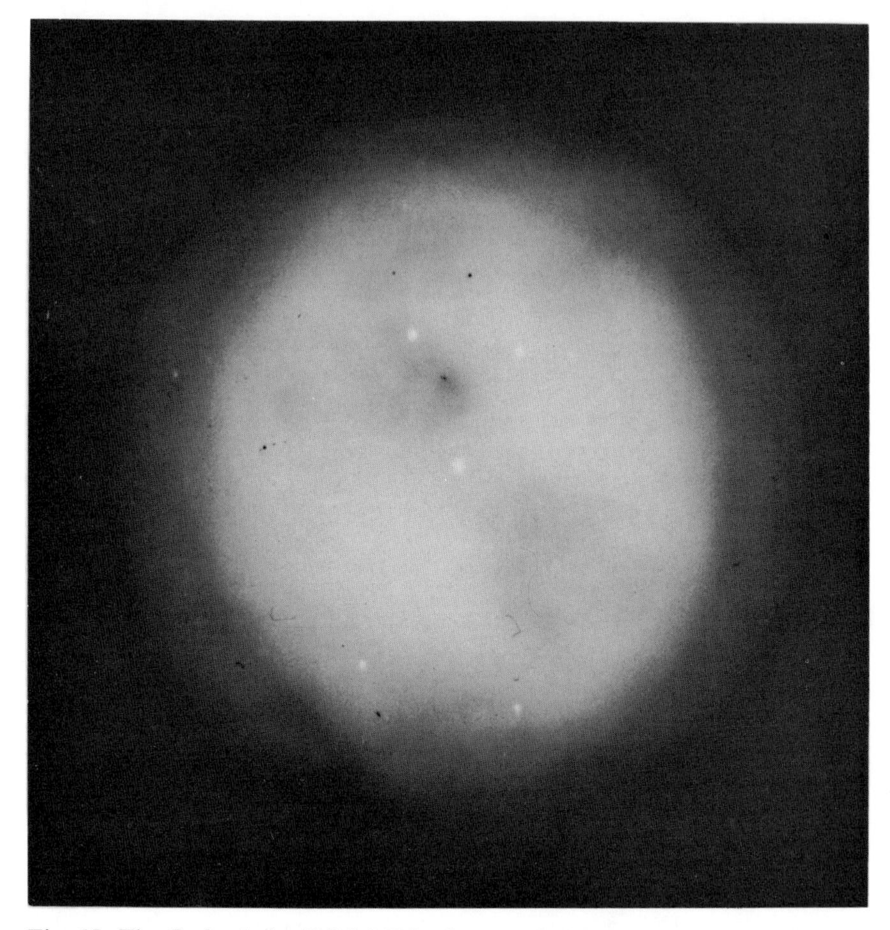

Fig. 65. The Owl nebula, NGC 3587, photographed by R. Minkowski with the 200-inch telescope. Compare the smooth appearance of this nebula with that of NGC 7662, in Fig. 66.

Observing Gaseous Nebulae

Very early spectroscopic work demonstrated that the radiation of gaseous nebulae was concentrated in discrete bright lines with a relatively weak continuum that is difficult to observe in all except the very brightest objects. (Diffuse galactic nebulae such as η Carinae or the Orion nebula may show a continuum that is partly due to background stars and light scattered by "dust" particles or solid grains.)

Ordinary direct photographs tell only part of the story, since images of several different ions, for example, ionized nitrogen and hydrogen, may overlap. Hence photographs with ordinary plate and filter combinations are "composites" representing contributions of radiations of several different ions.

By use of a very narrow-band-pass filter, that is, one that lets through radiation over only a very narrow range of color, it is sometimes possible to isolate the radiation of one or two strong lines. In this way, Minkowski has secured some striking photographs of various planetary nebulae (Figs. 65–67). A similar technique has been used to observe rarefied diffuse nebulae of very low surface brightness. With red-sensitive plates and narrow-band-pass filters whose transmission is a maximum for the red hydrogen line Hα, 6563 Å, it is possible to observe many faint nebulae in the Milky Way. A control observation obtained with a yellow filter and yellow-sensitive plate excludes all nebular radiation and records only background stars and starlight scattered by interstellar grains.

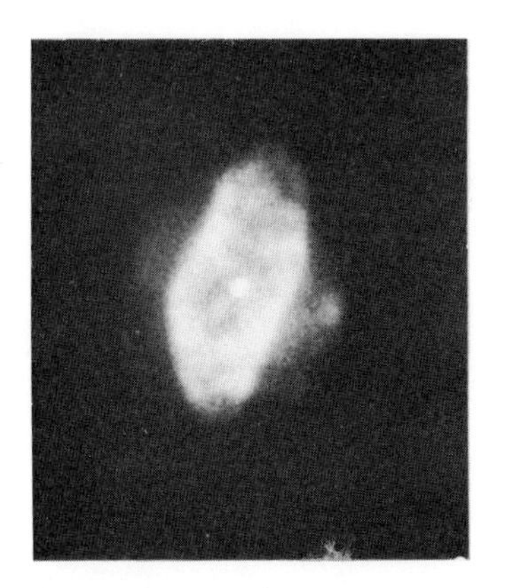

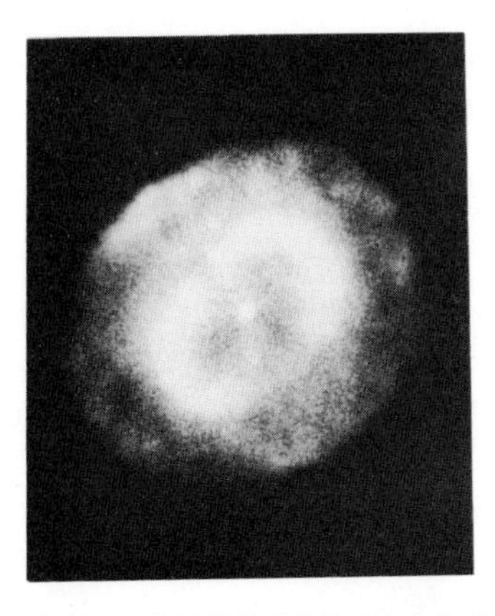

Fig. 66. The ring nebulae NGC 7009 (*left*) and NGC 7662 (*right*). Notice the central star and the bright inner rings. The outer rings are fainter and in NGC 7662 are broken up into a number of separate knots or condensations. (Courtesy R. Minkowski.)

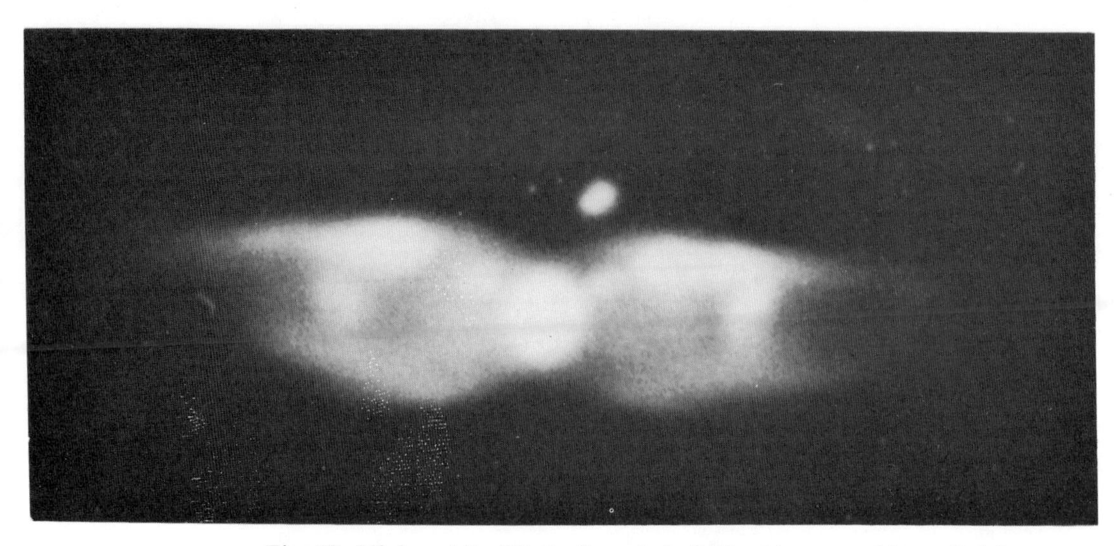

Fig. 67. Minkowski's "Butterfly nebula." The bizarre and beautiful forms exhibited by planetary nebulae are nowhere more aesthetically exhibited than in the nebula MHα 362–1. (Courtesy, R. Minkowski.)

Yet another method, practical only for nebulae of small angular size such as planetaries, is to remove the spectrograph slit and photograph the resultant "slitless" spectrum of the nebula. Figure 68 shows images of the small ring nebula IC 418 formed in this way in some of the strongest radiations of the nebula. Note that the image (*a*) due to the line of doubly ionized oxygen [O III] at 5007 Å is much smaller than the closely overlapping images (*f*) from the lines at 3726 and 3729 Å of [O II]. The radiation from doubly ionized neon (*e*) is closely confined to the center, whereas that of singly ionized nitrogen (*g*) is concentrated in the outer ring.

For most purposes, however, one employs a slit spectrograph. Since most spectral lines of interest represent a very low surface brightness, it is necessary to employ very fast spectrographs or to use electronic image converters. Figure 69 compares the spectra of several nebulae, photographed with the nebular spectrograph at the prime focus of the Lick Observatory 120-inch reflector. Figure 70 shows several spectra obtained at the coudé focus with the Lallemand electronic camera.

Quantitative work on nebulae requires accurate measurements of their surface brightnesses in their characteristic radiation. For this purpose the astronomer employs a photoelectric spectrum scanner—a spectrograph in which the photographic plate is replaced by a photocell behind a narrow slit.

Extended gaseous nebulae and the brighter planetaries can be observed in radiation with wavelengths such as 4, 10, 20, 80 centimeters, that is, short radio waves. Most gaseous nebulae, planetaries in particular, radiate as theory predicts that a hot ionized gas should. There are, however, some

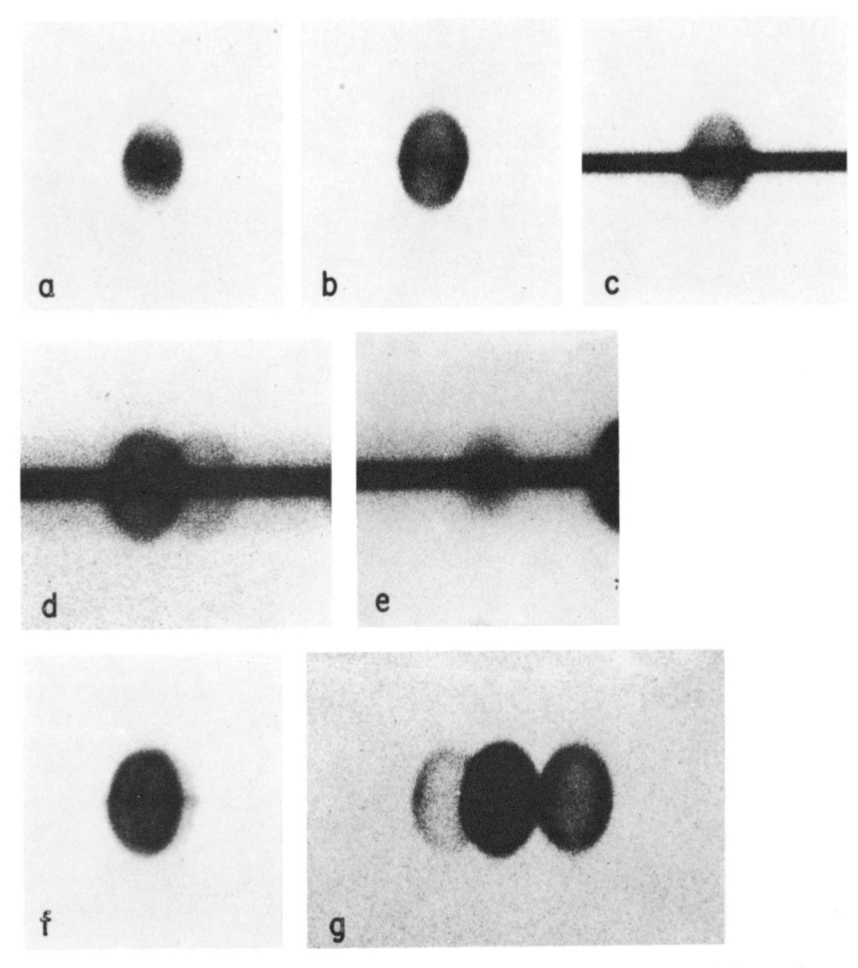

Fig. 68. Slitless-spectrograph images of IC 418 (negative prints). These photographs were obtained by using the coudé spectrograph of the 100-inch telescope in a slitless form. Each spectral line is replaced by an image of the nebula in that particular radiation. The horizontal streak in images (*c*), (*d*), and (*e*) is the image of the central star. (*a*) 5007 Å, [O III]; (*b*) 4861 Å, Hα; (*c*) 4471 Å, He I; (*d*) 4068, 4076 Å, [S II]; (*e*) 3868 Å, [Ne III]; (*f*) 3726, 3728 Å, [O II]; (*g*) 6548, 6584 Å, [N II]; the central (burned-out) image is due to Hα at 6563 Å. (Courtesy O. C. Wilson.)

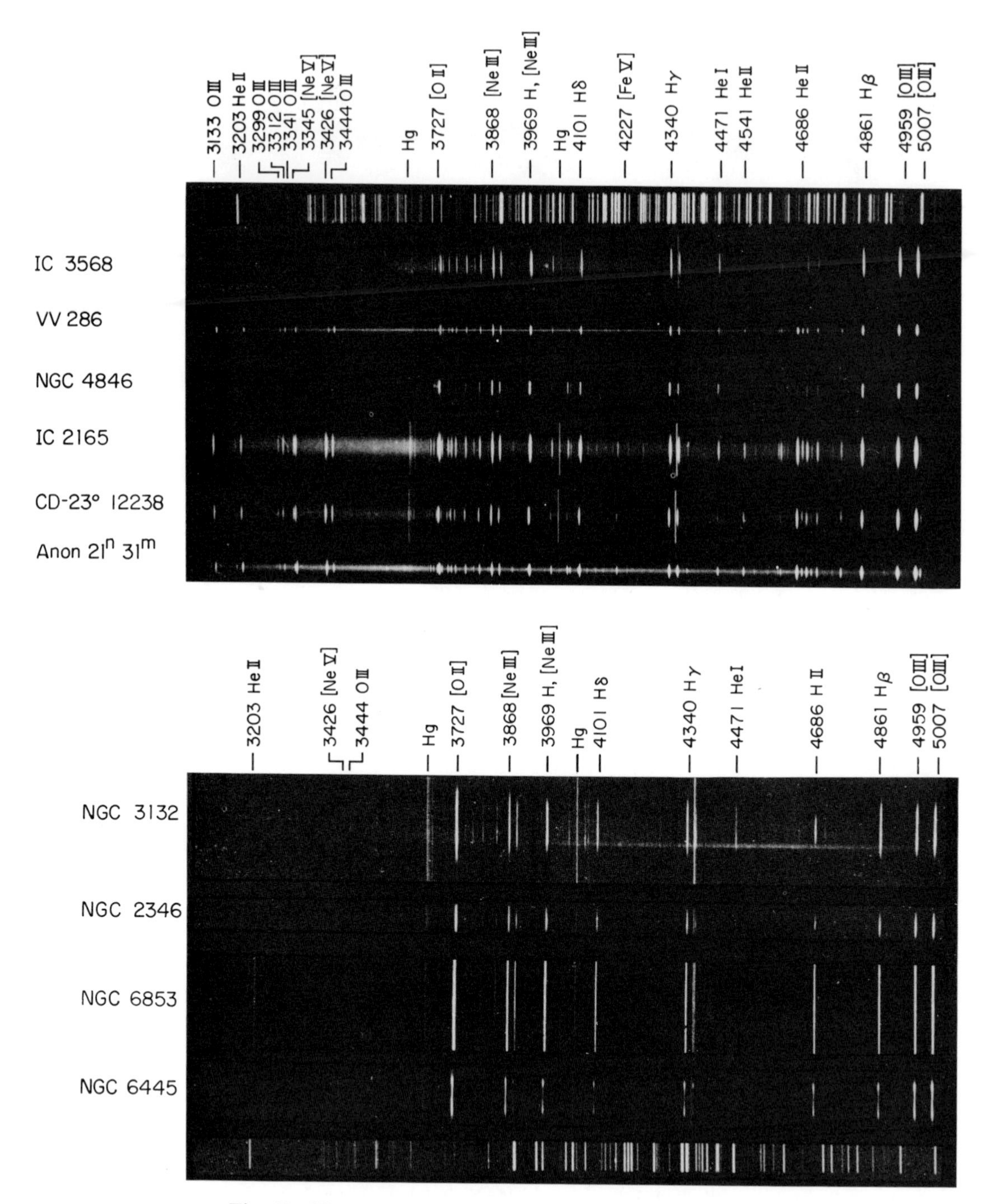

Fig. 69. Slit spectra of planetary nebulae. The principal nebular emissions are marked. Notice the high intensity of the green lines of doubly ionized oxygen at 5007 and 4959 Å and the continous spectrum at the head of the Balmer series. The Balmer continuum, which is especially strong in IC 2165, starts just to the left of the [O II] line at 3727 Å. The lines marked Hg fall on top of this continuum; they arise from mercury-vapor lamps. (Photographed with the nebular spectrograph at the 120-inch telescope, Lick Observatory, University of California.)

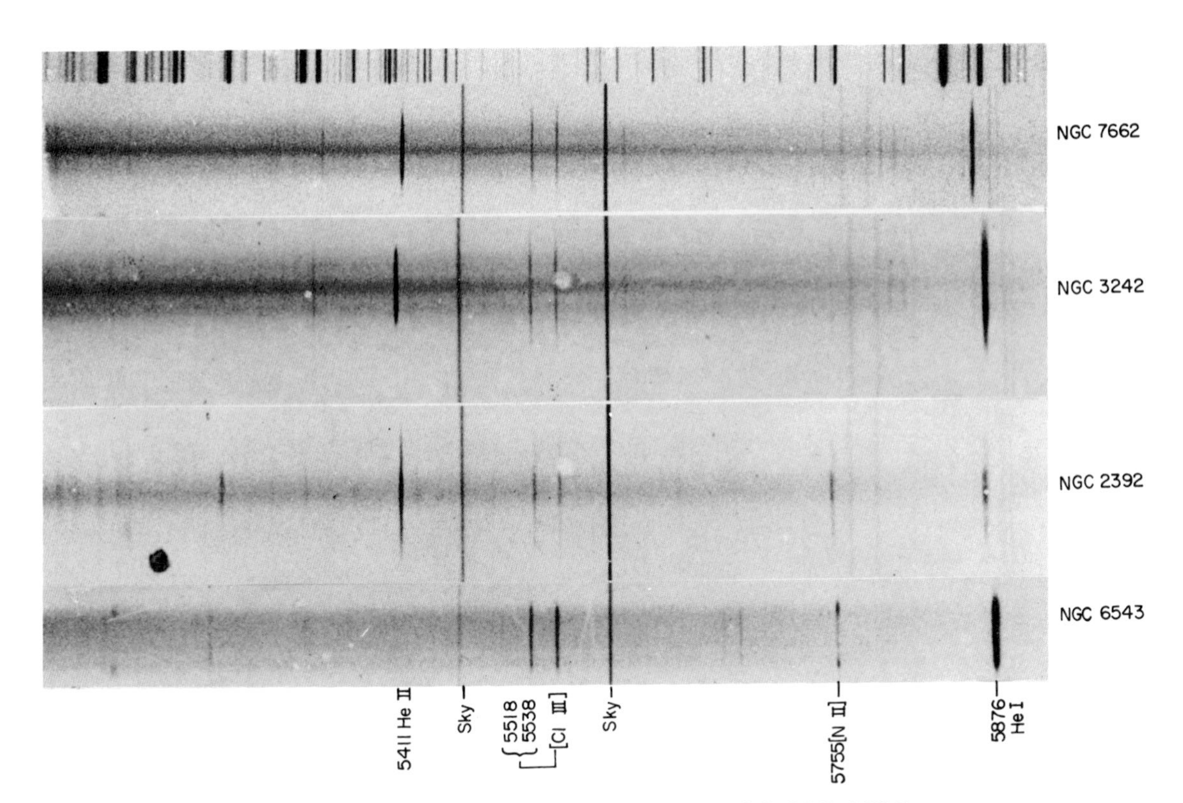

Fig. 70. Spectra of planetary nebulae in the green region. NGC 7662, NGC 3242, and NGC 2392 are high-excitation nebulae; NGC 6543 is one of moderate excitation. Notice the strong lines of both neutral and ionized helium (5876 and 5411 Å, respectively) in the three high-excitation nebulae, and of neutral helium in NGC 6543. The [Cl III] lines are prominent in all nebulae; the [N II] line at 5755 Å appears in NGC 6543. (Photographed with the Lallemand tube at the coudé spectrograph, Lick Observatory, University of California; courtesy M. F. Walker.)

exceptional nebulae whose radio emissions indicate directed motions of charged particles (see Chapter 11). Nebulae can also be observed in the infrared and, with the development of satellite techniques, in the ultraviolet. Some very important spectral lines fall in these regions.

The Mystery of Nebulium

Although gaseous nebulae display the well-known Balmer lines of hydrogen and familiar lines of helium, their strongest spectral lines, in the green at 4959 and 5007 Å and ultraviolet at 3726 and 3729 Å, have never been observed in any laboratory. Believing themselves well acquainted with the spectra of all common elements, astronomers at first were inclined to ascribe the unexplained nebular lines to a mysterious element unknown on the earth, which was called nebulium. Advances in physics and chemistry left no room for such a hypothetical element, and the aforementioned strong lines turned out to be due to oxygen. Other unidentified lines were shown to come from nitrogen, neon, argon, sulfur, and other elements shining under physical conditions not readily attainable on earth. Let us see how these lines were identified.

As we found in Chapter 3, spectral lines arise when an electron jumps from one energy level in an atom to another. The transitions that are most probable, that is, easiest for the electron to accomplish, normally give rise to strong lines; those that are highly improbable, that is, difficult, result in weak lines. The rules governing transitions are relatively simple, and are so restrictive that the number of spectral lines is far less than the number of possible combinations of pairs of energy levels. When he begins to analyze a spectrum, the physicist knows only the wavelengths, and hence the frequencies, of the spectral lines, and that these frequencies result from differences between atomic energy levels. The process of deducing the energy levels from the observed frequencies, which is somewhat like solving a jigsaw puzzle, is illustrated in Fig. 71. The observed lines are (1), (2), (3), (1′), (2′), and (3′). From the fact that the differences

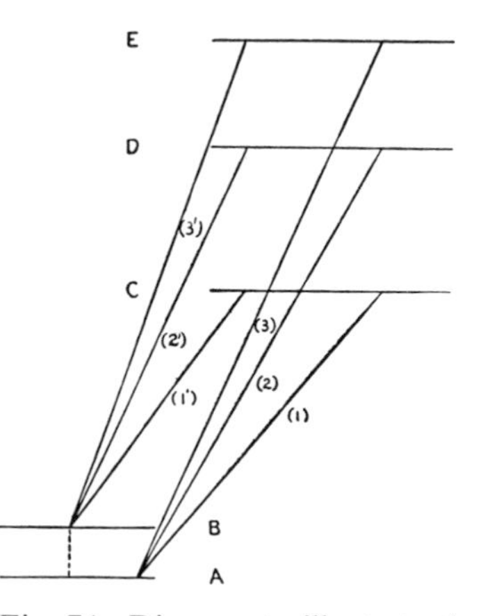

Fig. 71. Diagram to illustrate the identification of energy levels. Lines (1), (2), and (3), corresponding to jumps from levels *C*, *D*, and *E* to level *A*, and lines (1′), (2′), and (3′), corresponding to jumps from *C*, *D*, and *E* to *B*, are observed. From these observations the existence of energy levels *C*, *D*, *E*, and *B* is inferred, although the line corresponding to the jump from *B* to *A* is not observed.

in frequency between the lines (1) and (1′), (2) and (2′), and (3) and (3′) are constant, we infer the existence of the pair of energy levels *A* and *B* with the same frequency difference. Likewise, we find the levels *C*, *D*, and *E*, always bearing in mind that the final pattern of levels must be consistent with the frequencies of all the observed lines. In spite of the fact that the line *AB* is not observed, because the rules of the game demand that this transition be highly improbable, we can still discover the levels *A* and *B*. Once an atom gets into level *B*, perhaps by collision with an electron, it must adopt a circuitous route to return to level *A*. It may, for example, absorb energy of the right wavelength to take it up to *C*, *D*, or *E*, after which it can come back to *A* by radiating the lines (1), (2), or (3).

By similar reasoning, physicists were able to deduce that the lowest energy levels of doubly ionized oxygen, O III, formed the pattern shown in Fig. 72, even though transitions between these levels had never been observed in the laboratory. In pondering the origin of nebulium, I. S. Bowen noticed, in 1927, that the energy differences between *B* and A_1 and between *B* and A_2 correspond exactly to the frequencies of the pair of intense green nebular lines at wavelengths 4959 and 5007 Å. Also, the difference between *C* and *B* agreed with the wavelength of another nebular line at 4363 Å.

Bowen's discovery revealed the remarkable nature of physical conditions in gaseous nebulae, as the following arguments show. According

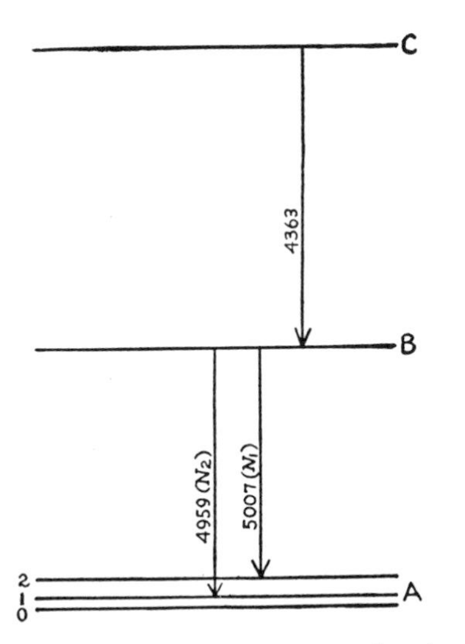

Fig. 72. The lowest energy levels of doubly ionized oxygen, O III. Transitions between these levels give rise to the 5007- and 4959-Å (green nebular) lines and the 4363-Å line.

to modern theory, an O III atom may jump from level B to level A, or from C to B, but its chances of doing so are exceedingly small—about a hundred million times less than the chance that a hydrogen atom will emit a line of the Balmer series. Another way of putting it is that, although an atom will linger only a hundred-millionth of a second in an ordinary level, it will remain in levels like B or C, so-called metastable levels, for seconds or minutes before returning to the ground level. For this reason, transitions of the type CB or BA_2 have been called "forbidden," although actually they are only highly improbable. Forbidden lines are generally indicated by brackets around the symbol of the ion. Thus the forbidden violet line of doubly ionized oxygen at 4363 Å is denoted by λ4363 [O III].

Why, then, do the forbidden lines dominate the spectra of planetary and many other gaseous nebulae? The answer is that the normal, or so-called permitted, lines are very hard to produce under the conditions existing in these objects, while the forbidden lines are not.

In the discharge tube, atoms are excited and de-excited by collisions with fast-moving electrons. An oxygen atom that happens to land in level B, for example, would have about 1 chance in 100 of emitting a forbidden line in a second. Since collisions with other atoms and electrons occur at the rate of several million per second, only a tiny fraction of the total number of oxygen atoms excited to level B will produce a forbidden line.

In a typical gaseous nebula, the electrons are not moving fast enough to excite atoms to the normal levels (whose excitation potentials are often as high as 10 or 20 volts). On the other hand, the free electrons are moving sufficiently fast to excite atoms from the ground level to one or the other of the metastable levels that are close to the ground level. The nebular density is so low that once an oxygen atom is in one of these metastable levels it is almost certain to radiate a forbidden line. Once a quantum of forbidden radiation is created, it is sure to escape from the nebula, since the probability that it will be reabsorbed is negligible. In the laboratory or in the nebula, the rate of emission of forbidden line quanta per unit volume and time will always be the number of atoms in the upper level multiplied by the transition probability expressed in reciprocal seconds, for example, [O III] λ4363 1.6N(C), and λ5007 0.021N(B). In the laboratory, even with O^{++} ions present, this radiation would be overwhelmed by permitted line and continuum radiation and possibly light from the discharge-tube walls. Although the forbidden radiation per atom per second is actually somewhat weaker in the nebula than in the discharge tube, the nebula is so vast (of radius about 10^{12} kilometers or greater) that the forbidden lines produced there may attain a considerable intensity. The emission per unit volume in normal lines is weakened enormously and that in the forbidden lines relatively little in going from a vacuum tube to a gaseous nebula.

The O II ion is of particular interest because it can remain in a metastable level for hours before radiating the close "3727" [O II] pair at 3726.1 and 3728.8 Å (Fig. 73). The relative intensities of these two lines

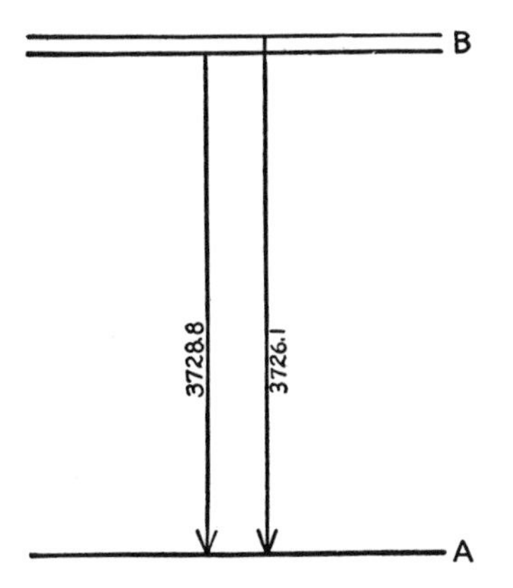

Fig. 73. The origin of the "3727-Å" pair of [O II] lines.

are sensitive indicators of the density at the low densities found in typical gaseous nebulae. At such high densities as are found in novae, atoms get in and out of the metastable levels by collision. In most instances, at low densities, they get in by collision and out by radiation. Hence the 3726.1/3728.8 intensity ratio is a valuable tool for probing the densities of gaseous nebulae.

Progress in this field depends not only on excellent observational material but also on theoretical calculations of collisional effects, lifetimes of excited levels, and so on. Such calculations are being carried out by R. Garstang, M. J. Seaton, S. J. Czyzak, D. E. Osterbrock, and their associates.

Fluorescence in Gaseous Nebulae

Fluorescent rocks are among the most fascinating of all minerals displayed in museums. We enter a windowless room and see a case filled with specimens, most of which shine dully by reflected white light, but, on pressing a button, we work an almost miraculous transformation. The white light vanishes, and suddenly the rocks glow in a sparkling array of colors. What has happened? When we extinguished the white light, we also turned on a source of ultraviolet radiation. Although invisible to the eye, the ultraviolet light is absorbed by the rocks and reradiated in visible colors; each ultraviolet quantum is split up into two or more quanta of longer wavelength.

A similar process is at work in gaseous nebulae. In spite of the fact that the nebulosity must derive all of its energy from the illuminating stars,

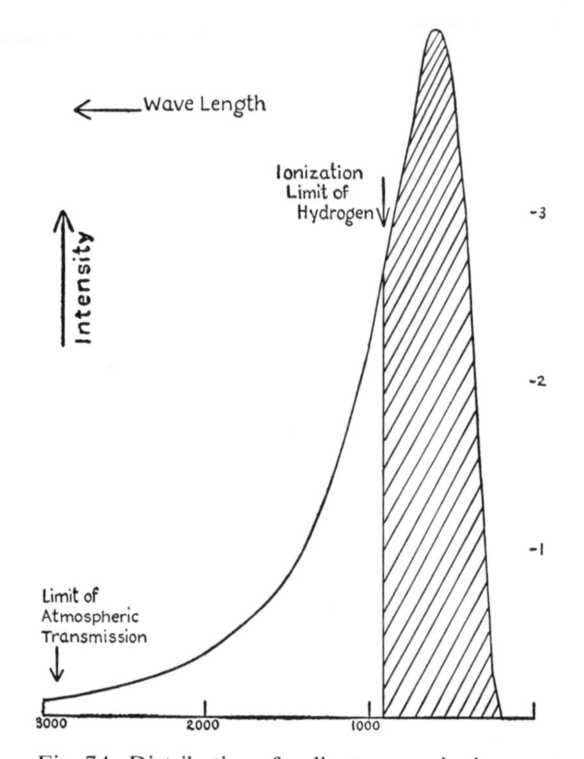

Fig. 74. Distribution of radiant energy in the spectrum of a star that radiates as a black body whose temperature is 50,000°K, plotted for the region beyond 3000 Å. Note that the energy maximum is in the far ultraviolet. The hatched area corresponds to the fraction of the total energy available for the ionization of hydrogen. Virtually all of the energy we observe in planetary nebulae comes ultimately from radiation in the far ultraviolet. More recent detailed calculations, such as those by K. B. Gebbie and M. J. Seaton, by D. Hummer and D. Mihalas, and by K. H. Böhm and his associates, show that photoionization of H, He, O, Ne, and other atoms produces noticeable distortions of the energy distribution similar to that produced at the limit of the Balmer series in hydrogen in hot stars (see Fig. 43). The qualitative character of the energy curve is unchanged, however, at least for very hot stars.

the total amount of visible light radiated by a planetary nebula may be as much as 40 or 50 times greater than that emitted by the central star itself. The explanation may be traced to the fact that the central star is so hot that most of its energy is given out in the form of invisible ultraviolet light, as illustrated in Fig. 74. After the invisible light energy has been absorbed by the nebular atoms, it is re-emitted in visible form. Hydrogen, by far the most abundant constituent of the stars and nebulae, is the atom that is mainly responsible for the transformation of the unseen into the seen.

Let us see what happens when the ultraviolet radiation of the star falls upon a shell of hydrogen gas. A quantum of wavelength less than 912 Å possesses sufficient energy to tear the electron away from a hydrogen atom. Such an electron, detached from an atom, may wander about in space until it is recaptured by a proton. Although the electron was torn away from the smallest orbit, it may, when it is recaptured by some other proton, land in any of the orbits, the higher ones as well as the lower ones (Fig. 75). If the free electron is captured in the lowest orbit, a quantum similar to the original quantum from the star will be reborn, and this quantum, escaping from the atom, may ionize yet another hydrogen atom.

Now an electron captured in one of the higher orbits may jump to any

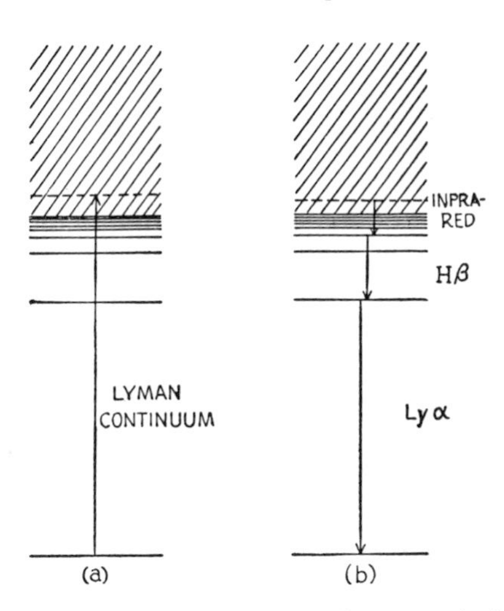

Fig. 75. The origin of hydrogen emission lines in the spectra of gaseous nebulae. (*a*) An atom in the ground level of hydrogen absorbs an ultraviolet quantum of energy and the electron is ejected from the atom. (*b*) The electron may be recaptured by the fourth level with the emission of an unseen infrared quantum. Then it may cascade to the second level with the emission of Hβ, which is observed, and finally to the ground level with the emission of unobservable Lyman α.

one of the lower levels, radiating as it jumps, or in principle it may absorb another quantum of starlight and leap to a still higher orbit. But conditions in gaseous nebulae do not favor this latter process. The nebula is so enormous compared with the star that the starlight is spread over a vast area and its intensity at any point is very low. When radiation is so "diluted," an excited hydrogen atom has little chance of absorbing another quantum, because it remains excited for only a hundred-millionth of a second, whereas it might have to wait 20 years for a quantum of just the right frequency to come along. Also, as we have already seen, the density is so very low that the prospect of a collision with another particle is remote. Hence the electron has no alternative but to return to the ground level, either in one jump or by cascading down by stages.

Each electronic jump is of course accompanied by the radiation of a light quantum. Thus if the electron is captured in the second level, the atom will radiate energy in the near ultraviolet, beyond the limit of the Balmer series. The exact wavelength is determined by the energy of motion of the free electron. From the second level the electron drops to the lowest level and the atom emits the first line of the Lyman series. Many of the electrons caught in high levels, however, will fall to the second level and thereby produce the bright Balmer lines so prominent in the spectra of planetary nebulae. Eventually, all of the stellar radiations of wavelength shorter than 912 Å are converted into light of lower frequencies, a large percentage of which falls in the visible region of the spectrum.

We saw in Chapter 2 that captures of free electrons in the second hydrogen level produce a continuous range of emission that starts at the Balmer limit and decreases in intensity toward the ultraviolet. A similar continuum is observed at the head of the Paschen series. In addition, there is also a curious continuum produced by atoms escaping from the second hydrogen level to the ground level. Normally we would expect the atom to radiate the Lyman-α line. Actually the $n = 2$ level is split into two groups of energy states of very nearly the same energy. Jumps from one group to the ground level are permitted; jumps from the other group are forbidden. The atom caught in one of the second group of states may jump to a virtual (fictitious) level between the $n = 2$ level and the ground level and then from the virtual level to the ground level. Since the virtual level may lie anywhere between the $n = 2$ level and the ground level, the result of the effects of many atoms will be a continuous spectrum filling the normally observable range. This two-photon continuum may produce most of the visual continuous spectra of many diffuse nebulae, and play an important role in planetaries as well.

Ordinary laboratory lines of carbon, nitrogen, neon, and oxygen in several stages of ionization are observed in high-density nebulae. They also originate by ionization followed by recapture.

The observed radiations of neutral and ionized helium originate in the same way as for hydrogen, by ionization and recapture, but the radiations necessary to remove one and two electrons from helium lie very far in the ultraviolet, beyond 506 and 228 Å, respectively. The central stars of planetary nebulae that show lines of ionized helium are therefore among the hottest stars known, with temperatures in excess of 100,000°. Zanstra and Menzel independently showed that, if all the stellar radiation emitted beyond the limit of the Lyman series were absorbed by hydrogen atoms in the nebula, it would be possible to estimate the temperature of the central star. If the nebula is so thick that there are a great number of absorptions and re-emissions, each quantum of ultraviolet energy eventually becomes broken down into a quantum of Lyman radiation and one of Balmer radiation. The latter escapes at once from the nebula; the former is repeatedly absorbed and re-emitted until it escapes. The number of quanta emitted by the whole nebula in the Balmer series may be observed (the energy in all Balmer slitless images divided by $h\nu$; see Fig. 68) and compared with that radiated in the ordinary photographic region of the star. Since the number of Balmer quanta is equal to the number of quanta beyond the Lyman limit radiated by the star, we can find what proportion of energy is radiated by the star in the far ultraviolet, and hence determine its temperature from the radiation laws (see Chapter 4).

Independent evidence for very high temperatures is offered by the spectra of the nuclei of planetary nebulae. Some display Class O spectra with weak emission lines; others show absorption lines, or may have continuous spectra with no features at all. Yet other planetaries show nuclei of the Wolf-Rayet type (Fig. 76), with diffuse emission lines superposed on a continuum. Special mention should be made of the nucleus of NGC 246, whose spectrum was found to contain lines of five-times-ionized oxygen, O VI. Since then, Greenstein has observed several other similar stars, all nuclei of nebulae of low surface brightness.

The Thermostat Action of Nebular Lines

An interesting sidelight in connection with the forbidden lines of oxygen, nitrogen, and neon is that their production acts as a thermostat to regulate the temperature of the nebular gas. Let us suppose that the nebula is composed entirely of hydrogen. The temperature of the gas is measured by the speeds of its atoms and electrons, which in turn depend on the

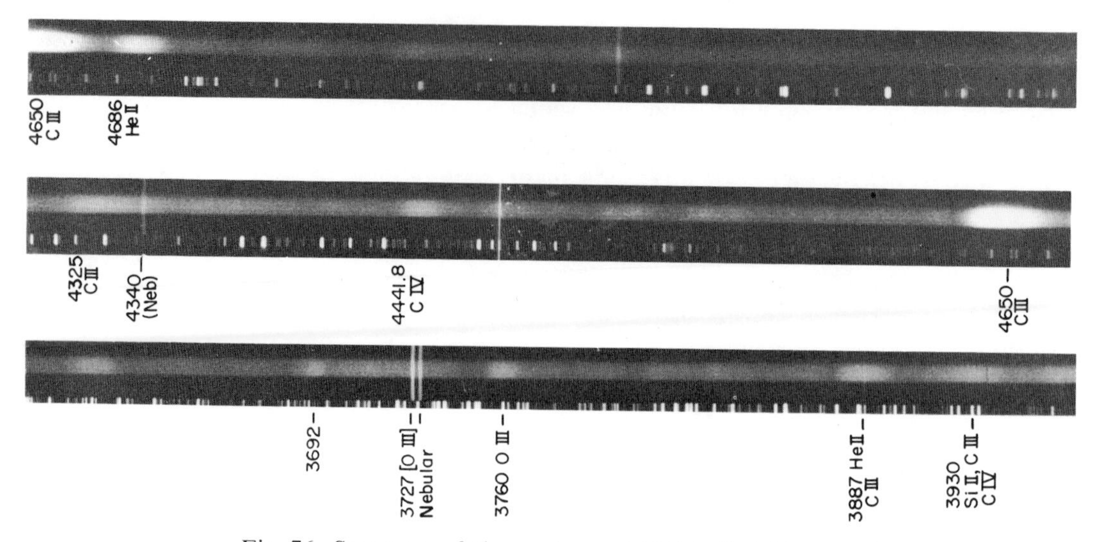

Fig. 76. Spectrum of the nucleus of NGC 40. This central star of the fairly bright northern planetary nebula NGC 40 has a spectrum of the Wolf-Rayet type, with strong, broad emission lines of helium, carbon, and oxygen in various stages of ionization. Other planetary nebulae have central stars whose spectra are continuous, have weak emission lines (with or without faint absorption lines also), or are Wolf-Rayet stars of an unusual type. Some of them show strong emission lines of O VI. (Lick Observatory, University of California.)

temperature of the central star. If the star is hot and therefore rich in high-frequency radiation, electrons will be torn away from their hydrogen atoms at high speeds. These electrons dash about and if they collide with neutral hydrogen atoms they bounce away without loss of energy unless they are moving very fast indeed, about 200 kilometers per second. At this and higher speeds they possess energy enough to excite hydrogen atoms from the ground level up to the first excited level. In a pure hydrogen nebula, the nebular temperature would depend strongly on that of the star until the temperature of the gas rose to about 25,000°, when excitation of the second and higher orbits of hydrogen by collisions with electrons would become important. At this point so much energy would be lost by the electrons in collisions that a further temperature rise would be inhibited.

Suppose now that we introduce small amounts of oxygen, nitrogen, and neon into the hydrogen nebula. These atoms possess metastable levels 2 or 3 electron volts above the ground level, and the energies of the free electrons become dissipated in exciting these "foreign" atoms to metastable levels. The energy radiated in forbidden lines, which arises at the expense of the energies of motion of the free electrons, is forever lost to the nebula, because the forbidden radiations cannot be reabsorbed. Thus the nebular gases are effectively cooled and although the temperatures of the central stars range from about 30,000° to 100,000° the nebular gas never gets much hotter than about 15,000–20,000°.

On the Importance of Coincidences

In addition to the forbidden lines of oxygen and nitrogen, which originate from metastable levels, some lines from normal levels have also been observed. Their origin is a fine illustration of the strange whims of nature. Figure 69 shows the strong lines in the ultraviolet spectrum of several high-excitation planetary nebulae.

One line, at 3203 Å, is due to ionized helium, and a pair arises from [Ne V], but our interest lies chiefly in the other strong lines, which are radiated by O III. These lines are perfectly normal in that they are commonly observed in the laboratory. The puzzling circumstance of their appearance is that other equally intense laboratory lines are absent.

The apparent favoritism has been demonstrated by Bowen to result from a remarkable coincidence. Bowen noticed that all the observed ordinary lines of O III in planetary nebulae originate from electrons cascading down from a single excited level. Furthermore, the wavelength of the radiation required to excite this level is 304 Å, which coincides almost exactly with the strongest line of ionized helium. The 304-Å line corresponds in ionized helium with the first line of the Lyman series in hydrogen. When a doubly ionized helium atom captures an electron, becoming singly ionized, the final stage in the cascading process is often the transition from level 2 to level 1. Hence this He II line, although invisible, probably attains great strength in the planetary nebulae. As shown in Fig. 77, the O III ions in the lower level absorb the plentiful radiation of He II, are excited to the high level, and then return by successive stages with the emission of the observed lines. The missing lines are not observed because they originate from other levels that do not have fortuitous sources of energy. Even more remarkable is the circumstance that as the O III ion finally returns to the lowest energy level it emits radiation at 374 Å, which is of just the right wavelength to generate a similar cycle in producing certain observed lines of doubly ionized nitrogen, N III, near 4640 and 4100 Å. Bowen's explanation is supported by the fact that the permitted

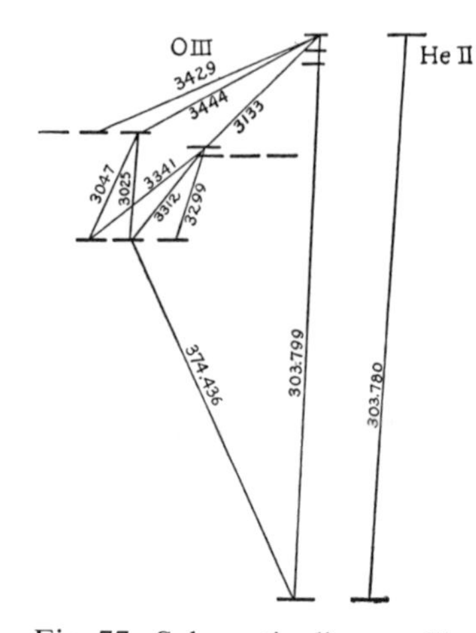

Fig. 77. Schematic diagram illustrating the Bowen mechanism for the excitation of the ultraviolet O III lines. The 303.78-Å line emitted by ionized helium is absorbed by the O III ion, which is raised to the upper level. The O III ion then cascades downward with the emission of the observed lines marked on the diagram.

lines of O III appear only in those portions of the nebula where the observable lines of ionized helium are strong, and therefore where the 304-Å line also is presumably intense.

Some Problems Presented by Planetary Nebulae

We have stressed planetary nebulae largely because their physical state can be understood both qualitatively and quantitatively, but it would be most unfortunate to give the impression that the basic problems of these nebulae have all been solved, particularly as regards their structure and internal motions.

If the distance of a gaseous nebula is known, its density may be found from its surface brightness as observed in $H\beta$ or in the continuous spectrum at the head of the Balmer series, provided we have an estimate of the electron temperature. Now the intensity ratios of certain forbidden lines, for example, 4363/(4959 + 5007) in [O III] (Fig. 72) and similar ratios for [O II], [S II], [Ne III], and so forth, will also depend on the electron density and temperatures. As far as these radiations originate in the same regions in the nebula, we may use them to derive electron density and temperature. When this is done, we find that for some nebulae both surface-brightness and line-ratio methods give the same electron densities but in others the forbidden lines give higher densities, indicating the presence of condensations. Such results are hardly surprising for objects such as NGC 7009 or NGC 7662 (Fig. 66), but blobs are sometimes indicated in nebulae for which there is no direct evidence of inhomogeneity. Other objects, such as the Owl nebula, NGC 3587 (Fig. 65), seem to be perfectly smooth. Almost any process we can contemplate would tend to smooth out irregularities. What produces the ones we see? We do not know; no mechanism so far proposed seems adequate.

Some planetary nebulae show no symmetry at all; others show symmetries of such a character as to suggest the presence of large-scale magnetic fields (Fig. 67). These fields may be of the order of 100,000 times smaller than the earth's field, yet enormous when the low density of the material is considered.

The nebular shells do not sit motionless in space; they are slowly expanding. If the slit of a spectrograph is placed across the telescopic image of a nebula, as Moore and Campbell did many years ago, the lines are often observed to be bowed or doubled, indicating an expanding shell. Olin Wilson employed a multislit spectrograph, that is, one with a set of strictly parallel slits, each of which would cross a different section of the nebula. In this way a grid of the line-of-sight velocities can be obtained. Near the center of the nebula both approaching and receding strata are observed, while toward the edge only material moving perpendicular to the line of sight is seen.

Wilson found that different spectral lines gave different rates of expansion, the ion of highest excitation giving the lowest expansion rate and that of lowest excitation the highest expansion rate. Since high-excitation ions tend to be found closer to the central star than those of low excitation, a reasonable interpretation is that the shell is increasing in thickness in such a way that the outer radius grows rapidly with time while the inner radius changes slowly.

Each quantum of energy carries momentum as well as energy. When an atom absorbs a quantum of energy $h\nu$, it picks up momentum $h\nu/c$. Hence radiation falling upon matter exerts on it a pressure, which can become important in the highly attenuated gases of nebulae or envelopes of stars. In a planetary nebula, many quanta of Lyman-α radiation are created. As these quanta are scattered from atom to atom, momentum is transferred. As a consequence, an initially stationary uniform nebula would be forced to expand. Velocity differences would appear between one point and another in the nebula. A Lyman-α quantum emitted in one part of the shell could not be absorbed in a distant region, moving at a different velocity, because of the Doppler shift, and radiation pressure would drop off, leaving the nebula slowly expanding.

A planetary nebula starts out as a small, compact object of relatively high density. As time goes on, it gradually expands in size and decreases in density. As long as the nebula absorbs all the ultraviolet radiation from the central star, it remains fairly bright, but ultimately the density becomes so low that most of the radiation escapes through the nebula into space. By this time the planetary is an object of very low surface brightness. Thereafter it gradually fades away as the central star settles down as an intensely hot, small, white dwarf star.

By assuming that the rate of expansion is constant, Whipple estimated that the lifetime of a typical planetary nebula is about 30,000 years. Thus within a few tens of thousands of years—a mere moment in the life of a star—the nebulous fragments will have expanded into the "vacuum" of interstellar space. As they fade from the scene, new planetaries will appear to take their place. The immediate ancestors of planetary nebulae are not yet recognized, but it seems certain that the nuclei evolve into white dwarf stars. We return to this topic in Chapter 9.

Although some planetary nebulae may represent simpler physical situations than those existing in most galactic diffuse nebulae, with their extremely irregular, unknown geometrical patterns of the material, objects such as NGC 7027 present very difficult problems. Frank Low found this object to have unusually strong infrared radiation. Probably it arises from solid particles (Krishna-Swamy and O'Dell) rather than from atomic radiations (Goldberg).

The Interstellar Medium

Where the gas constituting the interstellar medium is dense and illuminated by bright hot stars it becomes ionized and fluoresces in a manner similar to that characteristic of planetary nebulae. The vast bulk of the interstellar gas, however, is at a very low density and often at a very low temperature as well.

The great extent of the gas cloud was first revealed by the spectroscope. Interstellar gas absorbs starlight in certain particular wavelengths in the same way that a stellar atmosphere absorbs light from a photosphere and thus imprints absorption lines of the familiar elements upon the continuous spectrum. So tenuous is the interstellar cloud, however, that an exceedingly long column of gas is required to produce an absorption line of sufficient strength to be visible in a stellar spectrum. The observation of interstellar lines is therefore confined chiefly to the more distant stars, and is best done with spectrographs of very high dispersion.

The first element to be discovered in interstellar space was ionized calcium, found by Hartmann in 1904 in the spectrum of δ Orionis, a close double star. When the line of sight is nearly in the orbital plane of a binary star, the revolving components will alternately approach and recede from the observer. The motion will be mirrored in the spectrum where, owing to the Doppler effect, the absorption lines will appear to oscillate in position, with the same period as the revolution (Fig. 78). Absorption lines produced by interstellar gas, which does not share the orbital motion, will, however, appear stationary. Since the *K* line of ionized calcium does not share the periodic motion of the other lines of δ Orionis, it must be produced in a cloud of gas detached from the star.

At Lick Observatory in 1919 Miss Heger discovered the stationary lines of interstellar sodium, but it was not until Theodore Dunham, Jr., developed the Mount Wilson high-dispersion coudé spectrograph that

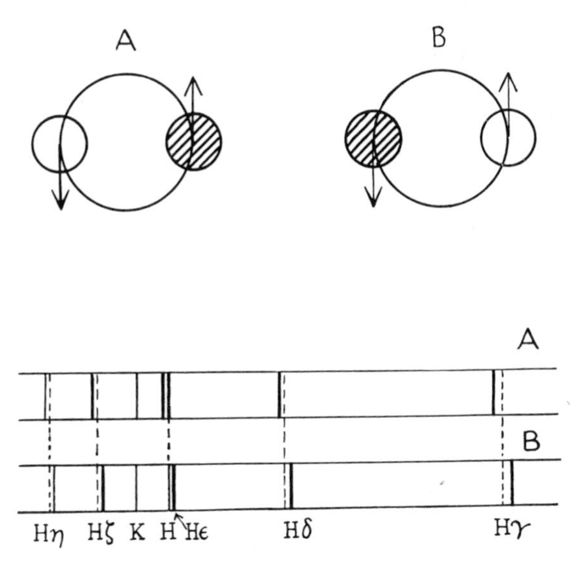

Fig. 78. How interstellar lines were discovered. In *A* the brighter star is approaching the observer; in *B* it is receding. Consequently, the absorption lines are shifted toward the violet in the upper spectrum and toward the red in the lower, but the stationary *H* and K lines, due to the interstellar cloud, remain unchanged in position.

great progress in detection of new lines was made. He and W. S. Adams were able to add potassium, neutral calcium, titanium, and iron to the constituents of interstellar space. Some unidentified lines observed by Dunham and Adams were shown by McKellar to be due to absorption by fragmentary molecules, CH, CH^+, and CN, in their very lowest rotational states. To this list of molecules must be added OH, detected by Barrett and his associates by radio-astronomy techniques, and also water, ammonia, formaldehyde, formic acid, and methyl alcohol (see Appendix I).

The optical-region absorption lines are best seen in the spectra of the hot Class B stars, where the stellar lines of molecules and metallic atoms are absent. It is evident that the intensities of the dark interstellar lines will depend upon the distances of the stars, for the light of a faraway star must pass through a longer column of gas than light from a nearby star.

The actual appearance of some of these interstellar lines in two distant stars is shown in Fig. 79. In the spectrum of HD 14143, which is a member of the distant double cluster in Perseus, notice that the *D* lines, al-

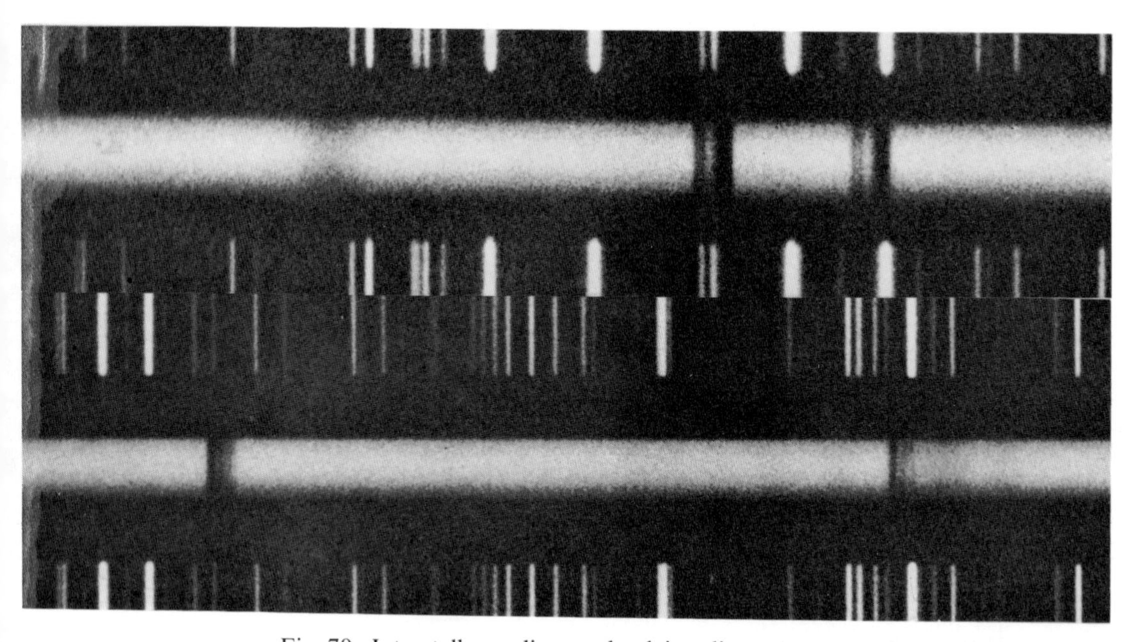

Fig. 79. Interstellar sodium and calcium lines: (*upper*) *D* lines in HD 14143 in the double cluster in Perseus; (*lower*) *H* and *K* lines in HD 173987. Notice the complex structure of the lines, indicating that there are several discrete clouds, each moving with a different radial velocity. (Courtesy G. Münch, Mount Wilson and Palomar Observatories.)

though multiple, are relatively sharp compared with the diffuse stellar 5876-Å line of He I. Figure 80 shows the multiplicity of the interstellar lines observed in the spectrum of ϵ Orionis. Each of these components reveals a distinct cloud with a different line-of-sight velocity.

When multiple lines are observed, the velocity of the strongest component usually corresponds to that expected from galactic rotation at a point half-way to the star. The others reveal the peculiar motions of distinct clouds, which have typical space velocities of the order of 14 kilometers per second. Some clouds, however, move with very high velocities. Typically, three or four clouds are found per 1000 light-years along a line of sight in the plane of the Milky Way. It is evident from the differences in intensities shown by different components that the clouds must have a huge spread in size or total material involved. Although the interstellar lines increase in strength and complexity with the distance of the background star, their total intensities do not provide a very precise measuring rod for estimating stellar distances.

If we could employ a cosmic steam shovel to reach out into space and scoop up a gob of the interstellar stuff, we would probably find that the sample contained most of the common elements that are found on the

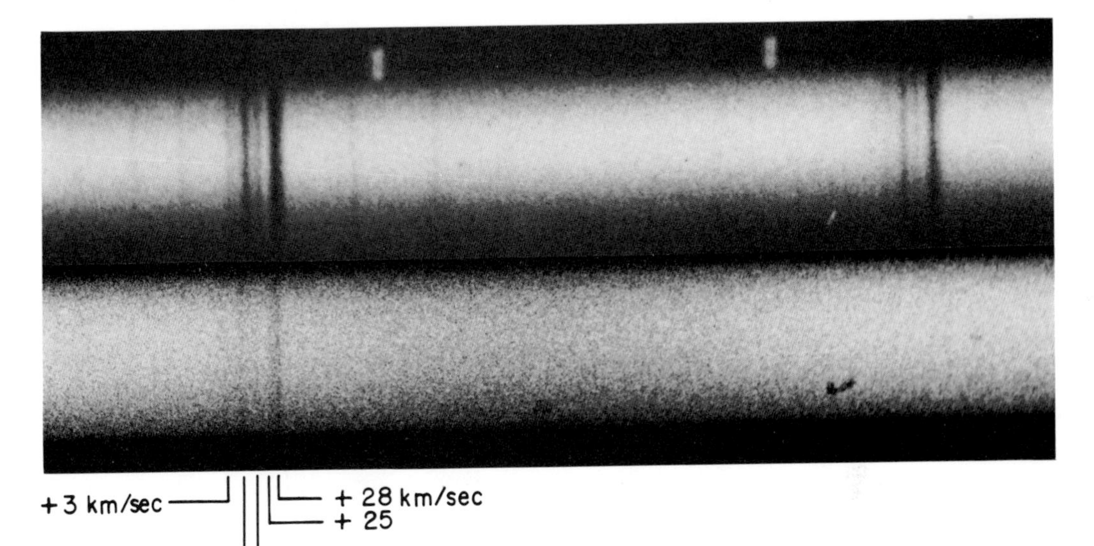

+ 3 km/sec + 28 km/sec + 25 + 11 + 18

Fig. 80. The multiple interstellar lines in the spectrum of ϵ Orionis. Notice the separate components, corresponding to discrete clouds with radial velocities of +3.0, +11.8, +18.3, +25.5, and +28.2 km/sec. (Courtesy G. Münch, Mount Wilson and Palomar Observatories.)

earth and in the stars. That the thumbprints of most of these elements have not been found may be traced to the physical conditions in interstellar space. Even the planetary nebulae are dense compared with the gaseous material of interstellar space; the gas is so exceedingly rarefied that only the most abundant elements and compounds can be detected. Nevertheless, there are some elements, notably Li, Be, and Al, whose lines should be expected if their abundances are at all comparable with their terrestrial values. Intensive searches, particularly by Spitzer, have failed to reveal any trace of lithium or beryllium. Either these elements are tied up in the solid grains or they are actually very much rarer in interstellar space than on earth.

With equipment flown in satellites and rockets it is possible to detect resonance (Lyman) lines of hydrogen and other elements that are abundant in the interstellar medium. Calculations predicted that not only would the Lyman lines be strong, but continuous absorption beyond the Lyman limit would extinguish all radiation from distant parts of the Galaxy, except in the x-ray region. R. Bless, A. Code, and their associates at the University of Wisconsin and D. C. Morton and E. B. Jenkins at Princeton have found the Lyman-α line to be weaker than predicted. The results can be understood if most of the absorption occurs in very cold regions, perhaps as cold as 20°K.

Ionization of the atoms takes place at a very slow rate because the ionizing radiation comes from great distances and is consequently highly attenuated. Nevertheless, once in a while a high-frequency quantum will come along and tear an electron away from an atom. Once ionized, an atom stays that way for a long, long time, because at the very low density its encounters with other free electrons are very infrequent. Consequently, most of the atoms of calcium and sodium, for example, are ionized.

Radiation capable of exciting atoms to higher energy levels is very weak, and collisions capable of lifting atoms from their ground levels to excited levels are rare. Consequently, an atom spends most of its time in the very lowest state of energy. Most atoms in this placid state, particularly those of the permanent gases, are indifferent to visible light, and will absorb radiation only from the invisible very short-wavelength region of the spectrum, which is not transmitted by the earth's atmosphere. Hence the majority of interstellar atoms evade discovery, even though they may be very abundant.

A case in point is hydrogen, whose interstellar Balmer lines have never been observed in absorption. Most of the hydrogen atoms near hot stars must be ionized; but occasionally they recapture electrons in the second, third, fourth, or higher orbits. As these electrons cascade to lower orbits they emit radiation. Now only atoms in the second energy level, that is, atoms whose electrons are in the second orbit, are capable of absorbing the Balmer series. Even though an electron may land in the second orbit, it will remain there but a hundred-millionth of a second before dropping to the lowest orbit. In order for a line of the Balmer series to be absorbed, a quantum of light of just the right frequency must come along during that time interval. The chance of that happening in interstellar space is extremely small.

It might be expected, however, that a downward cascading of captured electrons would produce a faint glow of light in the spaces between the stars. By means of a specially designed spectrograph, Struve and Elvey, at the McDonald Observatory in Texas, detected this faint light, in the form of bright lines of hydrogen and ionized oxygen, in large regions of the Milky Way (Fig. 81). They found, however, no emission lines in high galactic latitudes away from the Milky Way. The hydrogen emission regions are often sharply bounded and associated with groups of Class O stars. Struve estimated the diameters of these regions to range from 80 parsecs for a region in Orion to 250 parsecs for the Cygnus nebulosity. The 3727-Å line of [O II] is nearly always present in regions of hydrogen emission, and the green nebular lines of [O III] are also sometimes observed. (Jumps between very high levels in hydrogen, for example $n = 103$ and $n = 102$, produce lines in the radio-frequency range, and these have been detected in a number of regions in the Galaxy.)

As previously described, these regions of attenuated, glowing gas can also be observed directly by using red-sensitive plates and narrow-band-pass filters whose transmission is maximum for Hα, 6563 Å. In this way, Collin Gum discovered the largest gaseous nebula known, which covers many square degrees of the southern sky but is invisible to northern observers. The chief difference between these extended regions of low-density emission and objects such as the Trifid or η Carinae nebula would appear to be total mass and density. In "conventional" diffuse emission nebulae, the density amounts to 100–10,000 atoms per cubic centimeter, whereas in objects such as Gum's nebula it may be 1–10 atoms per cubic centimeter. Since for a given temperature the emission goes as the square of the density, the brightness of an object such as the Trifid or the Lagoon nebula will be enormously greater.

Strömgren worked out the theory of the ionization of hydrogen in the interstellar medium. He predicted that for a radius of about 30 parsecs

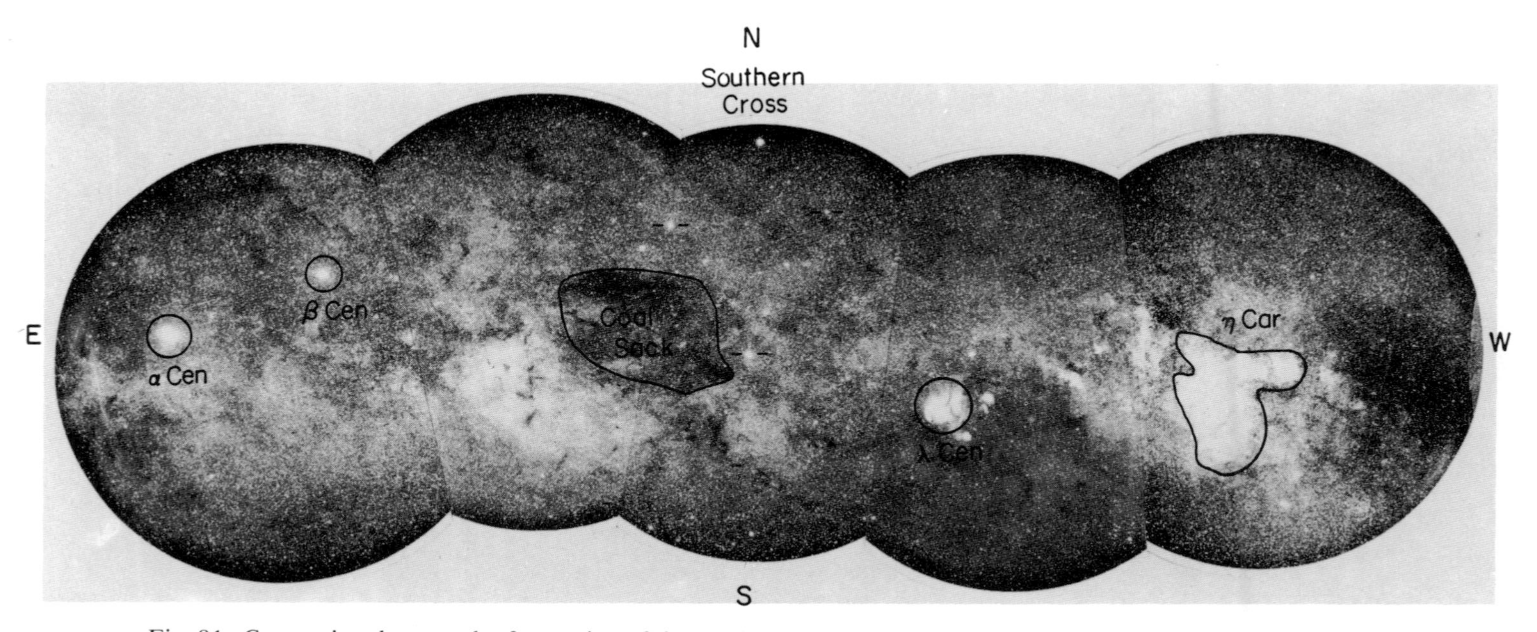

Fig. 81. Composite photograph of a portion of the southern Milky Way. The two bright stars at the left are the pointers to the Southern Cross, α and β Centauri. Note the dark area east and south of the Cross, the Coal Sack. Notice also the bright area near λ Centauri and the bright nebulosity associated with η Carinae. These are regions of ionized hydrogen (H II), which sometimes appear in roughly spherical form; see Fig. 83. (Courtesy Mount Stromlo Observatory.)

in the neighborhood of a Class O star hydrogen would be completely ionized if its density amounted to about 1 atom per cubic centimeter; then there would be an abrupt boundary and a few parsecs farther out all the hydrogen would be neutral (Fig. 82). Within the region of hydrogen ionization the emission lines would be produced as the electrons are recaptured and cascade to lower levels. Thus, if the density of the gas was relatively uniform in the neighborhood of a hot star, the latter would be surrounded by a disk of faintly glowing hydrogen. These Strömgren spheres have been observed on Hα photographs of the Milky Way and in external galaxies as well. There is strong evidence that objects such

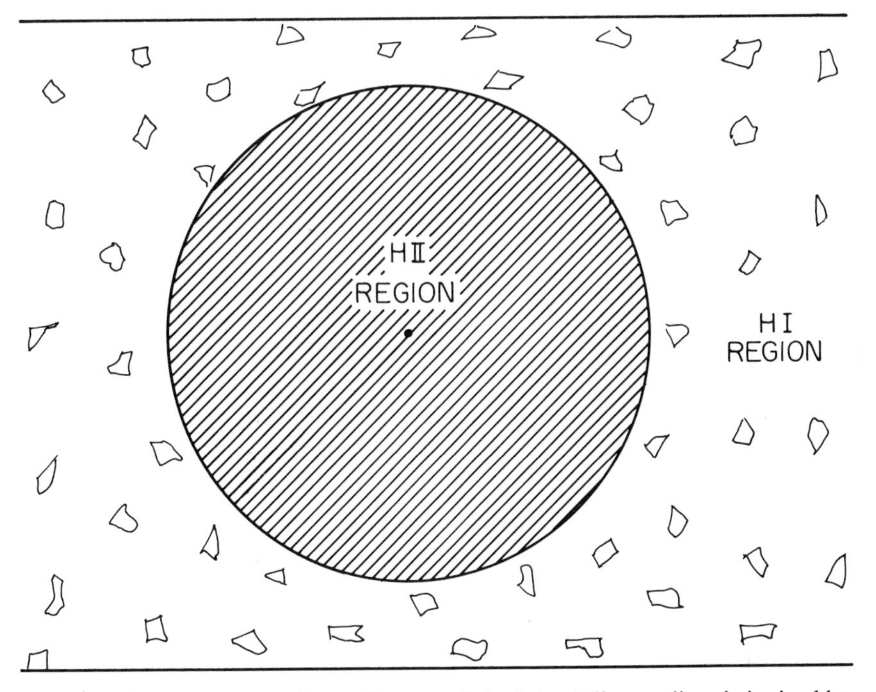

Fig. 82. The Strömgren sphere. The gas of the interstellar medium is ionized by the hot illuminating star imbedded within it, the temperature of the ionized gas being about 8000°K. The ionized region ends very abruptly and is surrounded by a relatively cool region of neutral hydrogen and solid grains. The unionized gas is very inhomogeneous, containing regions where the density is 10–20 particles per cubic centimeter, and the temperature is 100°K or less. The space between these denser clouds is filled by a hotter gas that occupies about 94 percent of the volume, and has a density of about 1 atom per 50 cm^3 and a gas kinetic temperature measured in thousands of degrees.

as the nebula in Orion represent Strömgren spheres in a denser medium. The theory may also be applicable to some planetary nebulae.

In the zone where the hydrogen is ionized there are 2 or 3 electrons per cubic centimeter; outside this zone, the number falls to 1 per 100 or 1000 cubic centimeters. Now the atoms of the metals may be easily ionized in either region since hydrogen has only a slight effect on the radiation available for their ionization. However, they may recover electrons in the region of ionized hydrogen much more easily than in the region of neutral hydrogen. Accordingly, most of the neutral sodium and neutral and singly ionized calcium probably exist in the former zone.

We might expect each Strömgren sphere to be surrounded by a zone of neutral hydrogen. In fact, the normal condition of the hydrogen gas in the Milky Way is to be neutral; only when a hot star is nearby does the gas become ionized. In the ionized-hydrogen or H II region the temperature is typically 6,000–10,000°K; in the outer H I regions where the electrons and protons are recombined it may be 100°–60°K or even lower in denser condensations. In addition to neutral hydrogen, molecular hydrogen, H_2, may also be present, but it is probably not abundant.

According to George Field, much evidence indicates that the H I region may be very inhomogeneous, consisting of cool clouds with densities of 20 atoms per cubic centimeter, with 1 electron every 25 cubic centimeters, at a temperature of 100°K or less. These clouds occupy about 4 percent of the volume. The background medium contains 1 atom per 50 cubic centimeters, but has a temperature that may be as high as 9000°K.

Neutral hydrogen absorbs Lyman lines in the far ultraviolet and we might expect that we could hope to observe this atom only from spectrographs flown in orbiting observatories. In this instance, nature is cooperative and has provided us with a powerful tool for study of the neutral hydrogen gas.

In Chapter 3 we described how ordinary atomic energy levels were split by the magnetic effects produced by the spin of the electron and its orbital motion. This interaction energy was usually small compared with the energy difference between normal levels. In addition to the spin of the electron, there is also a spin of the nucleus of the atom, but the magnetic effects of this spin are much smaller. Nevertheless, the interaction of the electron's magnetic effects with those of the nucleus produces a so-called hyperfine structure of spectral lines which is observed as a very small-scale splitting in lines of some atoms, such as manganese.

In a hydrogen atom the magnetic interaction of the spinning electron with the spinning proton is such that in one direction of electron spin the energy is slightly greater than in the other. The energy differences are small, about 2 million times smaller than the energy necessary to detach the electron from the atom. Thus the neutral hydrogen atom in its ground level can exist in one or the other of two states whose energies differ by 0.000006 electron volt. When the atom flips from one state to the other it absorbs or emits a quantum of energy whose wavelength is 21 centimeters. This radiation, which falls in the radio-frequency range, was predicted by van de Hulst in Leiden and first observed by Ewen and Purcell at Harvard. The probability of a flip is very low. If the atom is in the state of higher energy it will remain there for an average of 10 million years before flipping to the lower state.

The 21-centimeter radiation has turned out to be a very powerful tool both for galactic-structure research, described by Bart J. Bok and Priscilla F. Bok in *The Milky Way,* and for studying the properties of the interstellar medium. The displacement of the line tells the line-of-sight velocity of its source and the profile and central intensity of the line give the number of atoms in the line of sight and the temperature of the cool hydrogen gas. Note that the line originates in regions of neutral atomic hydrogen; it is not emitted in ionized-hydrogen regions nor where hydrogen exists in purely molecular form. Normally, the line is observed in emission (Fig. 83), but if the hydrogen is found in front of an emitter of intense continuous radio-frequency radiation, such as the Cassiopeia source (see Chapter 12), it may appear in absorption.

Effective studies of the distribution of neutral hydrogen require radio telescopes with large apertures, so that the angular resolution will not be so small that all details are lost. It is found that the interstellar medium as represented by clouds of neutral hydrogen extends over vast regions of the Galaxy. Investigations, particularly by Menon at Harvard, indicate that the nebula in Orion is only the hot, ionized neighborhood of the stars of the Trapezium and is surrounded by a cloud of cool gas, thousands of times as massive. The great 30 Doradus nebula (frontispiece), is surrounded by a much larger envelope of neutral hydrogen.

When the 21-centimeter line is very strong in emission, its central intensity may reveal something about the temperature of the gas. Interpretation of the measurements is complicated by the fact that neutral hydrogen gas may not be of uniform temperature. The denser portions of the interstellar medium are probably cold (20–100°K); the less dense

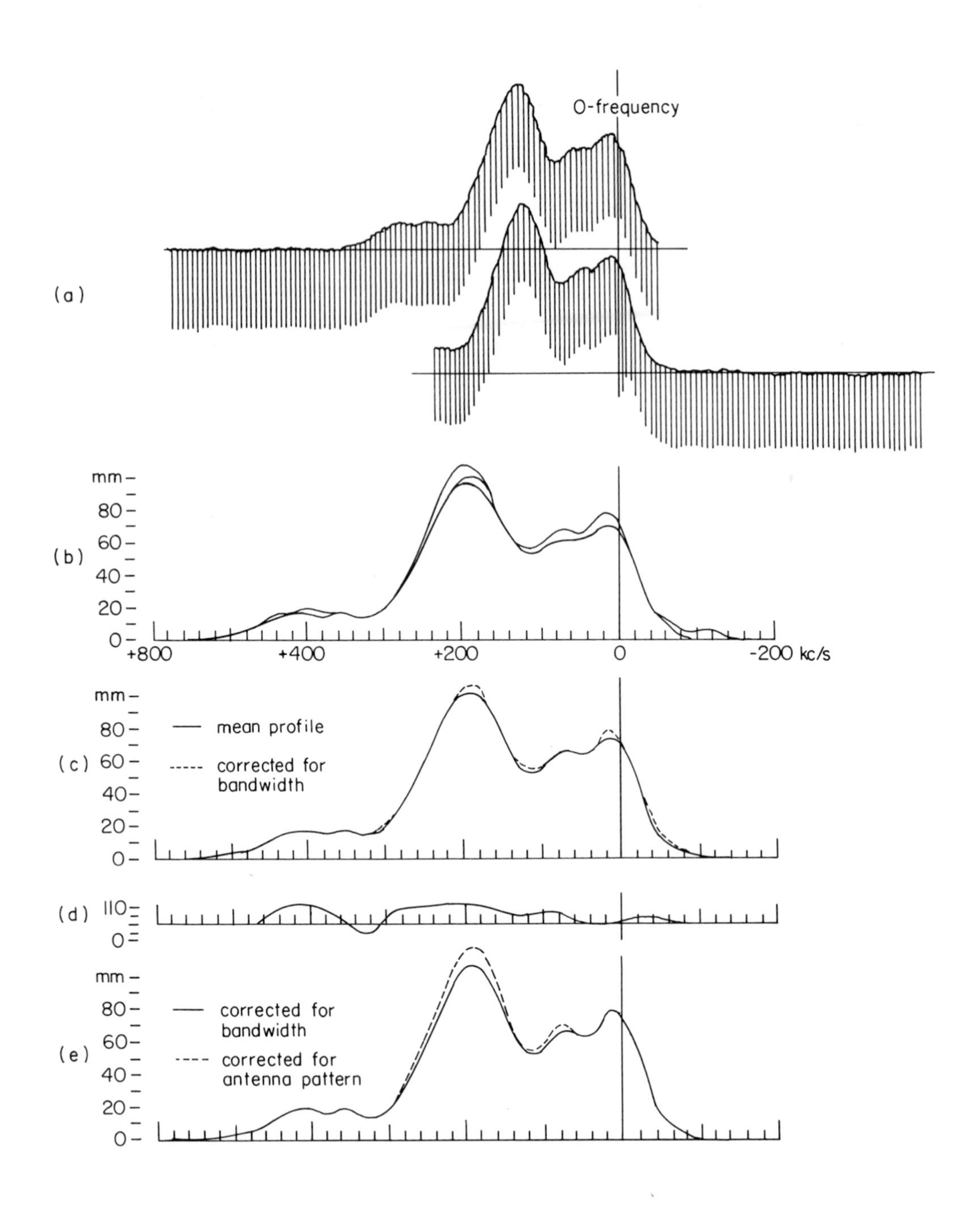

O-frequency
(a)
mm
80
60
40
20
0
(b)
+800
+400
+200
0
-200 kc/s
(c)
mean profile
corrected for bandwidth
(d)
110
(e)
corrected for bandwidth
corrected for antenna pattern

Fig. 83. A profile of the 21-cm line of neutral hydrogen. (*a*) The original records as obtained from the radio telescope and displayed by the receiving equipment. (*b*) Comparison of three separate measurements of the line profile, all on the same intensity scale. (*c*) The averaged profile and its correction for the spread in frequency of the receiving equipment. In ordinary optical spectrograms, each spectral line corresponds to a small range in frequencies. Analogously, a radio-telescope receiver accepts radiation over a small frequency range or bandwidth. Correction for this effect of the finite bandwidth steepens the line profile. (*d*) One must also correct for the "resolving power" of the radio telescope; that is, a point image would appear smeared out on the sky and any source of finite size appears fuzzier than it actually is. The curve gives the correction factor that must be applied. (*e*) The effect of the finite resolution (antenna pattern) is to smear out the profile. Hence, when this correction is applied, the profile is made narrower and steeper. The steeper, higher curve, then, is the final profile. (Courtesy of Gart Westerhout, *Bulletin of the Astronomical Institutes of the Netherlands 13* (1957), 211.)

portions may be much hotter. In the neighborhood of young, hot, very luminous stars, where the gas becomes heated and ionized, it appears as H II regions.

The Bright Nebulae

There seems to be a prominent tendency for the mixture of interstellar dust and gas to collect in small clouds and condensations. Often these clouds will not only dim but completely obscure the light from the stars behind. When the dark clouds chance to occur near one or more bright stars, however, they are illuminated and appear as bright diffuse nebulae. The Pleiades (Fig. 84) and the nebula in Orion are two typical examples of kinds of bright nebulosities that decorate the sky. An especially rich nebulous region is that centering on the Trifid and Lagoon nebulae, which are separated by only 1.5 degrees (Fig. 85). The fact that the entire region is strewn with faint nebulous streamers suggests that the two nebulae are physically associated and merely represent two condensations in an extensive cloud of material. It is important to bear in mind that, like planetary nebulae, bright irregular nebulae are not self-luminous but derive their brilliance only through the courtesy of associated stars. The spectroscope suggests that there are two principal ways in which these aggregations of dust and gas borrow energy from their stellar neighbors. E. P. Hubble of Mount Wilson first noticed that when the surfaces of the exciting stars are cooler than about 18,000° the spectra of bright nebulosities consist chiefly of dark lines. The similarity between the nebular spectrum

Fig. 84. Reflection nebulosity in the Pleiades. (Photographed with the Crossley Reflector, Lick Observatory, University of California.)

Fig. 85. The Trifid nebula (*above*) and the Lagoon nebula (*below*), two of the most striking gaseous nebulae in the whole sky. They are separated by only 1.5°, and they are also probably close together in space. Note the two open star clusters, NGC 6530 above and to the left of the Trifid nebula and NGC 6531 within the Lagoon nebula. (Photographed by F. S. Paraskevopoulos with the 24-inch Bruce Telescope at the Harvard Boyden Station.)

and the exciting stellar spectrum leaves little doubt that one is merely the reflection of the other. On the other hand, when the exciting star is very hot, the main features of the nebular spectrum are strong, bright lines of hydrogen and ionized oxygen. Such emission-line spectra are reminiscent of the spectra of certain low-excitation planetary nebulae like IC 418. The mechanism of light emission must also be similar to that in a planetary nebula, since we are again dealing with a gas of low density, excited by radiation from a hot, distant star.

An exception to the rule just stated must be noted for the small fanlike nebulae associated with variable stars of the T Tauri or R Monocerotis type. These nebulae are found only in regions such as Taurus where there exist extensive clouds of cool gas and obscuring material. The spectra of these nebulae are usually characteristic of low excitation, but they may contain forbidden lines of oxygen, nitrogen, and other elements, similar to those found in planetary nebulae.

Bright spectral lines are the trademarks of emission nebulae and have led to their discovery in all parts of the Milky Way and even in galaxies external to it. The most beautiful example of an emission nebula anywhere is the Tarantula nebula (30 Doradus) in the Large Magellanic Cloud (frontispiece). Notice the large, graceful arches of radiating material. The Large Magellanic Cloud contains a large number of these emission nebulosities (Fig. 86). Those in the Small Magellanic Cloud are less bright and conspicuous (Fig. 87).

Objective prisms used in conjunction with Schmidt cameras have revealed many emission nebulosities in external galaxies, such as NGC 6822 and the Triangulum (M 33) and Andromeda (M 31) spirals. Red-filter ($H\alpha$) photographs taken with large reflectors also reveal numerous emission patches in these objects.

We may visualize the stars as giant spotlights shining on a mixture of fluorescent material (atoms) and reflecting particles (dust). The continuous spectra of emission nebulae appear to be produced, not only by reflection from grains, but also by emission processes involving heavy atoms, as previously described. When the spotlight is rich in ultraviolet radiation, the gas fluoresces and bright lines appear. But when only relatively cool stars are present, there is no great store of ultraviolet radiation for the atom to absorb and re-emit in the form of visible bright lines. The starlight falling upon the nebula is then scattered by the dust particles, and the nebular spectrum is simply a reflection of the stellar spectrum. Thus Struve, Elvey, and Roach at the McDonald Observatory found the

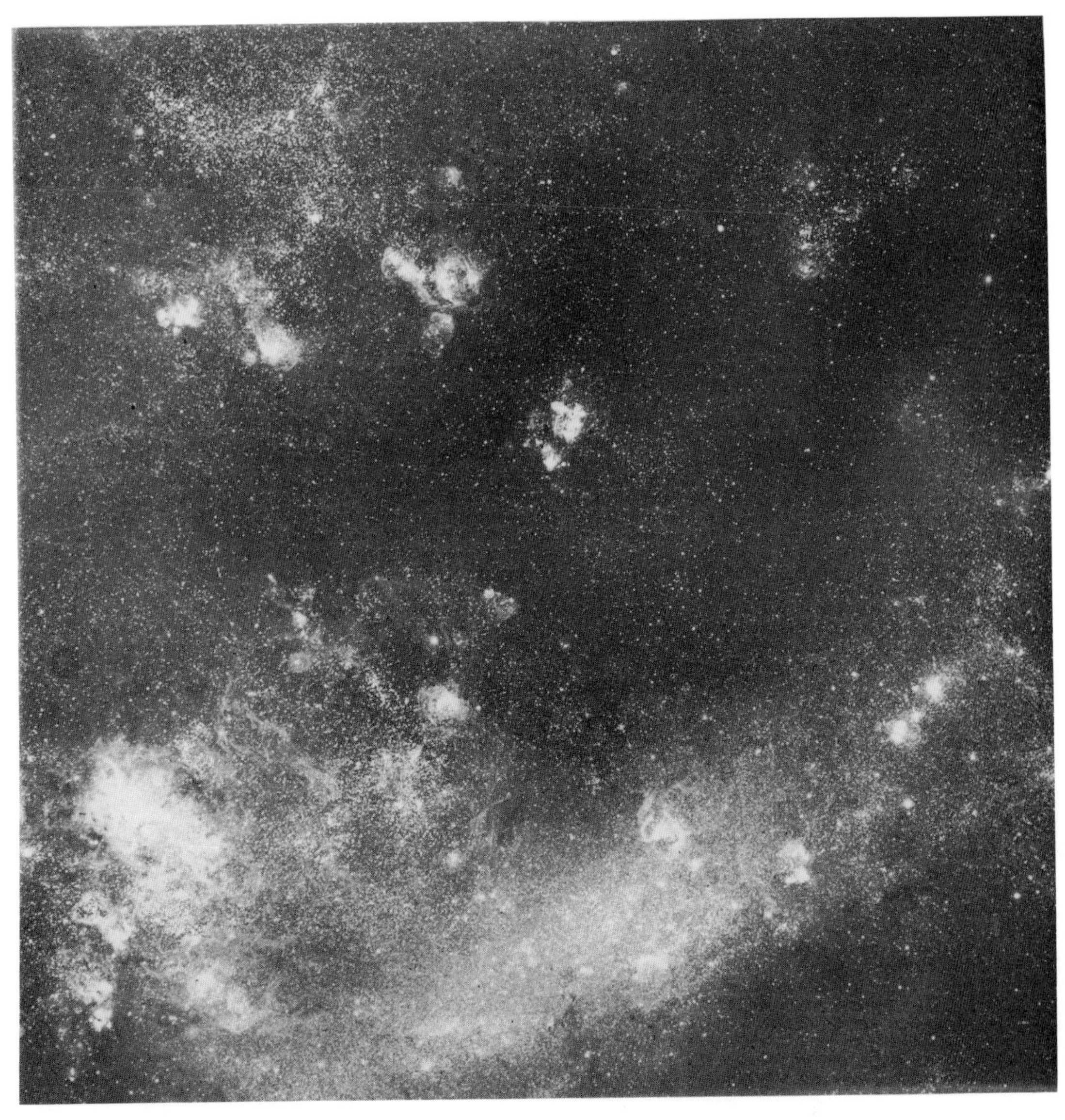

Fig. 86. The central region of the Large Magellanic Cloud showing a number of hydrogen emission nebulosities, the most spectacular of which is 30 Doradus (see frontispiece); photographed with the Uppsala Schmidt camera at Mount Stromlo Observatory. (Courtesy Bengt Westerlund.)

Fig. 87. The emission nebula NGC 346 in the Small Magellanic Cloud. (Courtesy Mount Stromlo Observatory.)

color of a large reflection nebula near the red supergiant Antares to be nearly the same as that of the star. The colors of other nebulae associated with blue B stars also were apparently about the same as those of the stars themselves. The Antares nebula covers about 1 degree in the sky and must be about 5 light-years in diameter.

Clouds of ionized hydrogen may be studied with radio telescopes, since these objects emit in the radio region. An electron passing near an ion is accelerated and radiates energy. Encounters of this type usually involve very small energy changes, which produce emission in the centimeter range. At the shorter radio wavelengths (for example, 4 centimeters) all diffuse nebulae are transparent; at longer wavelengths (for example, 83 centimeters) some of the more massive, thicker nebulae become opaque. For the nebula in Orion particularly, measurements at several wavelengths enable one to determine the total amount of material along the line of sight and also the temperature of the gas. Thus B. Y. Mills and P. A. Shaver found the Orion nebula to have an electron temperature of 8000°K, in good agreement with the value given by M. Peimbert's measurements of the [O III] lines (see Fig. 88).

The Interstellar Grains

Careful comparison of the color of an illuminating star with that of the surrounding reflection nebulosity gives some help in assessing the character of the grains that scatter and absorb starlight.

A significant clue to the nature of the particles responsible for "blacking out" distant stars comes from a study of the colors of stars that are only partially obscured. In Chapter 4 we saw how the color of a star is related to its temperature. The cool stars are red and yellow in color, whereas the hot stars are blue. The types of lines that appear in the stellar spectrum are a good indicator of the temperature, and therefore of the true color of the star. In many regions of the Milky Way, one finds stars that show spectral lines characteristic of high temperatures; yet these stars appear red. We may, therefore, surmise that the light reaching us from these objects has been not only dimmed, but reddened as well. The phenomenon is not unlike the appearance of the sun as it sets, reddened, in a dusty or smoky atmosphere.

The fact that some interstellar particles possess the ability to redden starlight tells us that they must be smaller than about 0.001 inch in diameter, for large obstacles, like meteoritic fragments, will simply block starlight without affecting its color. At the other end of the size scale, we find that the free electrons may also be ruled out as a factor in obscuring stars because they, too, are incapable of changing the color of light. On the other hand, particles of the size of atoms or ordinary molecules, about 1 hundred-millionth inch in diameter, are very potent reddeners. We have a daily example of this phenomenon in the blueness of the sky and in the redness of the rising and setting sun. The sky is blue only because the earth has an atmosphere. As the rays of sunlight pass through the atmosphere, they are deflected sidewise from their original paths by the molecules of air. The deflection of light by small particles is known as scattering, a term that we shall have occasion to use frequently. It so happens that blue rays are more easily deflected by molecules than red rays. Consequently, most of the sunlight that is scattered is blue and it is this diffused blue light from the sun that provides the beauty of the blue sky. Manifestly, the scattering process removes blue rays from the original solar beam, and the sun appears redder than it would be in the absence of the atmosphere. Also the reddening of the sun is most pronounced near the horizon, when its rays traverse a long column of the blue-eliminating atmosphere.

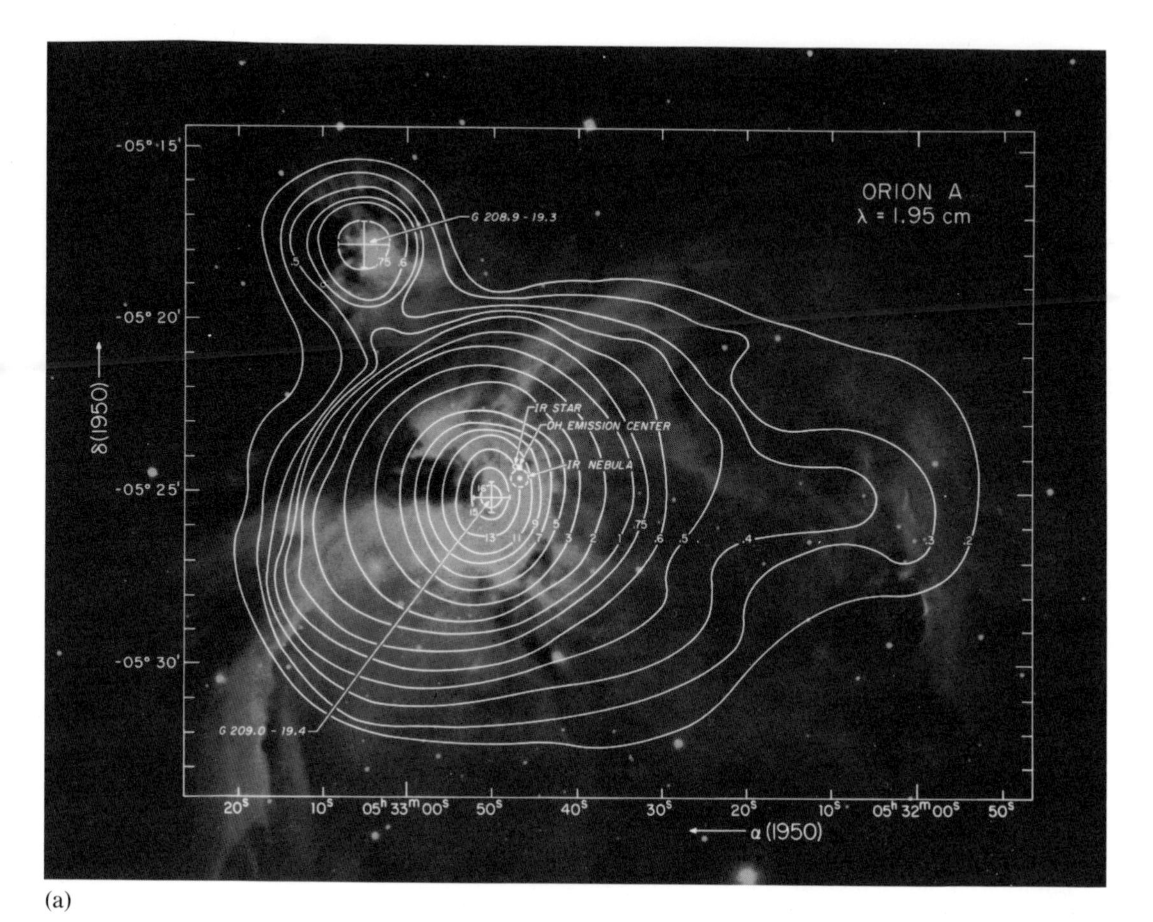

(a)

Fig. 88. Comparison of direct photographs and radio isophotes of the Orion nebula. (*a*) Radio-frequency measurements of the Orion nebula at 1.95 cm are shown by contours superimposed on a direct photograph. The radio-frequency observations were secured by Schraml and Mezger with the NRAO 140-foot telescope with an angular resolution of 2 minutes of arc. The OH source position has been observed by Raimond; the two infrared (IR) objects have been observed by Becklin and Neugebauer (star) and Kleinman and Low (nebula). (Courtesy P. G. Mezger, National Radio Astronomy Observatory.) (*b*) Radio-frequency measurements of the Orion nebula at 74 cm (frequency 408 MHz) with a resolution of 3 minutes of arc as secured with the Mills Cross at Molonglo Radio Observatory of Sydney University are compared with a Lick Observatory photograph. Notice that there are significant differences in detail between (*a*) and (*b*). At 408 MHz the radio telescope "sees through" the nebula, but at 2 cm much of the nebula is opaque and we "see" only the surface features in these regions. (Courtesy B. Y. Mills and P. A. Shaver.)

(b)

Yet, in spite of the fact that, as we have seen, interstellar space contains many atoms and molecules, these cannot be blamed for the dimming of distant stars. Atoms and molecules are simply too efficient as reddeners to be the cause of the dimming observed. Their scattering power varies inversely as the fourth power of the wavelength of the light that falls upon them, that is, they scatter ultraviolet light (3000 Å) sixteen times as efficiently as red light (6000 Å), whereas actual observations of the colors of stars show that the interstellar particles cut down the intensity of ultraviolet light only about twice as much as the red.

An important consideration is the ratio of reddening to total absorption. Suppose we measure the magnitudes, U, B, V, of a star in the three-color system of Johnson and Morgan. The "blue – visual" or $B - V$ color excess is defined by the equation

$$E_{B-V} = (B - V)_{\text{obs}} - (B - V)_{\text{true}}.$$

The method of determining the color excess for a star cluster is described in Appendix F. The ratio of total absorption to color excess, $A(V)/E_{B-V}$, is usually taken as about 3, although Harold Johnson has suggested that it may change from point to point in the Milky Way.

Theoretical, observational, and experimental studies, especially by Schalen of Uppsala, Sweden, by Greenstein and Henyey at the Yerkes Observatory, by van de Hulst of Leiden, Holland, and more recently by Wickramasinghe of NASA and by Greenberg of Rensselaer Polytechnic Institute, suggest that particles of intermediate size (from 0.001 inch to 0.00001 inch in diameter) will scatter light in such a way as to reproduce the observed degree of coloring of distant stars. Interstellar space would appear to be strewn with fine particles larger than molecules and atoms, yet so small that most of them could be seen only with a powerful microscope. Thus, grains so fine as to be invisible to the eye are responsible for the blacking out of distant stars.

What is the composition of these grains? Are they metallic (for example, iron or nickel), are they nonmetallic, or perhaps are they graphite? Obviously, we cannot yet get a sample of the interstellar material to analyze in the chemical laboratory. Fortunately, metallic, nonmetallic, and graphite particles behave differently in the ways they reflect, absorb, and scatter light. When light falls upon a nonmetallic substance such as sand, most of it is scattered, that is, reflected by myriads of tiny mirror-like surfaces. On the other hand, much of the radiation falling upon a metallic surface, say polished copper, may actually be absorbed, that is, converted into heat.

The reason for this behavior is to be found in the nature of light waves and certain differences between metallic and nonmetallic substances. Metals are good conductors of electricity, whereas nonmetallic substances like glass, silica, or ice are nonconductors, that is, insulators, or dielectrics. Metals contain large numbers of electrons that are not attached to atoms, but are free to wander to and fro. An electric current in a wire consists of a flow of these free electrons. A light wave has rapidly fluctuating electric and magnetic fields associated with it, and when it strikes a metal the electric field produces a rapid to-and-fro motion of the electrons. As the electrons rush about, they bump into atoms and lose energy, which appears as heat. The phenomenon is exactly the same as the heating of a wire by an electric current. The energy necessary to propel the electrons comes from the light beam, which is therefore diminished in intensity. On the other hand, the electrons in an insulating substance like sand are tightly bound and not free to move. Consequently when a light beam falls upon a nonmetallic, or insulating, substance, its energy is not dissipated by setting electrons in motion and the beam is reflected almost without energy loss. Thus the reflectivity, or albedo, of nonmetallic insulating materials tends to be high while that of metals is low.

The crystal structure of graphite involves a system of parallel hexagonal plates. Parallel to the plane of the plates graphite has a very high electrical conductivity, so the material behaves like a metal if the electric field of the impinging light waves causes the electrons to move parallel to the plates. Conversely, if the electric field is perpendicular to the plates, the material behaves like a dielectric.

Another proposal regarding the nature of the interstellar particles is that they are two-layer structures, with an inner core of metal or graphite and an outer shell of some dielectric. Soot flakes ejected from the atmospheres of carbon stars, or small particles of silicon dioxide from Class M stars, might serve as the "condensation nuclei" for such particles in interstellar space.

It is possible, as has been emphasized by Platt, that the absorption is not produced by small grains comparable in size with the wavelength of light but rather by much smaller particles containing only a few score atoms and absorbing at many different wavelengths. Since different particles are built up by different combinations of atoms, the net effect would be an extinction of starlight varying continuously with wavelength. Further observational tests are needed to distinguish between these hypotheses.

In any event, it appears not only that the grains are nonspherical, that is, elongated, but also that they are aligned parallel to one another over large regions of space, presumably by a magnetic field. If the light of a highly reddened star is observed through a piece of Polaroid, its brightness will appear to fluctuate by a small amount as the Polaroid is rotated. This means that the light is polarized, and hence that the obscuring particles scatter the light in a manner that depends on their orientation. To produce this observed effect, which was discovered by Hiltner and Hall, not only must the particles be ellipsoidal or needle-shaped, but they must also be lined up.

The manner of the orientation has been explained by Davis and Greenstein. In the interstellar gas, collisions of atoms with particles set them spinning rapidly. The most likely mode of spin is about the shorter diameter, like a pencil turning end over end. Unless the long diameter of the grain is perpendicular to the magnetic field, the grain will experience a change in magnetization as it rotates. This changing magnetization acts to dissipate the kinetic energy of the grain and causes it to line up with its short axis along the field. Then if the magnetic field is perpendicular to the line of sight, we will see many of the grains broadside. Under these conditions, they will produce polarization. If we are looking along the direction of the field we see the spinning particles tilted at every angle; there is no preferential direction and hence no polarization.

The existence of polarization tells us several things: (*a*) that there exist magnetic fields in space; (*b*) that the grains of interstellar dust must be cooler than the gas, for if they were at the same temperature or hotter they could not be lined up by the field; (*c*) that, since different types of grains would respond differently to magnetization and would have different polarization properties, we can eliminate certain types of particles. The actual amount of polarization is small, and it shows a slight dependence on the color of light.

Further evidence supplied by the colors of the reflection nebulosity in the Pleiades favors dielectric rather than metallic grains. The combined evidence from the dependence of extinction on color, colors of reflection nebulae, and polarization suggest that the dirty work of obscuring starlight is done not by metallic particles, although some may be present, but primarily by dielectric grains, presumably of frozen gases or dirty ices, that is, mixtures of water, ammonia, methane, and other gases rich in hydrogen, converted to a solid form by the low temperature, 100°K or less, that prevails in interstellar space. Water ice has a strong absorption band in the infrared, which should be observable from a balloon flown above the earth's atmosphere. Danielson, Woolfe, and Gaustad were unable to detect it in the spectra of highly reddened stars. Hence, the actual chemical structure of these substances may be more complicated than we have supposed. Also, the temperature fluctuations that occur within a given cloud in the interstellar medium may be of such a magnitude that ice cannot be the principal constitutent of the grains. A very low temperature would have to be maintained to prevent substantial losses by evaporation.

Presumably, grains grow by accretion, that is, by the gradual addition of one atom after another to an initial structure of some sort. If there were no processes acting to destroy them they would be much larger than they are now observed to be. Presumably they are destroyed by a number of processes; collisions between interstellar clouds, as envisaged by Oort and van de Hulst, would cause a rise in temperature of the gas and also of the grains, which could result in their destruction by evaporation or sputtering. Particles could also be destroyed by cosmic-ray impacts. The operation of these two competing processes would tend to maintain a fairly fixed distribution of sizes. There is some suggestion from variations in the reddening law that the particle sizes may vary from one part of the Galaxy to another, but the problem is not yet solved (see Fig. 89). Observations from the Orbiting Astronomical Observatory have enabled A. Code, R. Bless, and their associates at Wisconsin to extend the law of space absorption to the ultraviolet. The absorption curve does not rise monotonically as the wavelength decreases, but shows a strong maximum near 2200 Å, after which it descends to a minimum near 1600 Å and then rises again. The data can be explained by supposing that the particles are mixtures of materials like graphite, SiC, and $MgSiO_3$.

Magnetic Fields in the Galaxy

Large-scale electric fields probably do not exist in the Galaxy, since separation of electric charges cannot be maintained in an ionized gas. On the other hand, magnetic fields almost certainly do exist. We have already mentioned one piece of evidence—polarization of starlight, the Hiltner-Hall effect. Shajn noted that in the regions where polarization was observed, filaments of ionized gas tended to be lined up along the corresponding inferred direction of the magnetic field. Magnetic fields tend to constrain particles to motions parallel to the lines of force; hence a gaseous nebula would tend to deform along the direction of the magnetic field rather than perpendicular thereto.

Cosmic-ray physicists have postulated magnetic fields within the Galaxy in order to explain how high-energy particles are retained within

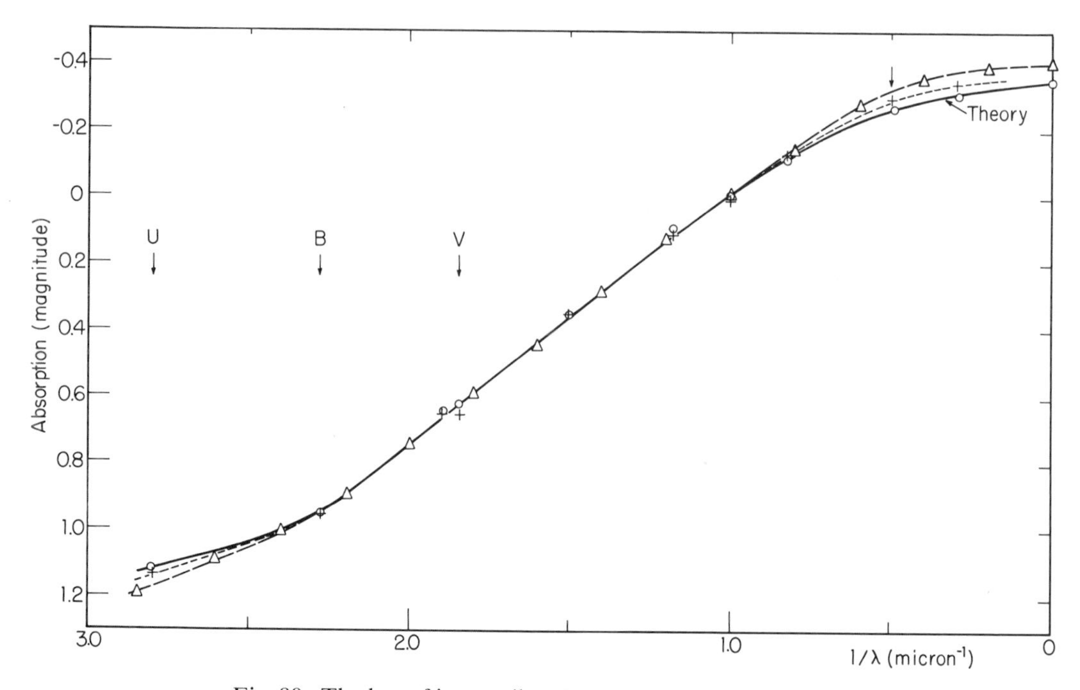

Fig. 89. The law of interstellar absorption. Absorption (in magnitudes) is plotted against the reciprocal of the wavelength (microns); 1 micron = 10,000 Å. The solid curve gives the theoretical values as deduced by van de Hulst, the short-dashed curve Johnson's data for NGC 6611, and the long-dashed curve, the mean values as compiled by A. E. Whitford. The symbols *U*, *B*, and *V* denote the effective wavelengths of the ordinary three-color system. Since the absorption is given in magnitudes, the rise of the curve at longer wavelengths corresponds to smaller absorption. More recently, the law of space absorption has been extended to the ultraviolet beyond 3000 Å.

it. Additional evidence is suggested by the presence of powerful emitters of radio-frequency radiation at very long wavelengths, the so-called non-thermal r.f. sources. This radiation is usually attributed to emission from very high-energy particles moving in magnetic fields.

One possible method for detecting interstellar magnetic fields might be from their Zeeman effect on the 21-centimeter hydrogen line. In the presence of a magnetic field, this line is split into three components, polarized according to the usual rules. Unfortunately, for a magnetic field more than 100,000 times weaker than the earth's field, the Zeeman splitting is extremely small; the profiles of the different components overlap almost perfectly and the field is therefore hard to detect. The problem is further complicated by the fact that the magnetic field may be oriented in different directions at different points along the line of sight.

The most promising method of detection appears to be the rotation of the plane of polarization by a magnetic field. More than a century ago, Michael Faraday noted that if plane-polarized light was passed through a transparent substance in a magnetic field, the plane of polarization was rotated. The effect was a maximum if the light ray was parallel to the magnetic field; it depended on the wavelength of light, the substance involved, and the strength of the field. The ionized gas of the Galaxy produces a distinct Faraday effect on the plane-polarized radio-frequency emission from distant galactic or extragalactic nonthermal sources, or from pulsars (see Chapter 12). Observations are difficult because some Faraday rotation may occur in the source itself, or in the earth's ionosphere, and the direction of rotation depends on the direction of the field. Thus if the radiation passes through a twisted tangle of magnetic fields, a considerable cancellation may occur.

The available evidence indicates that galactic magnetic fields lie between a millionth of a gauss and about 3/100,000 of a gauss, that is, it is about 0.000002 to 0.00006 of the earth's field. It is strongly variable and probably attains higher values only in a few dense regions.

The Gas Clouds and Their Motions

We may regard the interstellar medium as consisting primarily of clouds of hydrogen (and presumably also helium), with small quantities of other elements such as calcium present as "impurities." Small solid particles are also present. They contribute only a tiny fraction of the mass, but because of their absorbing properties they often obstruct the light of distant stars, and hence attain a considerable nuisance value in studies of galactic structure. Strömgren described a typical cloud as having a diameter of about 16 light-years, a density corresponding to 10 atoms per cubic centimeter and a total mass of about 400 suns. Such a cloud would contain enough solid grains to produce an extinction of about 0.2 stellar magnitude. In our local neighborhood about 7 percent of the volume is filled with clouds. Interstellar clouds show a vast range in size, mass, density, and extinction. Bok has described clouds ranging from small dense "globules," with a diameter of 0.2 light-year and a mass about 0.1 that of the sun, through "large globules" of about 3 solar masses, radii of 0.8 light-year and densities of 1600 atoms per cubic centimeter, to large clouds with 18,000 solar masses, radii of 65 light-years, and densities about 10 atoms per cubic centimeter. Figure 90, due to Donald Osterbrock, shows examples of some of these globules.

Between the clouds the densities appear to be of the order of 1 atom

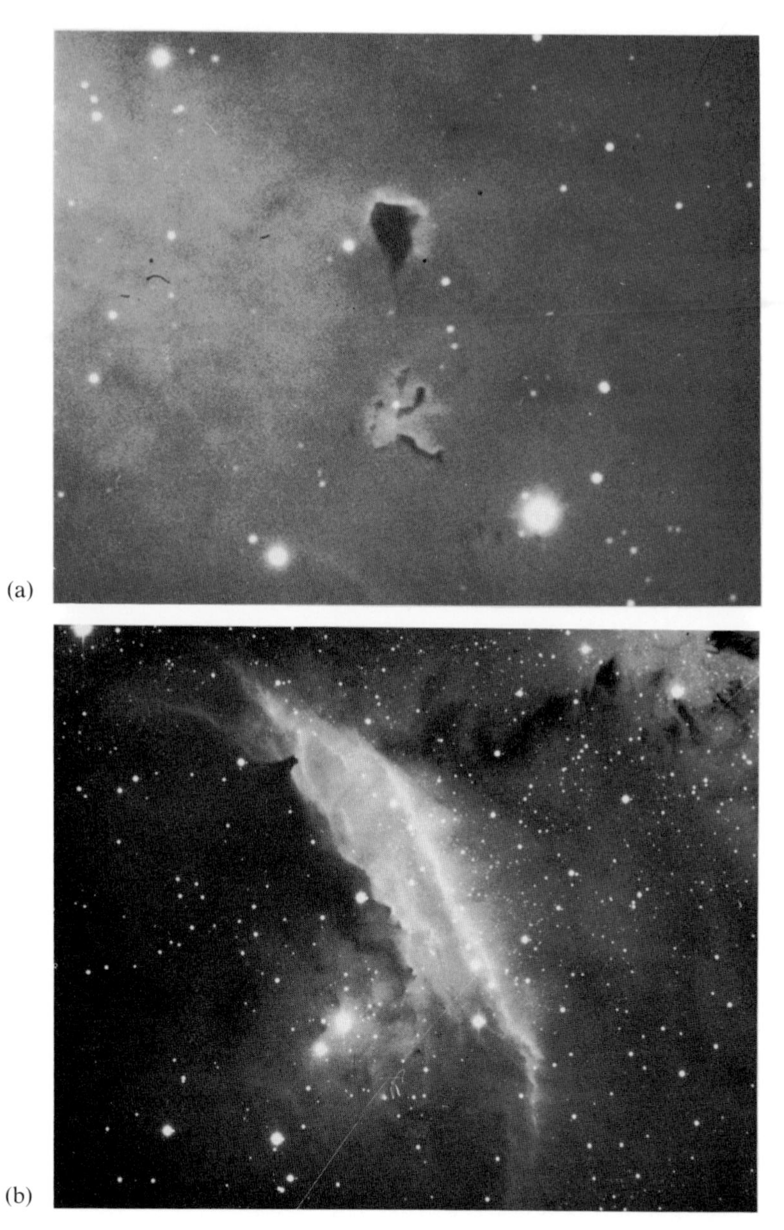

(a)

(b)

Fig. 90. Globules in the interstellar medium; globules are dense absorption features that are often surrounded or edged by ionized gas. (*a*) Globules in the diffuse nebula M 16. The largest one is 30″ × 20″; notice the bright edge toward the exciting star. (*b*) This H II region in S29 lies on one side of a dense cloud; the length of the front is 9′. (*c*) Density fluctuations just north of the central part of M 8; the area shown has dimensions 4′ × 5′. (*d*) A bright, very dense condensation near the center of M 8; the diameter of the hourglass is about 10″; the electron density, estimated from the 3729/3726-Å ratio of [O II], is about 5000 cm^{-3}. Photographed with the 200-inch Hale telescope, using a red filter which yields the radiation in H + [N II]. (Courtesy Donald Osterbrock.)

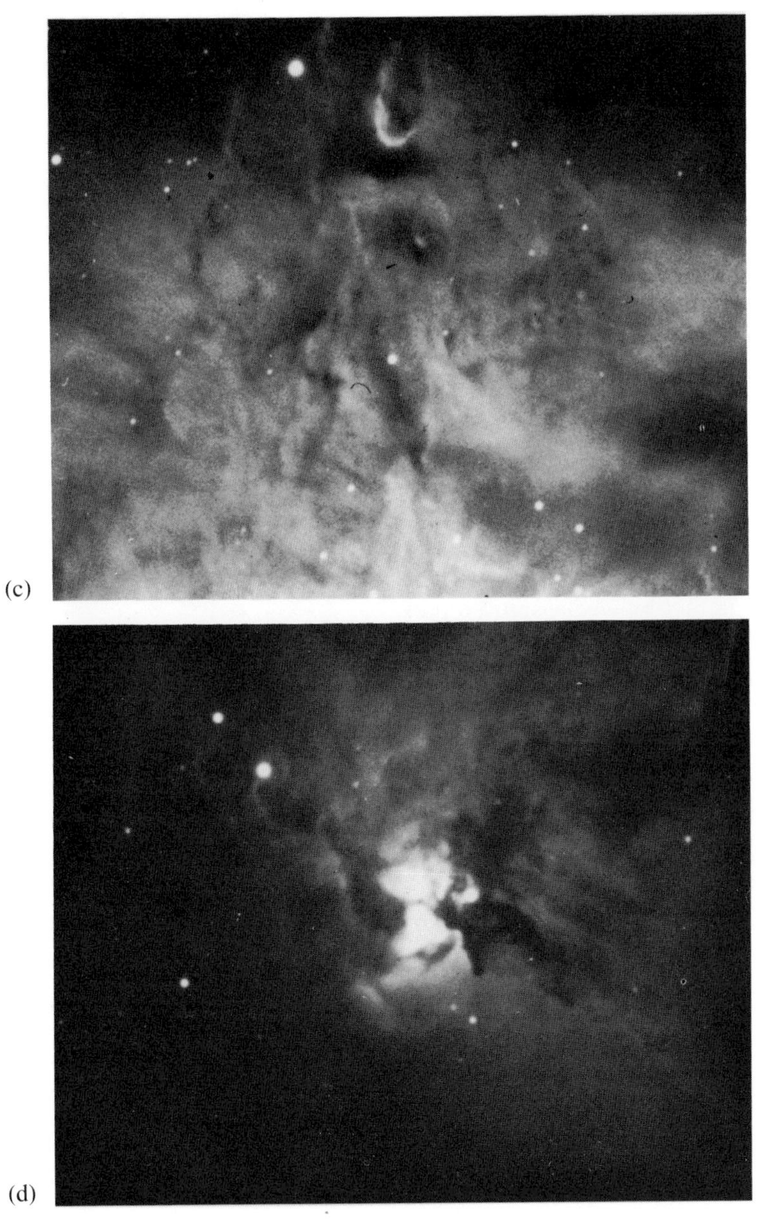

(c)

(d)

per 10 cubic centimeters. To say that the interstellar material is spread very thinly is a gross understatement; a good idea of the density of the stuff can be achieved by supposing an ordinary marble to be pulverized and the dust spread as evenly as possible throughout the volume of a sphere 500 or 1000 miles in diameter. Although the clouds are usually concentrated in the spiral arms of the Galaxy, they are found occasionally at large distances from the galactic plane. The multiple interstellar lines show that even within one spiral arm there are clouds with a large range of velocities, from 3 to 28 kilometers per second.

What controls the motion of the material? Why does it tend to concentrate in blobs and why does it not escape from the spiral arms? These questions have occupied the attention of many astronomers. We might expect radiation pressure to play an important role in the motions of the material. A hot star passing through a cloud would tend to repel the medium around it. Spitzer showed that grains in even a slightly ionized gas would tend to pick up a negative charge. Hence dust and gas would tend to be repelled together, since the ions and charged dust grains would interact by electrostatic forces.

Factors other than radiation pressure can become quite important, however. Suppose that a star of high luminosity is suddenly turned on within a dense cloud. It will immediately start to create a Strömgren sphere, but as the gas becomes ionized its temperature also rises and the pressure increases steeply. A shock wave is created and rushes through the surrounding medium.

Detailed examination of the problem indicates that the density near the newly born exciting star should fall rapidly. Thus, for the Rosette nebula in Monoceros, measurements of the density minimum suggest an age of 50,000 years. By the same argument, the fact that no density minimum is observed near the exciting Trapezium stars of the nebula in Orion indicates that this nebula cannot be more than 10,000 years old. On the other hand, from a comparison of 21-centimeter observations with theoretical predictions, Savedoff concluded that the present expansion velocity of 10 kilometers per second and radius of 195 light-years for the neutral hydrogen cloud indicates that the original cloud had a radius of 125 light-years and a density of 4.6 atoms per cubic centimeter, 2 million years ago. He suggested that the original star that initiated the expansion has disappeared. The total amount of outward-moving neutral hydrogen gas in the shell surrounding the Orion nebula region appears to be of the order of 60,000 suns. The total mass in the 30 Doradus nebula must be much greater than this.

Spitzer and his colleagues find that an expanding shell around a very luminous Class O star can accelerate large amounts of gas to velocities of the order of 10 kilometers per second; the process ceases when the density falls to around 1 hydrogen atom per cubic centimeter. How, then, can higher velocities be produced? Spitzer and Oort suggested that an isolated cloud in the neighborhood of a suddenly ignited O star might suffer a rocket effect. The side of the cloud toward the star would be subjected to a sudden stream of ultraviolet radiation that would ionize the gas. As the gas escaped, its mechanical reaction would propel the cloud outward, exactly as a rocket is propelled.

The interstellar medium is far from an equilibrium, static state. Any given portion of it is not even in a steady state. Two clouds, both composed of cold, neutral gas, collide. The temperature is raised rapidly, perhaps to 1000° or more, grains evaporate, and the gas cools by radiation. Energy is also supplied to gas clouds by supernova explosions (see Chapter 11) and by absorption of ordinary starlight. In an undisturbed cloud the grains may radiate more efficiently than the gas and reach a lower temperature.

One outstanding problem is why the interstellar medium fragments itself into clouds of greatly varying size and density. Why does not the material spread itself out more or less uniformly? The answer to this question is intimately related to the question of how the medium can evolve into stars, for the gas clouds are so tenuous that their internal gravitational attraction must be negligible. Only at a later stage, when the density has increased greatly, does gravitational attraction get a chance to pull the blobs together into actual stars.

The chemical composition of the interstellar medium is difficult to estimate. Only a few of the abundant metals whose resonance lines fall in favorable spectral regions are represented by absorption spectra. Perhaps the best estimates are those obtained from analyses of emission nebulae such as that in Orion. The theoretical methods are essentially the same as those used for studies of planetary nebulae. The composition of the Orion nebula, for example, seems to be about the same as that of the bright stars contained in it. This result is not surprising, in view of our current conviction that these luminous blue stars are formed from the interstellar medium (see Chapter 8).

The importance of the interstellar medium in astronomy is being increasingly appreciated. First, it has long been realized that the precise way in which the interstellar matter affects the passage of starlight must be learned before we can fully assess the large-scale distribution of the stars. Second, we have to learn the composition and structure of the solid particles, and why they, together with the gas, tend to collect in clouds which ultimately may form into stars. The interstellar medium is the great reservoir into which are poured the envelopes of dying stars and from which new stars are continually being formed.

Perhaps the most exciting development concerning the interstellar medium has been the discovery of the radiofrequency emission from the hydroxyl radical OH, ammonia NH_3, water H_2O, formaldehyde H_2CO, hydrogen H_2, hydrogen cyanide HCN, cyanoacetylene HC_3N, wood alcohol CH_3OH, and formic acid HCOOH (see Appendix I). The OH and H_2CO radiation originates in regions of very small angular dimensions, perhaps actually comparable in size to the solar system. Furthermore, the relative intensities of the several lines observed differ from source to source and change with time. The ammonia and water lines show anomalous intensities suggestive of maser action, that is, a tendency for molecules to be pumped into certain selected energy levels in numbers greatly in excess of what one would expect at the local temperature. Another curious feature is the tendency for reactions to favor production of organic molecules such as HCN, CH_3OH, H_2CO, HCOOH, CO, and CN rather than OH and NH_3. Even more complex organic molecules such as amino acids may be found in interstellar space. We can now understand, perhaps, why organic material has been found in certain chondritic meteorites, which are among the oldest objects in the solar system.

Is it possible that we have at last located actual regions of star formation, where even complex organic molecules are built up as the material becomes concentrated in ever denser blobs? Are we witnessing the first stages of the formations of stars and solar systems? Many astronomers feel that the answer is yes and that we may be on the verge of a solution to an ancient problem.

8 What Makes the Stars Shine?

In preceding chapters we have seen how the surface temperature, the atmospheric density, and the chemical composition of a star are learned from its spectrum. If we are lucky enough to find that this star is an eclipsing-binary system, we can also often get its dimensions, total mass, and average density. Since all sorts of stars are found in binary systems, we have a fair idea of the masses, luminosities, surface temperatures, radii, and compositions of our celestial neighbors.

From spectral studies we can learn much about the atmosphere of a star, but of the vast bulk of it, that is, its interior, we really have little direct information. We seek to know where originates the energy of the radiation that the stars are so generously pouring out into space, and through what kinds of processes it is produced. In order to answer these questions properly we shall have to consider rather carefully the ways in which astronomers and physicists have sought to understand the processes occurring in stellar interiors. Then we shall try to reconstruct the life histories of typical stars and shall show how the concept of stellar evolution leads to a fitting together of data from many diverse sources into a coherent story. Finally, we shall show how the origin of the elements themselves is to be sought in processes in stellar interiors.

Celestial Powerhouses

Our whole cosmogony, all our speculations on the history and future of the physical universe, depend upon the answer to the question: "What makes the stars shine?" Only when this question is answered is it possible to reconstruct the life history of a star such as the sun, and to interpret many of our basic data such as the mass-luminosity correlation and the Hertzsprung-Russell diagram.

It is not difficult to get a good estimate of the power output of the sun. As Pouillet did in 1838, one may observe the rate of heating of a blackened flask containing water and exposed to sunlight. The loss of heat by convection and radiation may be gauged by observing the rate of cooling of the heated flask when it is shielded from the sun, and the effects of atmospheric absorption may be roughly measured by doing the experiment at different altitudes of the sun. In a refined version of the observation, the heating effects and losses are more accurately assessed and the fact that atmospheric extinction depends on wavelength is also taken into account. Thus it is found that the energy received by each square centimeter outside the earth's atmosphere is 1.96 calories per minute. Since the distance of the sun is known, one can calculate the total power output of the sun; it amounts to 3.84×10^{23} kilowatts or 5.06×10^{23} horsepower. At the rate of 1 cent per kilowatt hour, this means that the sun, a fairly modest star, radiates a million million million dollars worth of energy every second. Such a figure does not convey much meaning. Statisticians have estimated the cost of the Second World War as a million million dollars. Hence the wealth dissipated in that conflict would run the sun for only a millionth of a second! Together, all the stars in the Galaxy radiate about a thousand million times as much energy as the sun.

By what kinds of processes do stars generate so enormous a quantity of energy? How long have the stars been shining and for how long may we expect them to continue to shine? So far as the sun is concerned, a partial answer to the second question comes, strangely enough, from paleontology. By methods of radioactive dating, geology tells us that the age of the oldest rocks in which fossils of primitive plants and animals are found is about 1000 million years. During all this time the sun must have been shining very much as it is now; for life is a fearfully fragile phenomenon, capable of existing over only a very small range of temperature between 0° and 100°C (except for spores, which under certain conditions may survive higher or lower temperatures). If the surface temperature of the sun were to change by as much as 10 percent, either upward or downward, life on the earth would probably be extinguished. Furthermore, yet older rocks may be dated by radioactive methods and in this way it is found that the crust of the earth has an age of 3500 million years. At that time the surface temperature of the earth appears not to have been drastically different from what it is now. What source of energy has enabled the sun to shine so dependably for more than 3000 million years?

An elementary calculation shows the hopeless inadequacy of any ordinary source of power, such as chemical combustion, that is, burning. Even if the sun were made of pure carbon, with just enough oxygen present to sustain combustion, it would have burned to ashes in a few thousand years. A more efficient, but still inadequate, source of energy is gravitational contraction. As a large, distended body contracts under the pull of its own gravity, the outer parts literally fall toward the center, and the energy of the falling material is converted into heat and light. Helmholtz suggested, nearly 100 years ago, that an annual contraction of the sun's radius by 140 feet would be sufficient to account for the observed rate of heat liberation. Further calculations show, however, that by shrinking from an almost infinite size to its present dimensions, the sun could shine at its present rate for less than 50 million years. Twenty million years ago the sun would have been at least as large as the earth's orbit, and at that time our planet presumably had essentially modern types of life on it.

One copious source of power, which looks very promising, is the conversion of matter into energy. Early in the present century, Einstein showed that mass and energy were related by the simple equation

$$E = mc^2,$$

where E (in ergs) is the energy that is obtainable from the complete conversion of m grams of matter, c being the velocity of light, 3×10^{10} centimeters per second. In order to keep the sun shining at its present rate, 4,200,000 tons of material would have to be transformed into energy every second. Yet the sun is so massive that its mass would thereby be diminished by only 0.1 percent in 15,000 million years.

What are the operations whereby stars may convert mass into energy? Several possibilities present themselves. First, as with radioactive substances, a small fraction of the mass may be automatically converted into energy. Second, as in certain laboratory experiments, some atoms may be transmuted with the transformation of roughly 1 percent of the mass into radiant energy. The third possibility, that of the conversion of all the matter of some atoms into energy, seems unlikely. In fact, there is no experimental evidence that justifies a belief in the total annihilation of matter, so this possibility can be excluded.

The first and most obvious suggestion is that stars continue to radiate because they contain great quantities of radioactive material. Experiments in the laboratory show that the disintegration of uranium into radium and eventually into lead is accompanied by a release of considerable amounts of energy in the form of high-speed particles and radiation. The rate of conversion is slow; a piece of pure uranium will be converted into equal parts of lead and uranium in about 4000 million years. But the rate of disintegration is always the same, whatever the nature of the surroundings; we would therefore expect the luminosity of a star that is dependent on its radioactivity to be directly proportional to its mass. The mass-luminosity relation (Chapter 6) shows, however, that the luminosities increase more rapidly than the masses. A star twice as massive as the sun is about sixteen times as bright. It is highly improbable that the more massive stars would have been stocked with greater sources of radioactive materials. Furthermore, even a pure uranium sun would not provide enough energy to maintain its observed rate of radiation. We might imagine it to contain super-radioactive elements, but experimental nuclear physics shows that materials with the required properties do not exist. It would be extremely difficult to construct a star of uranium, thorium, and other radioactive elements that would supply anything near the required energy without having it explode like an atom bomb.

The second hypothesis, that of a transmutation of elements, with a bit of the masses of the interacting atoms consumed to provide energy, seems much more promising. Accordingly, we shall look into this possibility.

The Anatomy of a Star

As is frequently true in science, the answer to our question "What makes the stars shine?" depends upon the answers to other questions, those relating to the structure of stars and atomic nuclei. The physicist may probe into the nuclei of atoms in the laboratory, but the astronomer can penetrate only the very outermost skin of the star, its atmosphere. The task of exploring the interior of a star is not as hopeless as it might seem, however, for the physicist has supplied us with the necessary tools.

Our problem is as follows: given the mass, luminosity, chemical composition, and radius of a star, and certain laws of nature such as those of gravitation, radiation, and gases, what are the densities, pressures, and temperatures at various depths within the star? Let us suppose that the star is rotating so slowly that we can neglect the effects of rotation on its structure. The luminosity of a stable star will equal the total rate of energy output in its interior. With a specified radius and luminosity, the surface temperature will adjust itself so that the surface area times the amount of energy radiated per unit area will equal the total amount of energy generated. If a bright star is relatively small, it will have a high surface temperature; if it is large, it will have a low surface temperature. Let us, for the moment, suppose that the star is of uniform chemical composition throughout. This is true for all stars at the beginning of their lives and we shall see later how it is possible to allow for changes in the hydrogen-to-helium ratio brought about by evolutionary effects, that is, the aging of the star.

To illustrate, let us suppose not only that we know the mass of a star, but also that we have a very good idea about how the mass is distributed through the interior, that is, how the density varies from point to point. There is good evidence from the data of eclipsing binaries (see Chapter 1) that the density increases very rapidly toward the center. The orbits of certain eclipsing binaries are not circular but elliptical, so that the stars travel at different speeds at different times in their orbits. Consequently the secondary minimum shown in Fig. 6 is not exactly centrally placed between the two primary minima unless the long dimension (major axis) of the orbit is pointed toward the observer.

Two spherical stars would attract each other like point masses and, unless there was a third star in the system, the orbit would remain pointed for all time in the same direction in space. In reality the stars are distorted by their own rotations and tidal effects on one another, they no longer attract like spheres, and the orbit rotates in space. Consequently, if the light curve is repeatedly observed, the position of the secondary minimum will shift back and forth as the orbit turns around in space. The rate of turning of the orbit depends on the separation of the stars, the period, and the concentration of density toward the stellar center. The greater the rise of density toward the center, the slower will be the turning of the orbit. The observed rate of orbital turning (apsidal motion) is in harmony with the high central density concentrations predicted by the theory of stellar structure.

A star can be stable only if, at each point in its interior, gas and radiation pressures, which depend on the temperature and chemical composition of the material according to known laws, exactly balance the weight of the overlying layers. As we go deeper into the star the weight of the upper layers, which we may compute from the law of gravitation and the way in which the mass is spread throughout the star, increases and so also does the pressure of gas and radiation, which rises as the temperature increases. We see, therefore, how the observed mass, radius, and luminosity of the star may enable us to evaluate the pressure, density, and temperature everywhere in the interior. From the observed luminosity of the star, we can easily calculate the rate at which it is generating energy, for if a normal star is to remain stable the rate at which energy is radiated at the surface must equal the rate at which it is being released in the interior. If the liberated energy does not escape, but is stored up in some fashion, the mounting pressure of heated gas and radiation will soon cause the star to explode.

Of great importance to the luminosity of a star is its central temperature, since, as we shall see shortly, this temperature determines the rate of generation of energy. The chemical composition of a stellar interior also plays a decisive role, for three reasons: first, because the chemical composition largely determines the transparency, and hence the ease with which energy flows to the surface; second, and more important, because the central temperature depends on the composition; and third because the rate of generation of energy depends intrinsically on the chemical composition of the material involved.

The pressure p exerted by a gas is proportional to the temperature and the number n of individual particles per unit volume in accordance with the expression

$$p = nkT,$$

where k is the Boltzmann constant (Appendix C) and T is the absolute temperature. For a gas composed of neutral hydrogen atoms, the average molecular weight (the total mass divided by the total number of particles) is 1. We speak of the molecular weight of a gas even though we are dealing with atoms, for it measures the mass divided by the number of particles, the mass of the oxygen atom being taken as 16.00. (In ordinary usage, the molecular weight of a gas is the number of grams of the substance contained in 22,400 cubic centimeters at atmospheric pressure and 0°C.) But if the hydrogen is ionized, as it is in a stellar atmosphere, there are twice as many free particles—nuclei plus electrons—with no change of mass, and the molecular weight becomes 0.5. A completely ionized carbon atom, with mass 12, yields seven particles, six electrons and a nucleus, so its molecular weight is $^{12}/_{7}$, or 1.72. Each unit mass of hydrogen contributes two particles, each unit mass of carbon $^{7}/_{12}$ of a particle. A carbon star and a hydrogen star that were alike with respect to size, mass, and density variation would differ in internal temperature, because the particles of the carbon star would have to work harder, that is, move faster, than the more numerous hydrogen particles to support the weight of the overlying layers.

Thus the star composed of heavy elements is hotter inside than the hydrogen star. Unless there is a deficiency of the necessary energy-producing fuel, energy will be generated more rapidly in the star with the hotter interior, and it will shine more brightly. As long as its supply of fuel lasted, a metal star would be about 100 times as bright as a hydrogen star. If the sun were composed of pure hydrogen, its central temperature would be somewhere near 10 million degrees. The central temperature would be about 40 million degrees for heavy atoms, whereas a pure helium composition would require a temperature in the neighborhood of 27 million degrees. Therefore, the chemical composition of its interior will profoundly affect the structure and total luminosity of the star. If the star is not homogeneous in composition, its central temperature will differ from these figures. Early workers who were unaware of the overwhelming preponderance of hydrogen and helium assigned central temperatures that were much too high.

To determine the structure or "model" of a given star the astronomer proceeds by trial and error. He adopts the relative proportions of hydrogen, helium, and heavy elements from an analysis of the star's atmosphere. With the aid of the known mass of the star, the gas laws, the radiation laws, and the law of gravitation, he calculates how the pressure, temperature, and density increase toward the center of the star. The march of temperature and density with depth will depend on the manner of energy transport through the layers in question.

In stars of normal density, energy may be transported in a stellar interior either by radiation, each quantum being passed on from one atom to the next, or by large-scale mass motions, commonly called convection. Let us consider these processes in turn.

Throughout most of the interior of a star, the temperature is well above a million degrees. Hence most of the radiation occurs in the form of x-rays, and most atoms are stripped of all electrons down to the inner shells. Absorption of radiation occurs as photoelectric ejection of electrons occurs; the phenomenon is similar to the photoionization of hydrogen from the second level which produces the absorption at the head of the Balmer series in the spectra of the hotter stars. At yet higher temperatures all electrons may be stripped off and the chief hindrance to the outward flow of radiation may be scattering by free electrons. Since the lighter atoms lose their inner electrons more easily and heavy atoms such as iron may retain their innermost electrons at very high temperatures, the ability of the material to block outgoing radiation, that is, its opacity, will depend on its chemical composition. For any assumed mixture of elements, the opacity depends in a complicated way on temperature and density. Extensive calculations have been carried out, for example by Arthur Cox and his associates, for all temperatures and densities likely to be encountered in stars.

Convection is a familiar process for the redistribution of heat, in which the energy is carried by matter in mass motion. A stove heats a room primarily not by radiation but by warming a mass of air in its neighborhood, which then rises and moves across the top of the room. Meanwhile, its place is taken by cool air which in turn becomes heated by the stove. In a similar fashion, deep within a star, a blob of heated gas rises toward the surface, while cooler material from a higher layer sinks downward to replace it; the process is a continuous one. These large-scale convection currents may be orderly, completely disorderly (turbulent), or something in between.

Energy flows outward in a star because there is a steady decline of temperature from the center to the surface. The rate of radiation of energy by the star is given and this energy outflow must be provided by radiation or by convection currents. If the rate of fall of temperature (temperature gradient) required for radiative flow is too great, convection will set in. Convection is a more efficient mode for moving energy outward than is radiation but it can occur only in certain regions of the stellar interior where the ionization or temperature gradient is just right. It is usually easy to decide whether a given layer in a star is characterized by one kind of energy transport or the other, but if convection currents are at work it is trickier to estimate the rate of energy flow and the change of density and temperature with depth. In the visible layers of the sun's atmosphere the outward flow of energy takes place almost entirely by radiation, but just below the photosphere a convective region starts and extends downward about a fifth of the solar radius. In yet deeper regions, the flow of energy is entirely by radiation. In cooler, less massive, main-sequence stars the convection zone becomes more important, and in faint red dwarfs it appears that throughout the whole star the energy is carried outward by convection currents.

Proceeding to more massive main-sequence stars, we find that somewhere near spectral class F5 the outer convection zone suddenly shrinks to a small size. An intrinsically very luminous main-sequence star, such as Spica, develops a convective core, although throughout the vast bulk of the star energy is carried outward by radiation. These differences in the internal structures of stars have important consequences on their life histories.

To calculate a model for a star like the sun we would start with a chemically homogeneous body of the same mass and calculate its internal structure, including the central temperature and density, and predict its luminosity and radius. If such a program is carried out for the sun, using the best atmospheric estimates of the total hydrogen content and hydrogen-to-helium ratio, one finds the wrong radius and a luminosity that is too low. As we shall see in the next section, normal stars shine by converting hydrogen into helium. Our chemically homogeneous object corresponds to a star of zero age. Hence one must allow for the depletion of hydrogen and repeat the calculations until the computed radius and luminosity agree with the observed ones. Such calculations have been carried out by Sears and Brownlee, among others. Speaking in general terms, the model of a star must fit physically reasonable limitations. We cannot have a star with a hollow core or a core whose density increases without bound, a temperature decreasing with depth, or a stellar surface under a substantial pressure. Any physically possible model fulfills these conditions, of course, but this alone does not guarantee that it is acceptable.

Any normal star must be stable. That is, if it is subjected to a small disturbance, for example, the gravitational attraction of a nearby star, the star must rebound to restore the status quo. Oscillations may take place, as indeed occurs in the Cepheid variable stars (see Chapter 10), but these oscillations must remain finite. Thus, any star model that pretends to correspond to reality must be stable; this condition excludes many otherwise possible models.

The theory of stellar structure has advanced greatly in recent years. On the one hand, there have been striking advances in our knowledge of the underlying basic physics; on the other, improved computers have made it possible to carry out calculations that were wholly beyond our capabilities until quite recently.

Figure 91 displays the internal temperature and density distributions in a solar model computed numerically a few years ago by Ray Weymann. Although slight improvements are now possible, the essential features remain unchanged. The vast bulk of the sun has a density less than that of water and a temperature in excess of a million degrees.

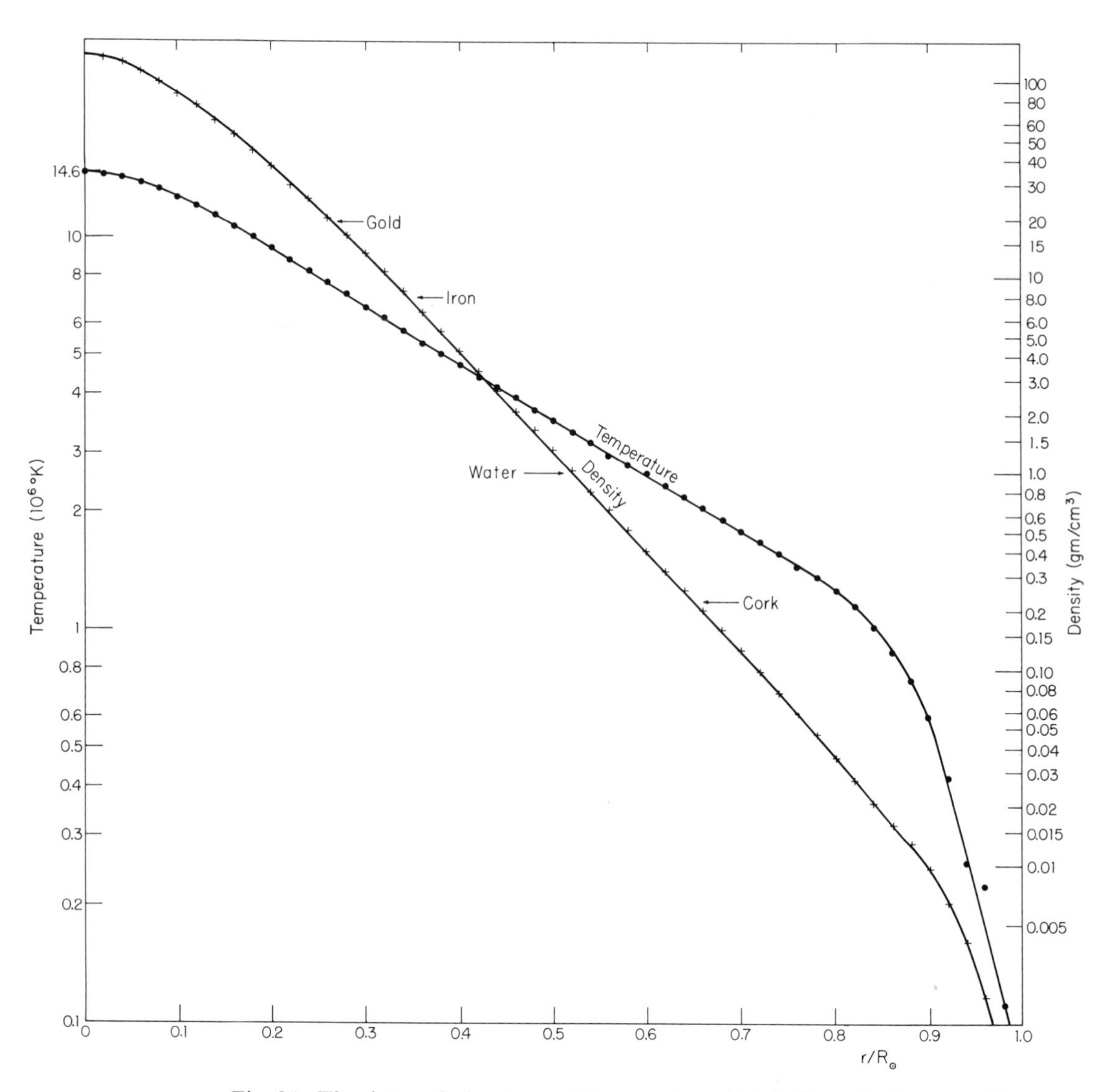

Fig. 91. The internal density and temperature distributions in the sun. The density (in terms of that of water) and temperature are plotted against the distance from the center (in terms of the radius of the sun). The steep density drop indicates that the vast bulk of the sun has a density less than that of water. The temperature of most of the interior of the sun is greater than 1 million degrees. The outer zone, corresponding to a shell with a depth equal to 0.14 of the solar radius, is in convective equilibrium. (After R. Weymann.)

An important outcome of the early work was that, even without knowing the specific mechanism of energy generation (assuming only that it depended on the density and temperature), it was possible to derive a mass-luminosity law, on the assumption that all stars considered were built on the same model. With plausible chemical compositions and the known masses of main-sequence stars, astronomers were able to predict stellar radii and luminosities to within the legitimate uncertainties of the basic physics employed (for example, the absorptivity of matter for the x-rays prevailing at temperatures of millions of degrees can be obtained only by theoretical calculations). Hence one could have confidence that the temperatures (10–35 million degrees Kelvin) and densities (20–200 grams per cubic centimeter) predicted for the centers of main-sequence stars were physically meaningful, although accurate values could not be stated for any one star. It turns out that, at precisely these temperatures and densities, things begin to happen to the nuclei of light atoms such as carbon and nitrogen when they are placed in a medium of hydrogen. What happens is that atoms are transmuted and energy is liberated in just the requisite amounts to explain the observed radiation of these stars.

The Transmutation of Elements

We have seen how, guided by the observed masses, luminosities, and diameters of the stars, and by well-known laws of nature, astronomers have deduced the physical conditions obtaining within a star. In searching for processes that will convert mass into energy, physicists have looked for one or more that will liberate enough energy at the predicted central pressures and temperatures of main-sequence stars to reproduce their observed luminosities.

Years ago the suggestion was made that the stars shine by converting hydrogen into helium. The atomic weight of hydrogen is 1.00813, and that of helium is 4.00386. Therefore, if four hydrogen atoms could be converted into one helium atom, 0.02866 unit of mass, or $1/141$ of the original mass, would appear as energy. The stars of the main sequence and most giants, subgiants, and supergiants probably do shine by converting hydrogen into helium, but the process is not so simple as jamming four protons together to form a helium nucleus. In order to learn the conditions under which the transmutation of elements takes place, we shall turn to the studies of the physicist.

From Chapter 3 we recall that the nuclei of atoms are composed of protons and neutrons. The number of protons determines the charge of the nucleus and therefore the kind of atom; the number of neutrons determines the isotope of the element. We saw, for example, that an ordinary carbon atom, of atomic weight 12, has a nucleus consisting of 6 protons and 6 neutrons, while a carbon isotope of atomic weight 13 contains 6 protons and 7 neutrons. Chemically, the two atoms are similar.

The heaviest naturally occurring elements, uranium and thorium, spontaneously break down into less heavy atoms, such as radium and mesothorium, and ultimately into lead. But it is possible, by bombardment with high-speed protons, helium nuclei (alpha particles), or neutrons, to disintegrate other atoms and thus achieve the transmutation of the elements. Devices such as the electrostatic generator or the cyclotron enable the physicist to speed up bombarding particles to enormous velocities and to fire them at atoms.

What kinds of particles are most effective? Protons and helium nuclei are useful only in the bombardment of light elements. Positively charged atomic nuclei repel the similarly charged protons and helium nuclei, and, since heavy nuclei have very large positive charges, they strongly repel the incoming hydrogen or helium nuclei and drive them away before they can penetrate the nucleus. But the neutron possesses no charge and consequently may easily penetrate the heart of any atom.

One may obtain neutrons from a radioactive "pile" in an atomic-energy plant or from the heavy isotope of hydrogen, deuterium, whose nucleus consists of a single proton tightly bound to a neutron. When heavy water, that is, water made with heavy hydrogen, is bombarded with deuterium, each pair of colliding deuterium atoms is converted into one helium atom of atomic weight 3 and one neutron.

If we bombard nitrogen (atomic weight 14) with neutrons, we obtain boron (atomic weight 11) and helium (4). We may write the reaction in the form of an equation:

$$ {}_{7}N^{14} + {}_{0}n^{1} \rightarrow {}_{5}B^{11} + {}_{2}He^{4}, $$

where the superscript denotes the atomic weight and the subscript the charge. Similarly, iron bombarded by neutrons is transformed into a manganese isotope with the ejection of a proton:

$$ {}_{26}Fe^{56} + {}_{0}n^{1} \rightarrow {}_{25}Mn^{56} + {}_{1}H^{1}. $$

Reactions of this latter type appear to be of great importance in certain stars, not for the production of energy, since little energy is actually liberated, but rather for the building up of elements of higher atomic number. If a sufficient supply of neutrons can be produced, elements like titanium and iron can be built up by neutron capture into elements such as zirconium, barium, or even lead.

Several possibilities are open when a nucleus is bombarded by a proton. First, the proton may simply remain in the nucleus and produce a new nucleus with a mass and a charge each greater by one. Deuterium (heavy hydrogen) when bombarded by protons yields the helium isotope of atomic weight 3, plus radiant energy:

$$ {}_1H^2 + {}_1H^1 \rightarrow {}_2He^3 + \text{radiation}. \qquad (1) $$

Second, a proton colliding with a nucleus may be converted into a neutron, with the ejection of a positive electron, or positron, which has the same mass as a negative electron but a charge of the opposite sign. (Dirac predicted its existence from theory, and Anderson at the California Institute of Technology later found it experimentally.) The resulting nucleus retains the same charge but has a greater mass. For example, in the case of hydrogen, a proton-proton collision may form a heavy-hydrogen nucleus, consisting of a proton and a neutron:

$$ {}_1H^1 + {}_1H^1 \rightarrow {}_1H^2 + {}_1\epsilon^+ + \nu, \qquad (2) $$

where ϵ^+ stands for the positive electron that is created. The ejected positive electron then encounters an ordinary negative electron, the two annihilate each other, and the excess energy appears as radiation. The symbol ν denotes the neutrino, a neutral particle of negligible mass that is emitted with positive or negative electrons; it carries away momentum and energy. Hence, not all annihilated mass appears as energy; some is carried away by neutrinos. Third, the bombarded nucleus may break up into two or more parts, one of which is a helium nucleus. Lithium (atomic weight 7) bombarded by a proton breaks down into two helium nuclei:

$$ {}_3Li^7 + {}_1H^1 \rightarrow {}_2He^4 + {}_2He^4. \qquad (3) $$

The light nuclei of beryllium and boron are particularly vulnerable to proton collisions, as shown by the reactions

$$ {}_4Be^9 + {}_1H^1 \rightarrow {}_3Li^6 + {}_2He^4, \qquad (4) $$

$$ {}_5B^{11} + {}_1H^1 \rightarrow 3\,{}_2He^4. \qquad (5) $$

Note that the sums of the nuclear charges and of the atomic weights must be equal on both sides of each equation.

In Fig. 92 we show the tracks of a negative and a positive electron in a bubble chamber. Charged particles produce bubbles as they are slowed down in liquid hydrogen. These bubbles, illuminated and photographed,

Fig. 92. Tracks of negative and positive electrons, formed in a bubble chamber. In this device charged particles produce bubbles as they are slowed down in liquid hydrogen, and their tracks can be observed by proper illumination. This photograph shows one of the most important results of modern physics—the direct conversion of energy into matter. A gamma ray is converted in the field of a hydrogen nucleus into an electron-positron pair, which forms the upper V. The triplet consists of a similar pair and an additional electron, which was knocked out of a hydrogen atom when the gamma ray was converted in the field of the orbital electron. A strong magnetic field is impressed perpendicular to the plane of the tracks. Consequently, at the point of decay the electron spirals off to the left and the positron spirals to the right. Note the tightening of the spiral tracks, which is caused by the loss of energy by both positron and electron. As the velocity decreases the curvature of the path steadily increases. (Courtesy Lawrence Radiation Laboratory, University of California.)

show the paths of particles that are themselves much too small to be seen. A magnetic field has been applied, so that the path of the negative electron is bent in one direction and that of the positive electron in the other. Figure 93 shows how a particle of high energy, say a proton of 1000 million electron volts energy, may collide with an atom and produce a host of secondary high-energy particles; these in turn collide with other atoms and thus produce a shower of high-speed electrons and evanescent particles called mesons. In showers such as these, the nucleus that is hit is actually shattered and great quantities of energy (some of it in the form of short-lived particles) are liberated. Particles of such high energy are associated with the cosmic rays.

Reactions involving alpha particles are also of interest. Thus

$$\begin{aligned} {}_6C^{12} + {}_2He^4 &\rightarrow {}_8O^{16}, \\ {}_8O^{16} + {}_2He^4 &\rightarrow {}_{10}Ne^{20}, \end{aligned} \tag{6}$$

and so forth. These reactions require very high energies (or temperatures). At sufficiently high densities and temperatures two alpha particles may momentarily coalesce to form a nucleus of ${}_4Be^8$; this nucleus is unstable and disintegrates spontaneously, within 10^{-14} second, into two alpha particles, but if before it has time to fly apart another alpha particle is captured, a ${}_6C^{12}$ nucleus is formed.

With yet higher temperatures (energies), reactions such as

$${}_6C^{12} + {}_6C^{12} \rightarrow {}_{11}Na^{23} + {}_1H^1$$

or (7)

$${}_6C^{12} + {}_6C^{12} \rightarrow {}_{10}Ne^{20} + {}_2He^4$$

may occur. In the next chapter we shall find that some of these reactions can become important in stars in the late stages of their lives.

How Energy Is Produced in Main-Sequence Stars

Earlier in this chapter we saw that the internal temperatures of the stars amount to millions of degrees. At such temperatures, the constituent particles are moving about so rapidly that occasional encounters between protons and nuclei should be sufficiently violent to produce nuclear transformations. It is important to realize that the various nuclear transformations do not operate effectively at the same temperatures. Consequently, a nuclear process responsible for the generation of energy in one star may not work in another star with a different central temperature.

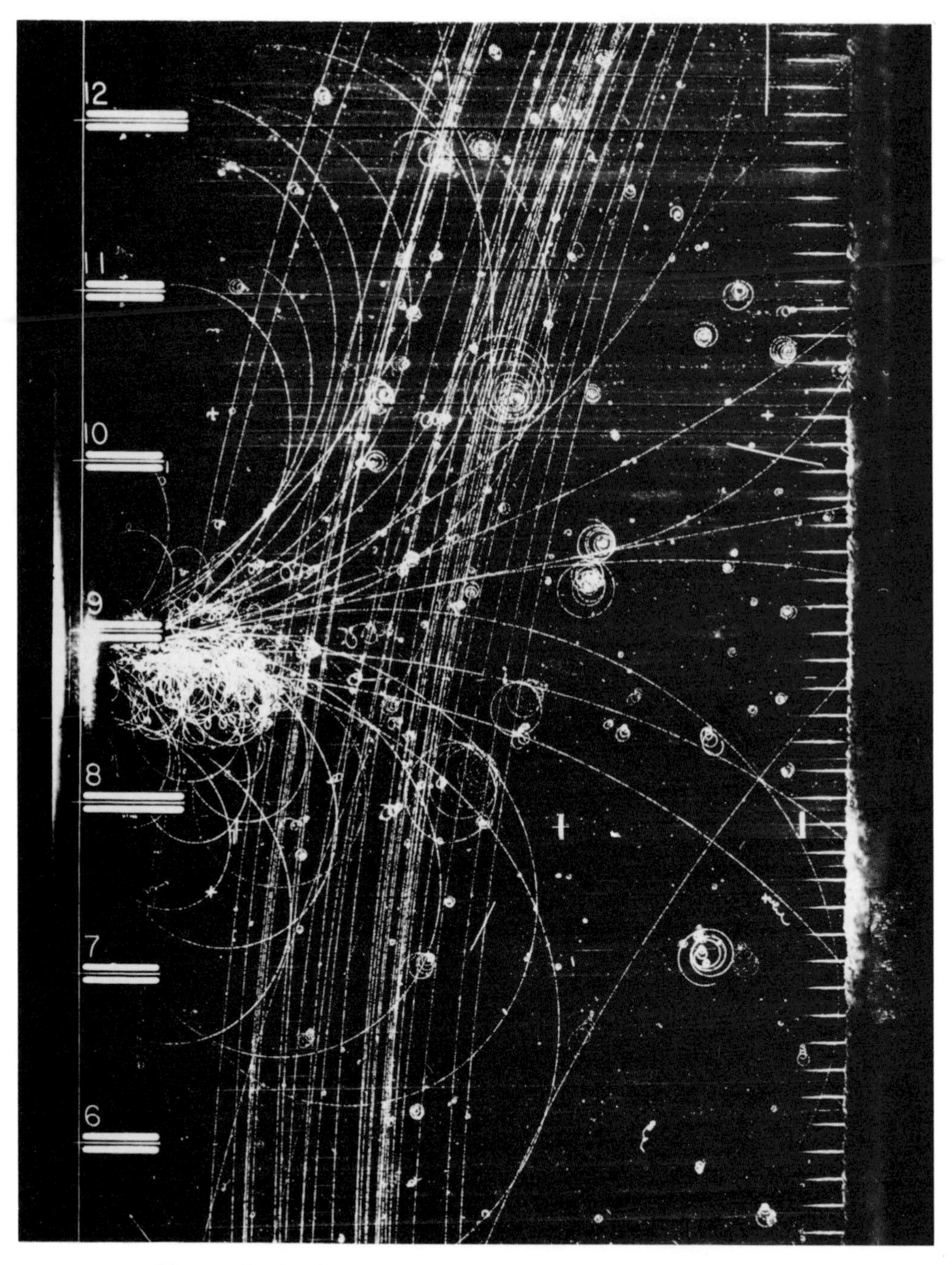

Fig. 93. Production of secondary particles by a high-energy cosmic-ray particle. This photograph, obtained with the 72-inch bubble chamber, shows what is called a cosmic-ray shower. The high-energy primary particle enters from the left, collides with an atom, and emits a shower of various particles. The impressed magnetic field (whose lines of force are perpendicular to the plane of the paper) causes the paths of the charged particles to be curved; the lower the speed of the particle the greater the curvature. (Courtesy of the Lawrence Radiation Laboratory, University of California.)

In stars such as the sun, whose central temperatures are in the neighborhood of 13–15 million degrees, the principal source of energy is the proton-proton reaction originally proposed by Critchfield and Bethe. In the first step, two protons collide to form a deuteron as in reaction (2); at the same time a positive electron and a neutrino are formed. Then the deuteron captures a proton to form a nucleus of $_2He^3$ by reaction (1) and two He^3 nuclei may collide to form a nucleus of He^4 and two protons:

$$_2He^3 + {}_2He^3 \rightarrow {}_2He^4 + {}_1H^1 + {}_1H^1. \tag{8}$$

Thus, in the course of events, four protons are destroyed to form an alpha particle; the two positrons that are also created are annihilated when they encounter ordinary electrons. The two neutrinos formed in reaction (2) steal parts of the energy and escape from the star. It is of interest that the first step in the reaction—the formation of the deuteron—has never been verified experimentally, but depends entirely on theory. Nevertheless, its validity seems well established. Alternative channels of the reaction are possible and have been discussed by a number of physicists. Fortunately, a direct experimental check on the predictions of theory is available.

Once a He^3 nucleus is formed, one of the possible reactions is

$$_2He^3 + {}_2He^4 \rightarrow {}_2Be^7 + \gamma;$$
$$_4Be^7 + {}_1H^1 \rightarrow {}_5B^8 + \gamma;$$
$$_5B^8 \rightarrow {}_4Be^{8*} + \epsilon^+ + \nu;$$
$$_4Be^8 \rightarrow {}_2He^4 + {}_2He^4.$$

That is, a He^3 nucleus collides with a He^4 nucleus to form Be^7 with emission of a γ-ray; Be^7 captures a proton to form B^8 and another γ-ray is emitted; B^8 decays first to an excited nuclear state of Be^8 with the emission of a positive electron and a neutrino, and the Be^8 promptly decays to two α-particles.

The neutrinos escape forthwith from the sun since they interact only very slightly with matter. It is precisely this weak interaction that makes them so hard to detect. A possible reaction is

$$_{17}Cl^{37} + \nu \rightarrow {}_{18}A^{37} + \epsilon^-,$$

that is, a Cl^{37} nucleus captures a neutrino to form an A^{37} nucleus which ejects an electron. As a detector, Raymond Davis used a vast quantity of carbon tetrachloride (CCl_4) and he made his observations in a deep mine to avoid extraneous reactions produced by cosmic rays.

The upper limit to the reaction rate, 3×10^{-36} per Cl^{37} nucleus per second, is about half the calculated rate, but perhaps further refinements in the nuclear theory (due to Bahcall), improvements in the theory of stellar structure, and assumed initial helium content may resolve the discordances.

In stars appreciably more massive than the sun, the principal source of energy is no longer the proton-proton reaction. When the central temperature approaches 18 or 20 million degrees, carbon is transformed and a remarkable process which leads to the production of helium from hydrogen takes place. The cycle is

$$
\begin{aligned}
{}_{6}C^{12} + {}_{1}H^{1} &\rightarrow {}_{7}N^{13} + \gamma; \\
{}_{7}N^{13} &\rightarrow {}_{6}C^{13} + \epsilon^{+}; \\
{}_{6}C^{13} + {}_{1}H^{1} &\rightarrow {}_{7}N^{14} + \gamma; \\
{}_{7}N^{14} + {}_{1}H^{1} &\rightarrow {}_{8}O^{15} + \gamma; \\
{}_{8}O^{15} &\rightarrow {}_{7}N^{15} + \epsilon^{+}; \\
{}_{7}N^{15} + {}_{1}H^{1} &\rightarrow {}_{6}C^{12} + {}_{2}He^{4}.
\end{aligned}
$$

A carbon-12 nucleus captures a proton and becomes radioactive nitrogen-13 with the emission of radiation (γ-ray). Nitrogen-13 disintegrates into carbon-13 with the ejection of a positive electron. The next proton collision transforms carbon-13 into ordinary nitrogen of atomic weight 14 with the emission of radiation. The nitrogen nucleus, struck by a proton, emits radiation and becomes oxygen of atomic weight 15. This nucleus is unstable, that is, radioactive, and disintegrates into nitrogen of atomic weight 15 with the ejection of a positive electron. When this heavy nitrogen nucleus captures a proton, it splits into an α-particle and the original carbon of atomic weight 12. By this cycle, four hydrogen nuclei have been converted into one helium nucleus, and the original carbon atom reappears. The carbon, which behaves like a chemical catalyst, can be used over and over again until all the hydrogen has been converted into helium.

This carbon cycle, discovered independently by H. Bethe of Cornell and C. F. von Weizsäcker in Germany, explains fairly well the luminosities of the stars along the brighter part of the main sequence and probably giant and supergiant stars also, that is, the majority of the naked-eye stars. The faint main-sequence stars appear to be explained adequately by the proton-proton reaction. In fact, Davis's neutrino experiment shows that the carbon cycle is responsible for less than 9 percent of the solar power output.

One very important point must be emphasized. As long as we require the stars to be chemically homogeneous or nearly so throughout their interiors, calculations of stellar models yield objects that fall on or near the main sequence. There is no possibility of explaining giant or supergiant stars by models of uniform chemical composition. This situation became apparent 50 years ago, when it was pointed out that giant stars constructed on the same model as the sun would have relatively low central temperatures. These stars must be constructed differently from main-sequence stars.

In fact, the only way giants and supergiants can be explained is by supposing that they have dense, hot cores that consist of some essentially inert material, principally helium. Surrounding this inert core is a thin shell where energy generation takes place. Outside this shell lies a vast outer "envelope" which accounts for the huge size of the star. How perfectly normal stars can develop such structures in the course of their lives will be described in the next chapter.

9 The Biography of a Star

The problem an astronomer faces in attempting to construct the life history of a star was described neatly more than a century ago in a classical essay by Sir John Herschel. Imagine a citizen of a megalopolis who had never seen a tree and was turned loose in a virgin forest for an hour and required to bring back an account of the life history of a tree. An observant man would quickly perceive a dead snag or a rotting log as the last chapter of a tree's history, but other stages, including the origin of trees, would be more difficult to establish. The analogy to the astronomer's problem will be evident.

The Course of Stellar Evolution

Within a period comparable with the age of the earth, many stars must show appreciable aging, or, as we usually call it, evolution. The term "evolution," which is generally applied to the history of a species, a society, a culture, or a civilization, but not of an individual, is probably an unfortunate choice. The universe or the Galaxy evolves, but not strictly speaking a star; its life history is determined by its mass, chemistry, and rate of rotation. However, the term is so ingrained that it cannot be dethroned.

If the lifetime of a typical star like the sun is measured in thousands of millions of years, it may appear difficult to recognize evolutionary effects. Certain stages of a star's evolution are passed over quickly, but aside from the catastrophic supernova phase, which may affect only a few stars in the course of their lives, most stars evolve so slowly that any effects that may occur could scarcely be recognized.

Throughout most of its life a star shines by converting hydrogen into helium. To keep on shining at the present rate, the sun, an unassuming dwarf, must convert 564,000,000 tons of hydrogen into 560,000,000 tons of helium every second. The sun has been shining for at least 4500 million years and should continue to shine for several thousand million more. On the other hand, Y Cygni, which is burning up hydrogen a thousand or so times as rapidly as the sun, seemingly cannot last more than about 100 million years longer, no matter how we juggle the hydrogen content. If the hydrogen content of Y Cygni is 80 percent, a rather high estimate, the star should be only about 35 million years old. But the universe has probably been in very much its present state for at least 10,000 million years. Therefore, if nuclear reactions are responsible for the energy generation in stars, objects like Y Cygni must either be recent creations or have been kept from shining for most of their lifetimes. Hence we may conclude that not only do stars evolve, but the processes of star formation may be going on before our very eyes.

Accordingly, let us describe, in broad terms, the life history of a typical star as it is pictured by a judicious combination of theory and observation. Our present views, which may be subject to considerable revision in details, envisage the stars beginning their lives as huge, tenuous, gaseous spheres, slowly contracting under gravitational attraction from condensations formed in the interstellar medium. At this stage the temperature is so low that nuclear reactions cannot occur. As soon as the temperature and density rise to the point where nuclear transformations can take place, enough energy is produced to raise the internal temperature and gas pressure sufficiently to stop the star from contracting further. The first energy-producing reaction to occur is the proton-proton reaction or the carbon cycle (Chapter 8). The star then has reached the main sequence and it remains there until the hydrogen in the interior has been seriously depleted. The further course of evolution of the star will depend on its mass. If it is about as massive as the sun or less so, hydrogen will be exhausted first in the center and then the hydrogen-to-helium conversion (hydrogen "burning") will gradually work outward in an energy-producing shell. The material in the interior of this shell will be inert.

On the other hand, the core of a massive star keeps itself stirred up by convection. Hence hydrogen is depleted at an almost uniform rate in a core having about a tenth of the radius and containing about a tenth of the mass of the star. In any event, the onset of an inert core has the following effect. Energy is produced by thermonuclear reactions only in a thin shell. The core continues to shrink in size even though it grows in mass, but the over-all size of the star increases. Its total luminosity may increase but the surface temperature falls. At this stage the star becomes a giant or, if it is sufficiently massive, a supergiant. The star continues to expand as more and more hydrogen is burned in the shell, until finally the outer layers become unstable and may even dissipate. Before this happens, the core itself may become so hot and dense that helium may be "burned" to carbon.

All theories of stellar evolution become somewhat vague at this point. The equations of stellar structure themselves indicate that the star becomes unstable, but they offer no help in tracking it through its last hours as an object shining by nuclear processes. The last stage of a star's life is well identified, however. It becomes a white dwarf, an extremely dense, small, faint object.

With this broad-brush picture as a guide, let us now examine the process in more detail. Extensive calculations have been undertaken by Schwarzschild, Hoyle, Henyey, Sears and Brownlee, Kippenhahn, Hayashi, Iben, and their associates.

Deferring for the time being a discussion of the earliest stages of a star's evolution, that is, its formation from the interstellar medium, let us consider it at the moment it reaches the main sequence and starts "burning" hydrogen into helium. The astronomer must not only calculate the initial model of the star (when it is chemically homogeneous throughout) but carry the calculation forward in time, allowing for the gradual change of hydrogen into helium. In practice, one divides the interior of the star into 30 or 40 concentric shells in each of which the temperature and density are known and the rate of conversion of hydrogen into helium can be computed. At the start, each shell has the same chemical composition.

Consider first a star of about one solar mass or somewhat less at time t_0. Throughout all of the energy-producing region, the gases are stagnant, that is, there is no mass motion, no mixing from one layer to another; the energy is carried outward by radiation. This model of the star we call $M_0(t_0)$.

We now wait for a time interval $t_1 - t_0$, which may be as short as a few million years or as long as 1000 million years, depending on the mass of the star and thus on the rate at which it burns hydrogen to helium. At the end of this period, the helium-to-hydrogen ratio in each of the shells will have increased, the change in composition being greater the closer the shell is to the center. Then with this new composition we calculate a new model for the star. If the changes between this and the preceding model are too large, it is necessary to take a smaller time interval and repeat the calculation. When a satisfactory model $M_1(t_1)$ is obtained, we repeat the calculation for a time interval $t_2 - t_1$ and get a new model $M_2(t_2)$, and so on.

If the mass of the star exceeds about twice that of the sun, its energy-generating core is a region where such vigorous convection occurs that out to about 10 percent of the radius of the star mixing is complete. One still calculates a succession of models for t_0, t_1, t_2, . . . but mixing of the material in the central regions causes the chemical composition of the core to be uniform throughout. The reason the massive stars have convective cores is that energy generation there takes place by the carbon-nitrogen cycle, which depends very steeply on the temperature. This means that a slight rise in temperature can cause a huge rise in energy output, so that the layers become mechanically unstable and vigorous convection sets in. On the other hand, stars of about the solar mass derive their energy from the proton-proton reaction, which depends much less steeply on the temperature.

Initially the star departs but little from its $M_0(t_0)$ model, but as time goes on, the hydrogen near the center will be more and more severely depleted. A star such as the sun develops a central inert core that gradually grows outward as the "helium rot" takes over one shell after another. The convective core of a luminous main-sequence star continues to burn hydrogen into helium until there is nothing but helium left.

When hydrogen is exhausted in the core, the star will continue to shine by burning hydrogen in a relatively thin shell around this inert region. The star is now built on a different model than before. It finally develops a convective layer outside the region of the shell, whatever the previous mechanical state of the material there may have been.

Now the outer part of the star expands. The core continues to shrink in size but its mass continues to grow as new material is added from the outer envelope. Thus the star begins to depart from the main sequence, becoming redder and more luminous as it evolves. At a certain stage it may move rapidly to the right on the Hertzprung-Russell diagram and become a red giant or supergiant.

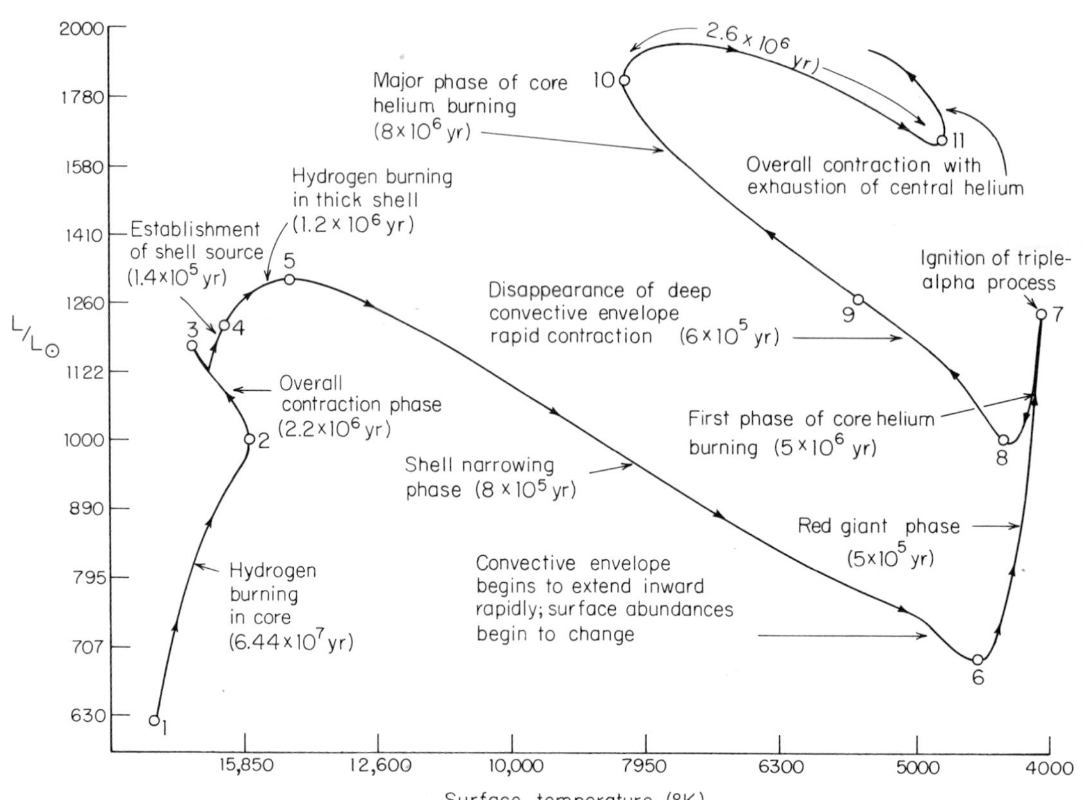

Fig. 94. The evolutionary path of a star of five solar masses, showing the times required for the star to travel between numbered points on the curve. The luminosity is given in terms of that of the sun (3.86×10^{33} erg/sec). (Adapted from Icko Iben, *Science 155* (1967), 786.)

The complicated paths that may be followed are shown in Figs. 94 and 95, which are due to Icko Iben. (Similar calculations have been carried out by Henyey and his associates.) As time goes on, the outer envelope of the star expands, and, although the total power output may increase (at least for the less massive stars), this increase cannot compensate for the vastly expanded area of the radiating surface, whose temperature therefore falls.

Note that stars of low mass evolve differently from those of large mass, because of the differences in their structures and energy-producing mechanisms. The surface temperature of a star more massive than 2 solar masses falls at first, then rises a bit before the star leaves the main sequence, and finally declines steadily. The surface temperature of a

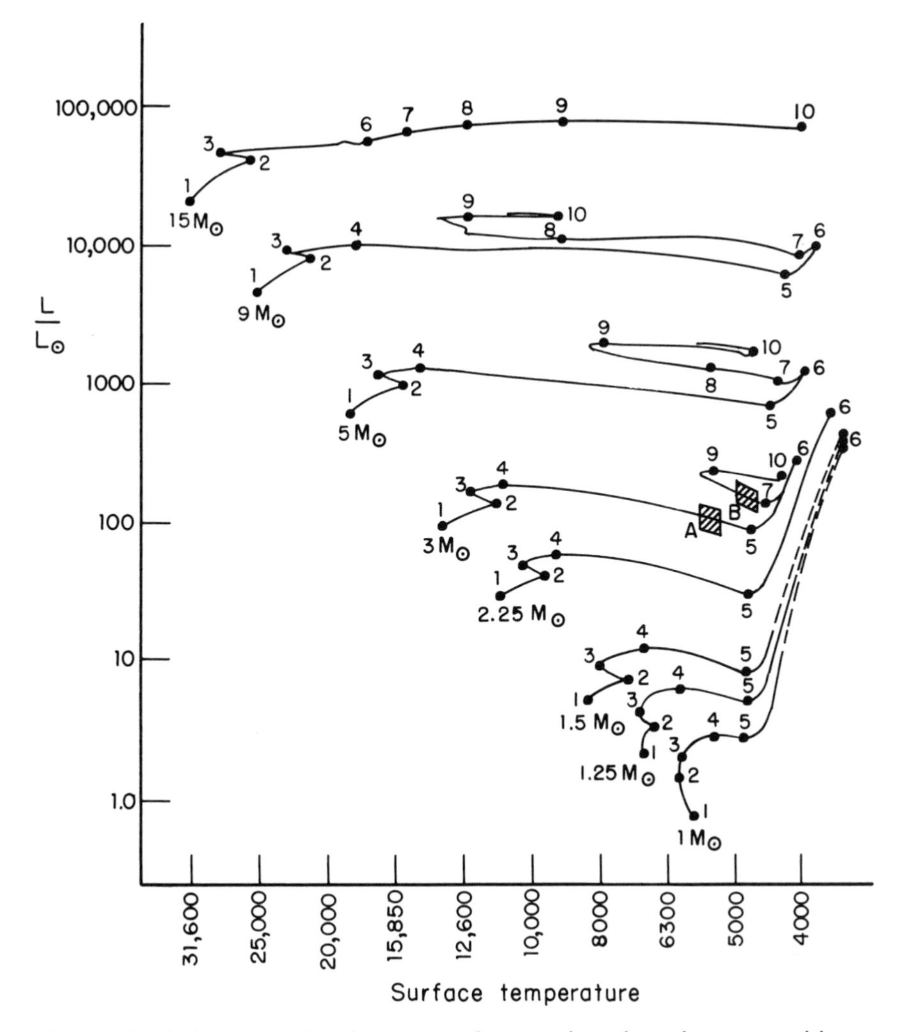

Fig. 95. Evolutionary paths for stars of approximately solar composition. Luminosity ratio is plotted against surface temperature for stars of 1, 1.25, 1.5, 2.25, 3, 5, 9, and 15 solar masses. Note that the shapes of the curves are quite different for stars of different masses. Stars of about 1 solar mass evolve slightly to the right and upward, while very massive stars evolve almost horizontally to the right (low-temperature) side of the diagram. The hatched areas refer to the components A and B of the star Capella. The numbered points refer to various stages in the calculations (the corresponding ages have been tabulated). (Adapted from Icko Iben, *Science 155* (1967), 788.)

star like the sun rises at first and then declines, but the luminosity increases considerably as the star becomes a red giant.

Exact calculations become difficult because we do not yet know enough about how energy is carried in an incandescent, wildly turbulent gas whose density increases rapidly with depth. As the helium core becomes increasingly dense, it approaches a physical state not unlike that encountered in a white-dwarf star, even in spite of a high temperature. If the star is sufficiently massive, helium can be burned into carbon and carbon into yet heavier elements.

If the mass exceeds 12 solar masses, the helium is burned before the star becomes a red supergiant, but if the mass is smaller than this the star evolves rapidly to the red-giant phase. It reaches maximum brightness where the entire outer region is convecting violently and helium starts to burn in the core. If the mass is 2.25 solar masses or less, helium burning may start suddenly and cause complications. In this stage of its life a star may have a helium-burning core and a shell where hydrogen burns to helium, and it may obtain some energy from contraction from time to time. Hence the evolutionary tracks can become quite complicated.

An interesting comparison is supplied by the two components of the spectroscopic binary Capella. According to theory, the more luminous star (which is the further evolved) is in vigorous convection outside the energy-producing shell whereas the fainter component is just starting out as a yellow giant and is not yet fully convective. This fainter component contains lithium, while the brighter component does not. In these stars the lithium was confined to a thin outer layer, since it would have been destroyed by thermonuclear reactions deeper in the star. When convection became established in the brighter star, most of the remaining lithium got carried to deeper layers and destroyed.

The Evolution of the Sun

The evolution of the sun is of particular interest since it is one star for which the radius, mass, and luminosity are accurately known. The chemical composition, save for the helium-to-hydrogen ratio, is also well established. Calculations by Sears and Brownlee, by Sears (1964), and by Demarque and Larson (1964) are plotted in Figs. 96 and 97. Figure 96 compares the present-day internal structure of the sun with its internal structure at the time it started out as a main-sequence star. Since then the core has contracted, both central density and temperature have risen, and

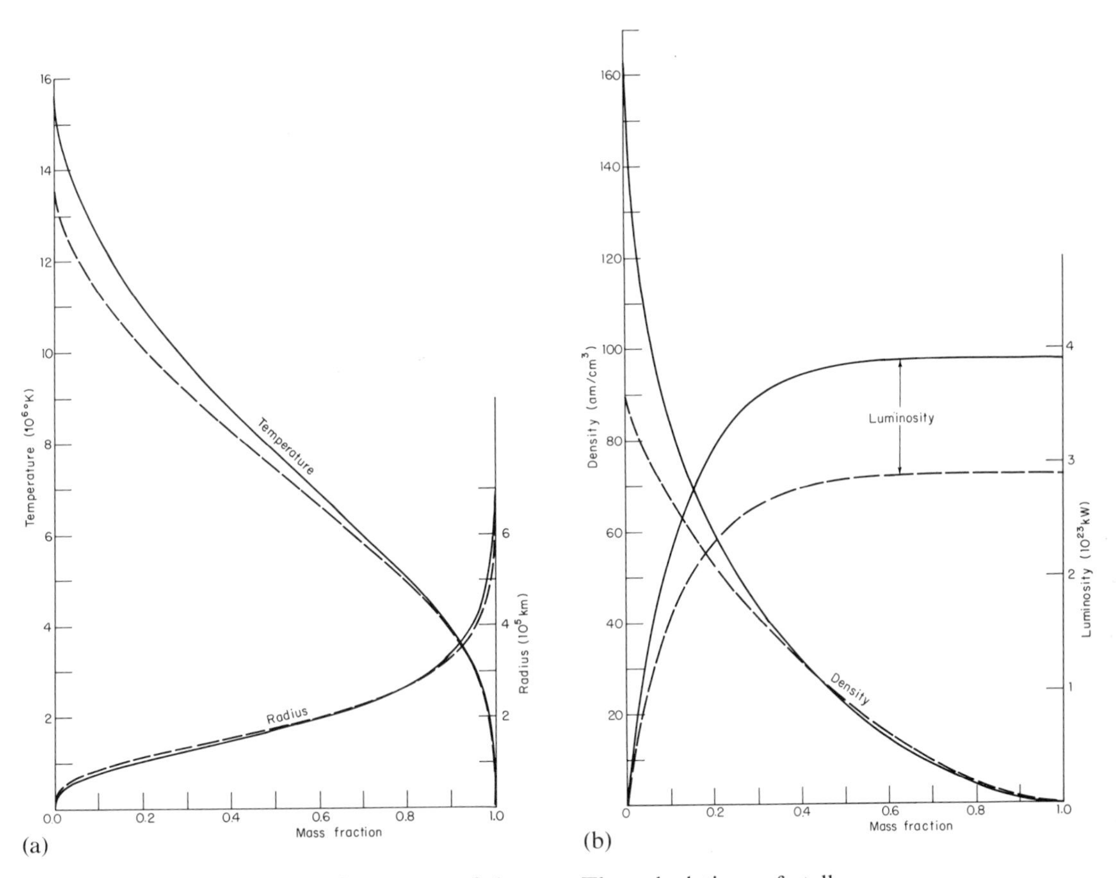

Fig. 96. The changing internal structure of the sun. The calculations of stellar models are carried out by dividing the star into shells of equal mass. Hence the variables are plotted against the mass fraction M_r, for times $t = 0$ (dashed curve) and $t = 4.5 \times 10^9$ yr (solid curve), the assumed age of the sun: (*a*) temperature and radius; (*b*) density and total energy generated within a given mass fraction. Note that virtually all the luminosity of the sun is produced by the inner half of its mass. Note also the contraction of the core of the sun, the rise in central temperature, the marked rise in central density, and the increase in total luminosity over the last 4.5×10^9 yr. (After R. Weymann and R. L. Sears, *Astrophysical Journal 142* (1965), 174.)

the central hydrogen content has been cut in half. All this has been accompanied by an increase in luminosity.

In Fig. 97 the sun's radius and luminosity (in terms of present values) are plotted over a period of 10,400 million years. During this period the radius and luminosity rise and the surface temperature changes more slowly. The lower limit to the earth's temperature for the next 5,800 million years is also plotted. Actually the temperature will exceed this

Fig. 97. Evolution of the sun. The gradually increasing radius and luminosity of the sun are plotted according to calculations by Demarque, together with the theoretical lower limit to the temperature of the earth; actually we can expect that the earth's temperature will rise faster than this.

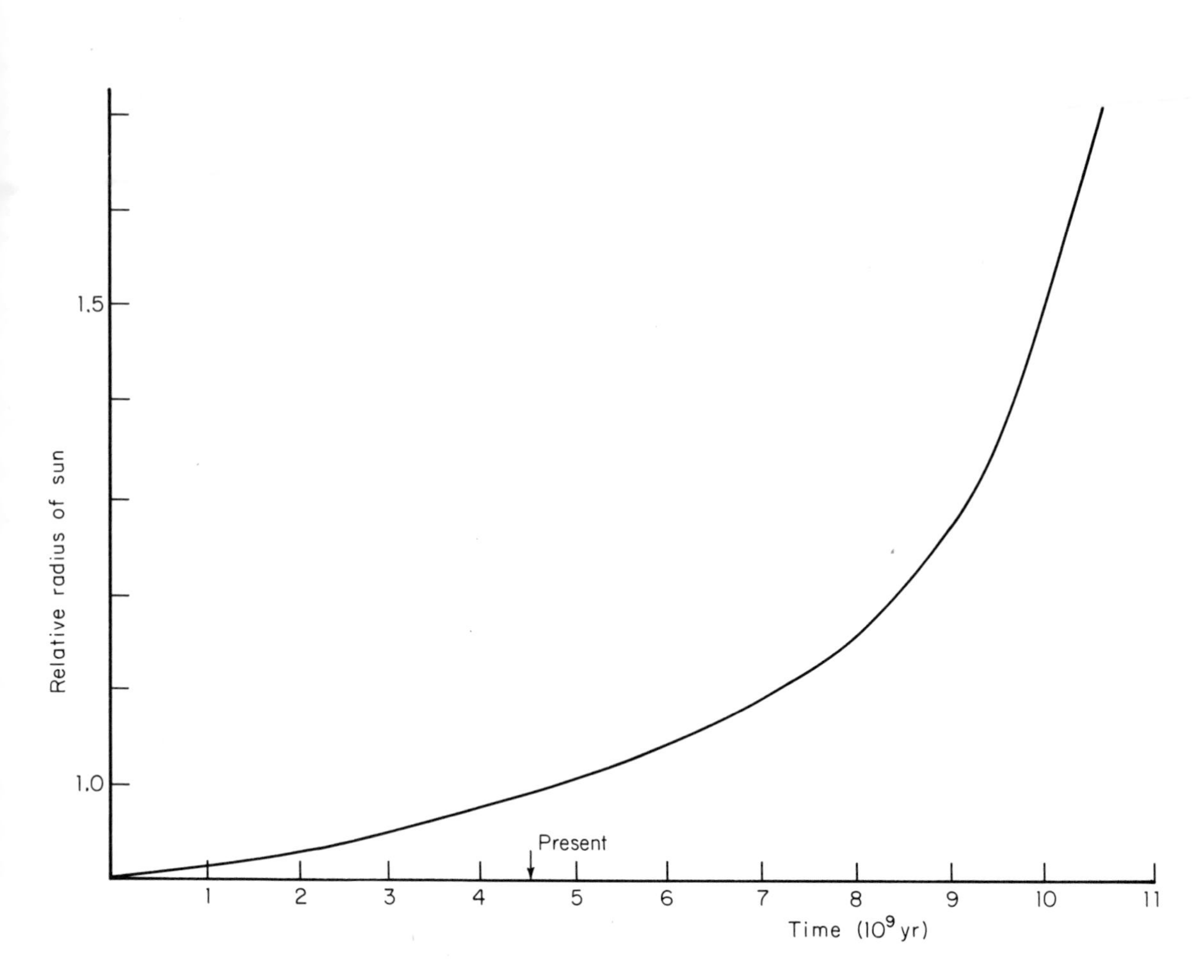

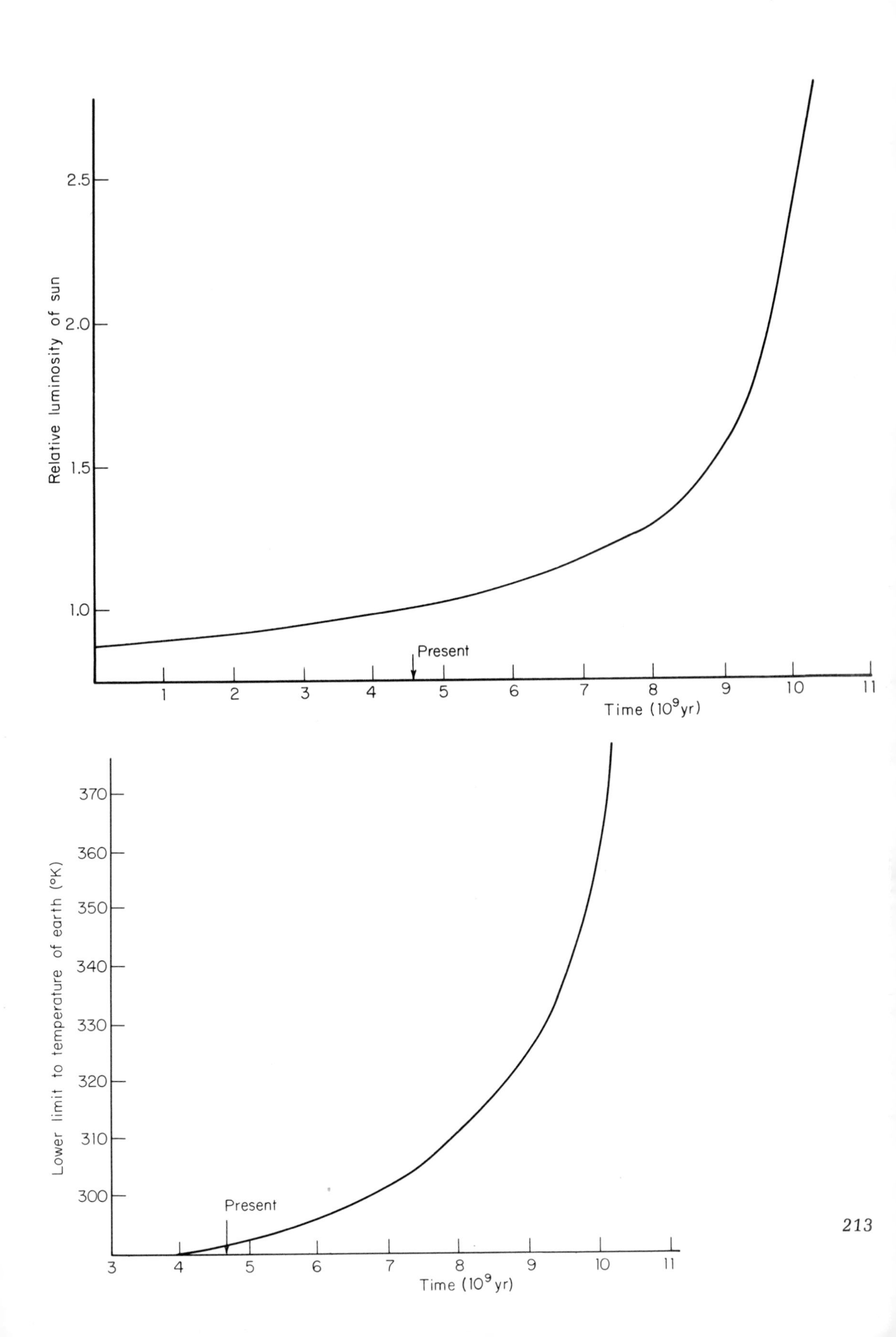

Relative luminosity of sun
2.5
2.0
1.5
1.0
Present
1
2
3
4
5
6
7
8
9
10
11
Time (10⁹ yr)
Lower limit to temperature of earth (°K)
370
360
350
340
330
320
310
300
Present
3
4
5
6
7
8
9
10
11
Time (10⁹ yr)

value. As the oceans are evaporated, the increasing supply of water vapor in the atmosphere will absorb infrared radiation ever more efficiently. Outgoing heat from the earth's surface will be turned back. As a consequence of this blanketing effect, the temperature will rise even more, more water will be evaporated, and the temperature will rise still further. Ultimately, the oceans are literally boiled away but the temperature will still rise as the atmospheric pressure rises and rocks become outgassed. Long before these dreadful things come to pass, life on the earth will have ceased to exist. The sun will continue to grow in size until ultimately it will engulf the orbit of Mercury and destroy that planet, but before it can reach the scorched cinder that had been the earth it will have expired and the core will have collapsed into a white dwarf.

Some Observational Evidence for Stellar Evolution

In Chapter 6 we saw that star clusters offered us a unique opportunity for studies of stellar evolution. Consider first the galactic clusters whose Hertzsprung-Russell diagrams are shown in Figs. 60 and 61. The heavy line bordering the left of the main sequence defines the positions occupied by homogeneous stars that have just started to burn hydrogen into helium. Thus it defines the zero-age main sequence ZAMS. As the helium content of the core increases, so does the mean molecular weight. Then, as we saw in Chapter 8, the temperature in the central regions must rise. When this occurs, the luminosity likewise increases and the star moves upward and eventually to the right in the diagram. The main sequence, therefore, has a finite width because it contains stars with cores of different helium-to-hydrogen ratios. If we were to construct an H-R diagram from field stars, that is, stars picked at random in the neighborhood of the sun, we would find a main sequence of substantial width because it would include objects of a wide range in age and helium-to-hydrogen ratio.

On the other hand, most of the stars of a given cluster probably were formed at very nearly the same time. An exception must be made for the faint red dwarfs, which are believed to condense from the interstellar medium rather slowly. In most clusters one observes only stars brighter than about absolute magnitude 7, and these may be regarded as having essentially the same age. Hence the main sequence in any one cluster is relatively narrow.

The lifetime of the star on the main sequence will depend on its mass. A star twice as massive as the sun will liberate energy eight or ten times as fast. Hence such a star will exhaust its hydrogen fuel much sooner than will the sun. A star of 10 solar masses will use up its hydrogen fuel even more rapidly. When the hydrogen is exhausted in the core, and the star continues to shine by burning hydrogen in a thin surrounding shell, it will evolve to the right on the diagram as a giant or supergiant.

These phenomena are well illustrated by various open clusters. All the stars of NGC 2362 fall on the main sequence. In *h* and χ Persei a handful of the most luminous stars have left the main sequence to become red supergiants, while the more luminous blue stars are beginning to depart from the main sequence. In the Pleiades there are no stars as bright as those in *h* and χ Persei, but the main sequence bends conspicuously to the right; there are no giants. The clusters M 11 and the Hyades represent a more advanced stage of evolution. Here the main sequence has been "rolled back" to stars like Sirius or Procyon, while there are now a number of red giants. In NGC 188 the main sequence includes only yellow and red dwarf stars, whereas the giant branch is connected to the main sequence by a continuous bridge across the subgiant region.

In the other clusters, from NGC 752 to *h* and χ Persei, there is a conspicuous gap between the main sequence and the red giants and supergiants which becomes wider the more luminous the star.

Thus, of the clusters depicted, NGC 2362 is the youngest, the Pleiades are older, the Hyades and NGC 752 are much older, and NGC 188 is the oldest of all. By noting the turn-off point and applying the results of stellar-evolution calculations which give the theoretical tracks for stars of different masses, one can estimate the ages of the cluster; *h* and χ Persei is but a few million years old, while the age of NGC 188 is about 10,000 million years.

The ages determined in this way become less accurate the more ancient the cluster. It must be emphasized that computed evolutionary tracks are sensitive to the assumed chemical composition; relatively slight variations can cause large changes in the predicted ages. Exact comparisons between theory and observation may be difficult for the brightest clusters because all stars may not have been formed at the same time.

Although open clusters show a huge spread in ages, the globular clusters are all very ancient objects; some of them have been assigned ages in excess of 10 or 12 thousand million years. (A possible exception must be noted for clusters such as NGC 1866 in the Large Magellanic Cloud, which contains great numbers of bright, blue, presumably main-sequence stars.) The theoretical tracks and derived ages depend on the assumed helium-to-hydrogen ratio. The differences between clusters in the observed H-R diagrams could be interpreted in terms of differing metal-to-hydrogen ratios.

The objects on the giant branches represent stars that have only relatively recently evolved from main-sequence stars not very much brighter or more massive than the sun (Fig. 98). Just how the horizontal branch fits into the evolutionary picture is not clear. Do the stars move to

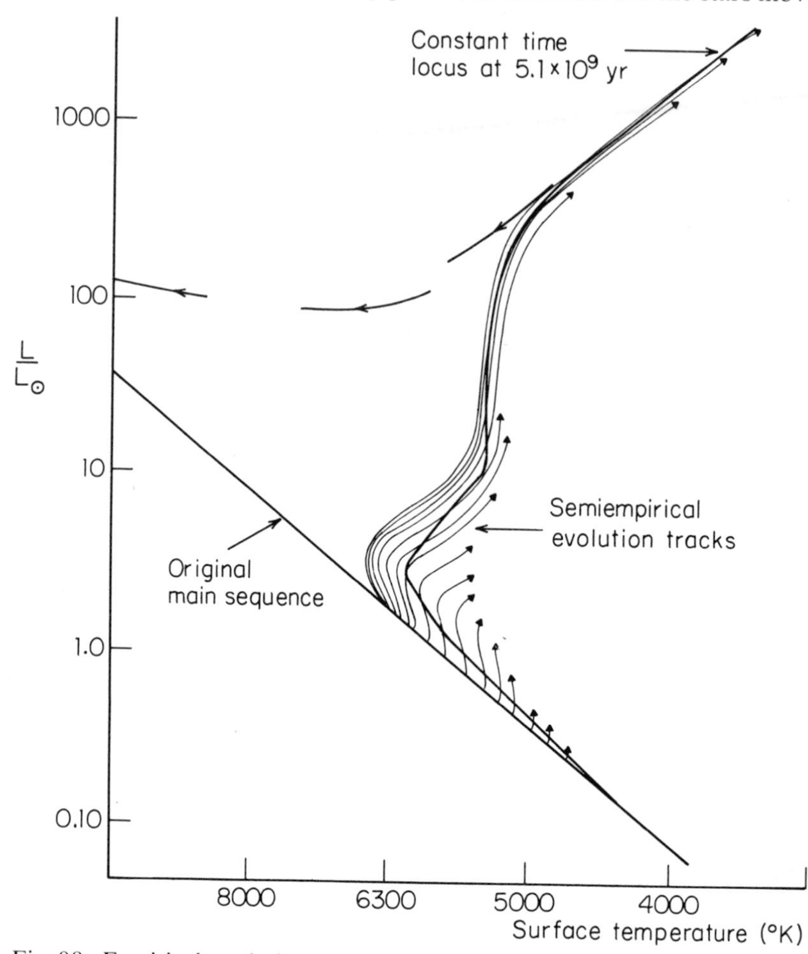

Fig. 98. Empirical evolutionary tracks in globular clusters. The luminosity (in terms of that of the sun) is plotted against the surface temperature for the stars of the cluster Messier 3. The arrows indicate the directions of evolution for stars on the main sequence. The curve leaving the main sequence and rising up to the giant branch then represents the points defined by stars of the same age but different masses which have evolved off the main sequence. The age of 5.1×10^9 years is probably too low; it has been revised upward by more recent theoretical calculations. (Adapted from A. R. Sandage, *Mémoires de la Societé Royale des Sciences de Liège 14* (1954), 266—The Liege Symposium on Nuclear Processes in Stars, September 1953.)

the right or to the left, since the bluest stars in and just below the horizontal branch are believed to be precursors of white dwarfs?

Most globular clusters have lower metal contents than galactic clusters. Yet giant stars in globular clusters attain substantially higher luminosities than giants in ancient open clusters such as M 67 and NGC 188. In these stars metals are not sufficiently abundant to influence the molecular weight, but they do affect the opacity. The material of which the globular-cluster stars are composed is more transparent. Hence it is no surprise that the globular-cluster giants are brighter than those of ancient galactic clusters.

Theoretical calculations of evolutionary tracks seem to agree at least semiquantitatively with these empirical color-luminosity diagrams so that the identification of giant and supergiant stars with objects that had been previously on the main sequence seems well established. Detailed evolutionary tracks of stars in the supergiant stage may become quite complicated.

Presumably, then, a star must evolve from a giant or supergiant to a white dwarf. How? One possibility is that the outer part of the star simply swells up until the atmospheric layers begin to escape continuously into space. Another possibility is that the star explodes. Direct observational evidence for the latter kind of event is supplied by supernovae (see Chapter 11). The amount of energy released in a supernova explosion is so great that no star could possibly survive such a catastrophe without serious modification in its structure. Yet very few stars must end as supernovae; those that do, have a profound influence on the surrounding interstellar medium and stars that are subsequently formed from it.

Direct observational evidence may be found for the loss of mass by supergiants, as A. J. Deutsch demonstrated for the binary α Herculis. This system consists of a Class M supergiant and a Class G secondary. In addition to the broad lines characteristic of a typical M star there appear a number of sharp displaced components that seem to come from an expanding envelope around the star. Deutsch found that these displaced components also appear superposed on the spectrum of the G companion. Evidently, the atmosphere escaping from the M supergiant envelops the G star as well. Deutsch found evidence for escaping envelopes and shells around other supergiant M stars. A classic example is Antares, which is surrounded by a small nebula showing bright lines of ionized iron excited by the high-temperature companion.

It appears that the outer parts of the star simply escape to the interstellar medium and the core settles down as a white dwarf. Once in a while we should be able to catch a star in the final stages of its transition from a giant or supergiant to a white dwarf. A number of puzzling "combination" variables that may provide significant clues are known. These stars show absorption M-type spectra upon which are superposed bright lines of hydrogen, helium, and sometimes ionized iron. There also appear in a number of these stars bright lines characteristic of gaseous nebulae and there is a bluish continuous spectrum in the ultraviolet. It has frequently been suggested that stars of this type, for example, Z Andromeda, CI Cygni, and BF Cygni, are binaries, but some of them may be giant or supergiant stars in the last stages of dissolution. The bright hydrogen and helium lines and the bluish continuum are produced presumably in the core; the M-type spectrum comes from the disintegrating envelope, while the nebular spectrum is excited in the extended cloud of gathering debris by the high-temperature radiation from the core. In any event, these stars will merit careful watching.

Very often, the blob out of which a star is eventually formed will be in slow rotation. Such a body will condense into a rapidly spinning star or perhaps split in two and become a binary. The great prevalence of rapidly spinning stars among the brighter main-sequence objects is easily understood. What is curious is the sudden disappearance of rapidly spinning stars near spectral class F5 and the slow rotation of stars of later spectral classes along the main sequence.

Figure 99, due to Olin Wilson, is a portion of an H-R diagram for field stars depicting the domains of rapidly rotating and slowly rotating stars. Rotational velocities in excess of 50 kilometers per second set in suddenly at F4 on the zero-age main sequence (ZAMS). Main-sequence stars that lie well above this line may show rapid rotation as late as spectral class G0—G5.

Wilson suggests that these stars may have evolved from stars that were initially near F4. Another curious feature is that stars with bright *H* and *K* lines, which therefore presumably possess extended chromospheres, all lie near the zero-age line. Rotation and chromospheric activity seem to be mutually exclusive. The sun belongs to the slow-rotation group.

The onset of rapid rotation seems to occur at the point where the outer convection layer becomes very shallow or disappears. It is suggested that stars with convection layers have chromospheres and coronas like the sun. The sun ejects material continuously; this is the phenomenon of the solar wind, which appears to remove little mass but does tend to slow down the solar rotation. Thus, on this hypothesis the stars in the "no-

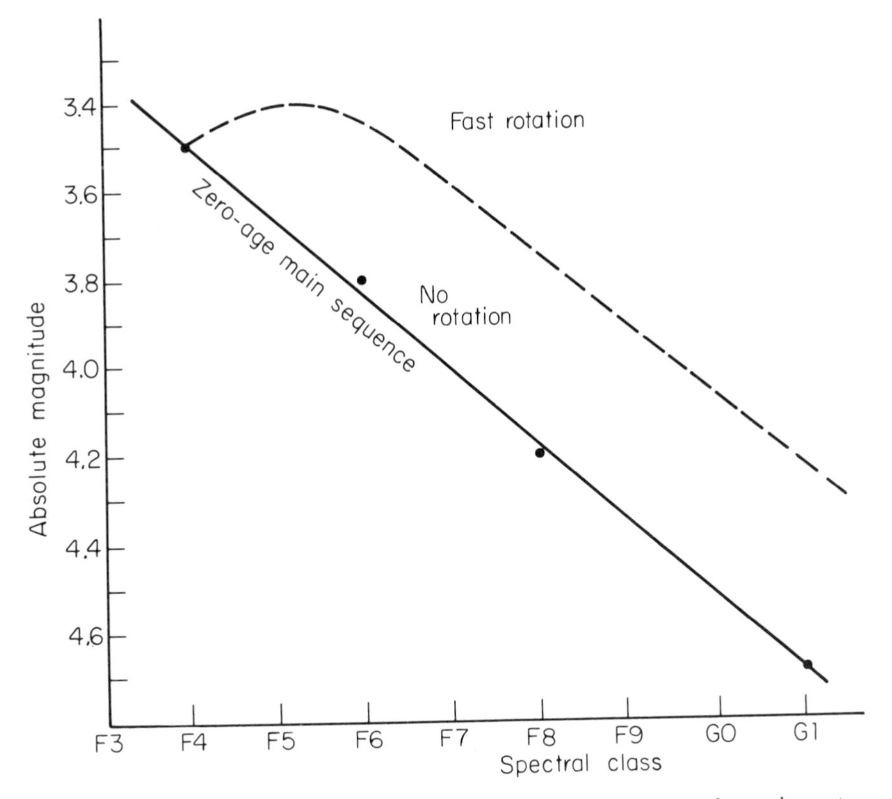

Fig. 99. Relation between absolute magnitude, spectral class, and rotation rates for main-sequence stars. Rapidly rotating stars lie above the dashed line; slowly rotating or nonrotating stars lie between the dashed line and the zero-age main sequence. (After Olin Wilson.)

rotation zone" would all have been slowed down by material ejection. One suggestion is that the cores of these stars continue to spin rapidly although the outer layers rotate slowly. This hypothesis has a number of attractive features; one is that the dependence of solar rotation rate on latitude now becomes intelligible. There are serious difficulties, however, in maintaining a big rotation difference between the core and the outer layer.

Often binary stars must be formed; if the stars are close together the evolutionary development of each will be modified by the presence of the other. These effects explain the unusual types of stars sometimes found in eclipsing-binary systems that have no counterparts among isolated stars. We explore this problem in our discussion of novae (see Chapter 11.)

The White-Dwarf Stars

The longest-established and most securely anchored fact in stellar-evolution studies is that the last visible stage in the life of a star is a white dwarf. The first of these objects to be discovered was the companion of Sirius. The existence of this star was predicted by Bessel from its gravitational effects, which produced a variation in the motion of Sirius across the sky. It was actually discovered in 1862 as an extremely dim object about 400 times fainter than the sun. Since it has about the same mass as the sun we would expect it to be a cool red star. Actually it is a blue star with a temperature of about 8200°K. Hence, since each unit area radiates four times as much energy as the sun, the ratio of areas must be 1600. Therefore, the diameter is about $1/40$ that of the sun and the volume is $1/(40^3) = 1.56 \times 10^{-5}$ that of the sun. The mean density must be 64,000 times that of the sun, hence about 90,000 times that of water.

Under such conditions, matter has very peculiar properties. Electrons are all or nearly all stripped from the atoms and so the mass can be compressed to a very high density. The electrons do not obey the ordinary gas laws at all, but are said to constitute a "degenerate" gas. Such material has very high electrical and thermal conductivity. Any excess energy could quickly flow from one region to another by simple heat conduction. The temperature no longer appears in the gas law; the pressure depends simply on the density. Once a star gets in this state it cannot get out of it. No energy is available to expand the star against its own gravitational attraction on itself. The star shines feebly, radiating away its last resources of heat energy, the nuclear sources being long since exhausted. The process may be slow, requiring several thousand million years as the star gradually blackens as a burned-out cinder.

Although they are called "white dwarfs" after the prototype Sirius B, these virtually defunct stars actually show a considerable spread in color from blue to red. Their spectra, which have been studied in greatest detail by Greenstein, show remarkable properties (Fig. 100). Spectral lines are invariably broadened because of the high densities. Most of them show hydrogen lines; most of the remainder display helium lines or a continuous spectrum in which all lines are washed out.

These spectra all support the hypothesis that stars convert hydrogen into helium, and sometimes helium into carbon or even eventually into heavier atoms. Carbon-rich stars, some of them showing molecular bands of carbon, are numerous. Evidently we are observing the residual cores of defunct red giants.

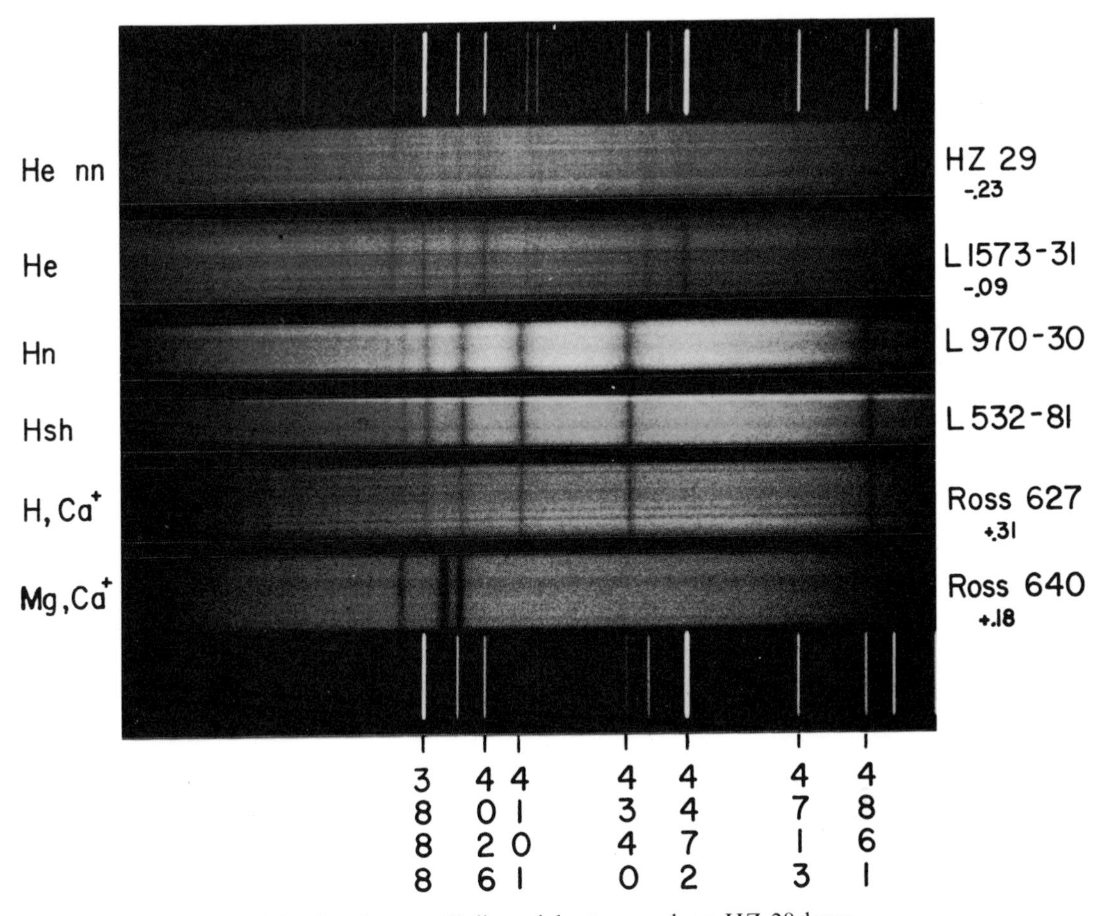

Fig. 100. Spectra of white-dwarf stars. Helium-rich stars such as HZ 29 have fuzzy-lined spectra; stars with sharp-lined spectra, such as L1573–31, L970–30, and L532–81, show prominent hydrogen lines. Ross 627 shows weak lines of Ca II, while Ross 640 is characterized by strong lines of Mg and Ca II, with no hydrogen. (Courtesy J. L. Greenstein.)

Theory shows that the radius of a white dwarf depends on its mass; the more massive such a star, the smaller the radius and the greater the surface gravity. A quantum of light escaping from the surface of a dense white dwarf has its frequency lowered, that is, the corresponding spectral line is shifted toward the red end of the spectrum. Measured "gravity shifts" in stars of known velocity indicate that white-dwarf stars have masses of the order of 0.6 or even 0.9 that of the sun. A body whose mass was one-fourteenth that of the sun (or less) would simply contract to a black dwarf without ever drawing on any nuclear energy sources. Kumar suggests that many faint red-dwarf stars may actually be contracting bodies that will never develop as hydrogen-burning objects. About 4 percent of the matter of the Galaxy is locked in white dwarfs of bluish color; the red ones may be even more abundant. They are found in open clusters, like the Hyades, that contain evolved stars (giants), but not in young clusters with no giant branch. They doubtless exist in profusion in globular clusters, where they are too faint to be seen. The best candidates for precursors of white-dwarf stars are the intrinsically faint nuclei of planetary nebulae, old novae, and the faint blue stars in globular clusters.

The Formation of Stars

Although the last stages of stellar evolution were easily identified, and the last flickers of a dying star are almost fully understood, the early stages of the formation of a star are still obscure. We have seen that the present existence of short-lived bright, highly luminous stars such as Y Cygni, μ Cephei, or Rigel requires that star formation be a continuing process, going on somewhere under our eyes. The basic difficulty is in recognizing it.

In some fashion the cold dust and gas of the interstellar medium must be gathered into a blob or cloud of sufficient density to permit it to pull itself together under the attraction of gravity. This first step in the formation of a star is extremely difficult to understand fully, and no suitable theoretical picture has yet been proposed. All theoreticians agree that once gravity has a chance to take hold the subsequent evolution of the star will be relatively rapid. The speed of evolution will depend on the mass; the more massive objects will contract relatively rapidly, whereas the less massive one will pull themselves together slowly. As the mass contracts, the interior becomes hotter and hotter until finally the object glows as a bona fide star. The star will continue to contract until the central temperature becomes high enough for nuclear reactions to occur. The central temperature stabilizes at a value that permits the star's entire energy output to be supplied by the conversion of hydrogen into helium. It continues to shine as a main-sequence star until hydrogen is exhausted in the core and its evolves away from the main sequence as a subgiant, giant, or supergiant. A body containing, say, only 1 percent of the solar mass would never develop a central temperature high enough to support nuclear reactions and would settle down as a cold planetlike object without ever having tasted the glory of being a star.

It would appear to be most promising to look for star formation among the dark clouds of the Milky Way. Nearly all young, bright, blue stars are found in or close to regions of high obscuration or areas showing evidence of interstellar material. We would not always expect to find bright stars in intimate association with dark clouds of absorbing material. Once a very luminous star is "turned on," its radiation produces a profound effect on the surrounding gas. The hydrogen in the immediate neighborhood of the star becomes heated to a high temperature and ionized. The expanding heated gas produces a shock wave that rushes through the cooler surrounding region and eventually causes the dissipation of much of the original interstellar cloud.

In our Galaxy astronomers have found a number of "associations," that is, groups of stars that appear to have been formed relatively recently in the neighborhood. Two types of association were recognized by Ambartsumyan, who called attention to the significance of the phenomenon. The so-called "O associations" consist of hot, luminous stars, and the "T associations" of relatively faint objects which appear to evolve into main-sequence stars not unlike the sun. The T associations owe their designation to the great number of variables of the T Tauri type—dwarfish stars immersed in vast clouds of smog and gas. These objects have been studied in great detail by Joy, Herbig, and Haro. All present evidence indicates that they are stars in the process of formation, not yet settled down to the main sequence.

Stellar associations are not clusters, although it is possible that an open cluster may form occasionally from an association. The stars are simply formed in a region; thereafter they spread apart as fast as their velocities will carry them. Now studies of the space motions of these stars, derived from their proper motions and radial velocities, give in many instances the rates of expansion of these associations; then, knowing their distances, we can determine their ages. In this way Blaauw found the Lacerta association to have an age of about 7 million years, while the Scorpio-Centaurus association appears to represent actually two or three associations with ages of about 20 million years. The Orion complex represents an especially interesting region. The expansion age for the belt is about 4 million years, but Strand finds indications that the Trapezium stars are younger than 1 million years. The stars in the Orion region contribute about 10,000 solar masses, but Menon, from studies of the 21-centimeter radiation of the cold hydrogen gas, finds a total mass of about 100,000 suns. Evidently, most of the material in the Orion region is in a disorganized state. Instructive clues are provided by color-magnitude diagrams for clusters or associations in which stars are be-

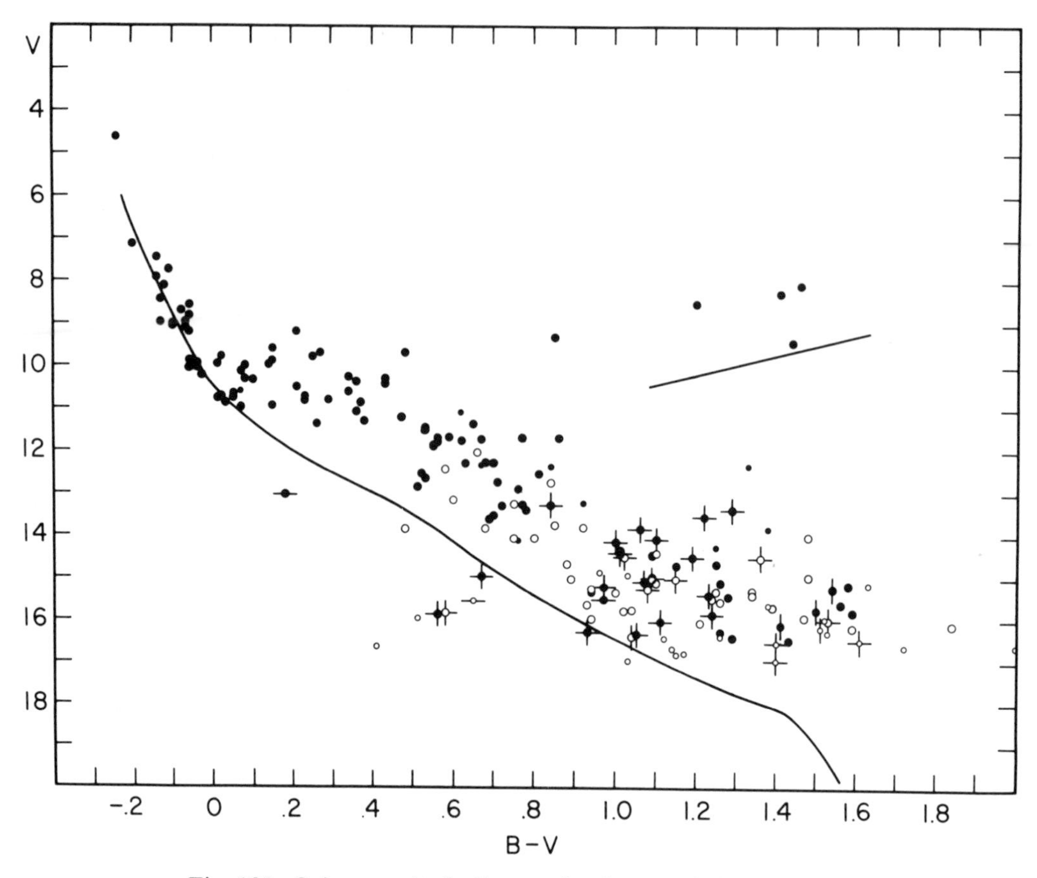

Fig. 101. Color-magnitude diagram for the association NGC 2264. The apparent V magnitude is plotted against the $(B - V)$ color index. (Courtesy M. F. Walker.)

lieved to be in the process of formation. One of the best examples is NGC 2264, which has been studied by M. F. Walker; his results are presented in Fig. 101.

The visual magnitude, or rather the more accurate photoelectric equivalent of the visual magnitude (the so-called V magnitude of Harold Johnson), is plotted against the blue magnitude minus the visual, the $B - V$ color index. The data plotted are the actually observed magnitudes and colors; no correction has been made for effects of space reddening. The dots represent photoelectric observations; the circles are photographic measurements, which are calibrated with the aid of the photoelectric measurements. Vertical lines through the points indicate known light variables, and the horizontal lines indicate stars in which Herbig has found Hα emission. The solid lines give the standard main sequence

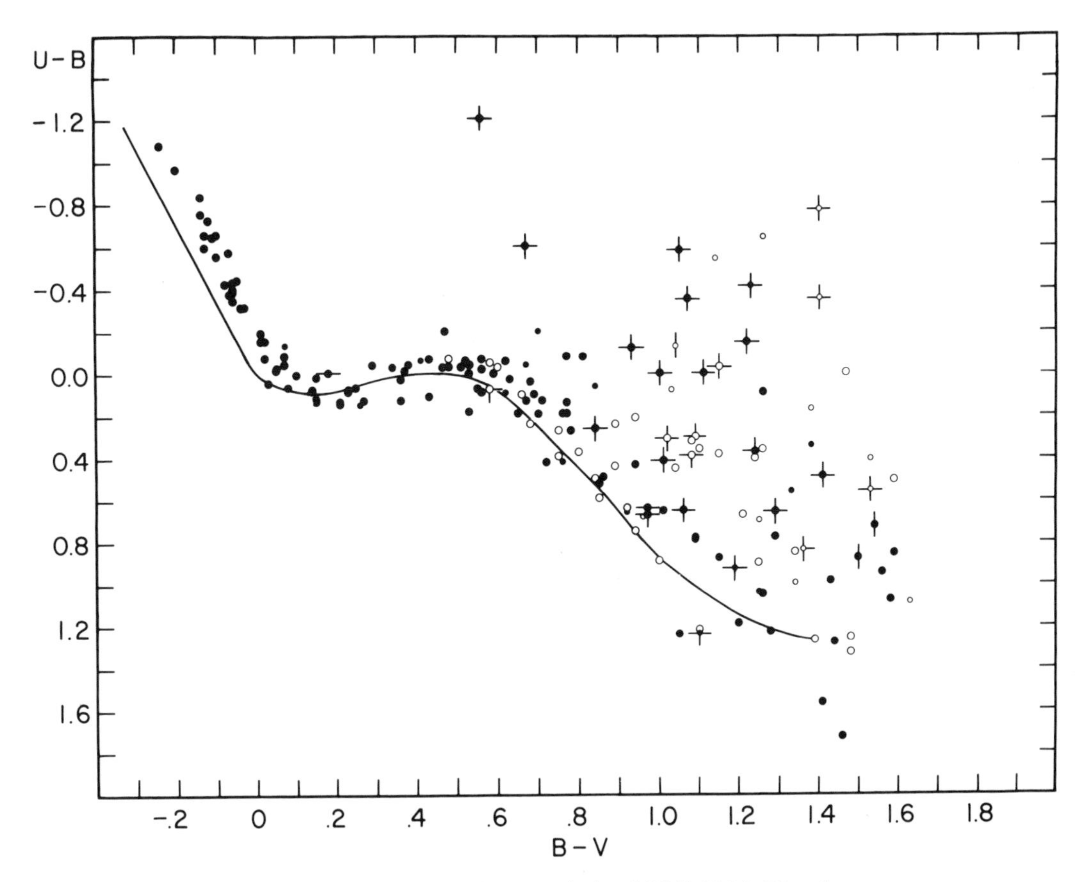

Fig. 102. Comparison of stellar colors in the association NGC 2264. The ultraviolet minus blue ($U - B$) color index is plotted against the blue minus visual ($B - V$) color index. (Courtesy M. F. Walker.)

(running from near the top left to the bottom right) and the giant sequence.

Figure 102 reproduces Walker's comparison of the colors of the stars in the cluster. He plots the ultraviolet minus blue $U - B$ color index against the blue minus visual $B - V$ color index. The solid line indicates the relation between $U - B$ and $B - V$ for normal unreddened main-sequence stars (see Appendix F). The small shift of the bluer stars from the line indicates a small amount of reddening. The effect of the reddening by space absorption is to make the $B - V$ color index about 0.082 magnitude. We call this space-absorption-imposed reddening the color excess. Assuming the ratio of color excess to total absorption to be the same as in other regions of the Galaxy, Walker finds a total absorption of about 0.25 magnitude. The distance of the cluster is about 975 parsecs.

Although the very brightest stars fit the main sequence defined by the normal stars in the neighborhood of the sun, the fainter stars show pronounced abnormalities. Examining the color-magnitude array we notice that beyond color index zero, corresponding to a Class A star like Sirius or Vega, nearly all the stars fall above the main sequence defined by the nearby normal stars. The fainter, redder stars are mostly variables with bright hydrogen emission, indicating extended chromospheres or involvement in nebulosity. Furthermore, the variable stars have an excess of radiation in the ultraviolet, as was pointed out by Herbig and Haro. On the $U - B$ vs. $B - V$ diagram they all fall above the standard main-sequence line, indicating an excess of radiation in the ultraviolet.

Evidently we are witnessing the formation of stars from the grains and gas of the great interstellar clouds of the Milky Way. As the stars pull themselves together by gravitational contraction they gradually move to the left in the V vs. $B - V$ diagram, but the evolution is evidently not a smooth process. Instabilities, evidenced by the marked variability and abnormal colors of these objects, abound. Theoretical attempts to handle the contraction of a star to the main sequence naturally assume an orderly development, so it is not surprising that the theory does not agree with observations in all details.

The early stage of a star's life, while it is contracting from a blob of interstellar material, seems to be a time of considerable instability. The star appears to form as a condensation in a cloud and material is probably both collected from the cloud and ejected into it. Gradually the star settles to some semblance of normalcy when the thermonuclear sources are ignited, but before that happens all sorts of interesting events can occur, such as the building of light elements, and even of systems of planets. Very often, a contracting blob must be in slow rotation and as it condenses it spins faster and faster until it breaks into two or more fragments. One of the fragments may undergo further breaking up into planetlike bodies. This stage in the star's evolution appears to be complicated by the presence of magnetic fields, which may do bizarre things to the involved, ionized gas and may even act as a brake on some of the spinning blobs. In Fig. 103 is shown the evolution of the sun as it approaches the main sequence, according to calculations by various investigators. Notice the rapid decline in brightness as the star shrinks and its temperature rises.

The formation of a star like the sun appears to be a process covering millions of years, so that we observe a number of such objects in our

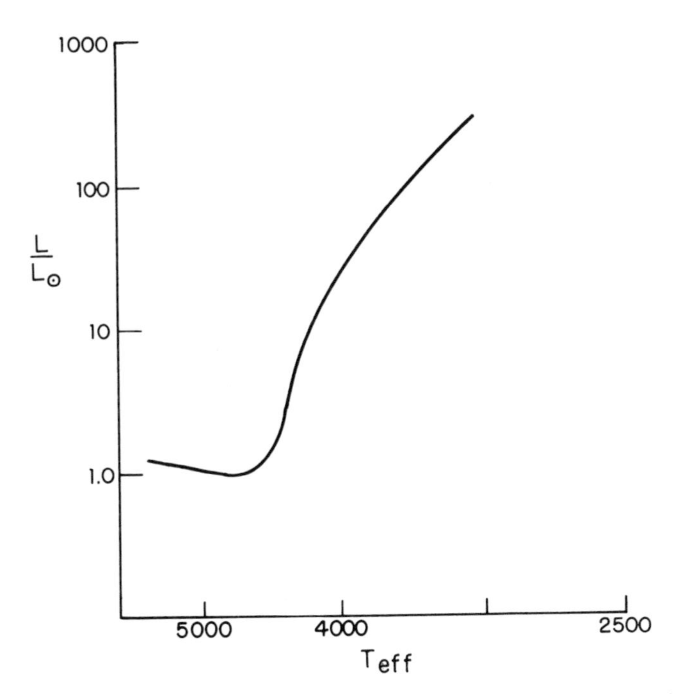

Fig. 103. The early evolution of the sun. The brightness of the sun is plotted against its surface temperature as it contracts and evolves toward the main sequence. (Adapted from calculations by Weymann and Moore; Hayashi, Höshi, and Sugimoto; Ezer and Cameron; and Upton; courtesy E. K. L. Upton, from *Stars and stellar systems,* G. P. Kuiper, ed. (copyright University of Chicago Press, Chicago, 1967), vol. 7, *Nebulae and interstellar matter,* B. M. Middlehurst and L. A. Aller, eds.)

neighborhood of the Galaxy, but the formation of highly luminous stars —or at least their bursting into incandescence—probably occurs over a very short space of time indeed. As we have remarked above, once a bright, hot star is formed, it will have a profound influence on the grains and gas in its neighborhood.

Occasionally, very luminous stars are found moving away from associations with very great velocities. Examples of such objects are AE Aurigae and μ Columbae, which appear to be escapees from the Orion region. These stars could not have picked up such high velocities from weak gravitational interactions with their neighboring association members, and Blaauw has made an interesting suggestion concerning their origins. Suppose that an extremely massive blob condenses into a normal star of 10 or 20 solar masses while the rest comprises a mass 200 or 300 times that of the sun. The pair form a double "star," revolving about their common center of mass while still contracting, but the more massive object never contracts into a stable one at all. The central temperature goes so high that the mass is blown to bits, leaving the normal star still moving with a speed of perhaps 100 kilometers per second. It therefore flies away with this great speed and leaves the association far behind.

Element Building in Stars

That chemical elements can be built in stars is one of the great generalizations established by modern astrophysics. The possibility that the required nuclear reactions could occur in stars had been suggested at various times, but only recently have various specific mechanisms been identified.

The chemical composition of the solar system, quantitatively as well as qualitatively, is not a meaningless datum. Properly interpreted, it constitutes a cosmic historical record of considerable significance. The first task is to obtain the abundances of the elements of the solar system. Analyses of rocks of the earth's crust are of limited usefulness. The chemical composition of the crust is certainly no more representative of the earth's average composition than is that of the slag in a smelter crucible representative of the composition of the original ore. In each instance, iron or nickel and the so-called noble metals like gold and platinum tend to sink to the bottom, while metals such as magnesium, aluminum, or sodium tend to ride to the top. Hence it is very difficult to deduce from the earth's crust what the original composition of the earth must have been. Moreover, during the process of its formation, the earth must have lost most of its hydrogen, helium, nitrogen, and other gases.

Ideally, the sun might be expected to give the most reliable data on the chemical composition of the primordial solar system. Unfortunately, determination of the solar chemical composition is difficult even for abundant elements because of uncertainties in the f-numbers, and weaker lines of less abundant elements are affected by blends with stronger lines of more plentiful metals (see Chapter 5). The solar abundance of many elements cannot be established at all; hence we must seek additional clues from other bodies of the solar system. It is generally believed that a particular class of stony meteorites called carbonaceous chondrites may give a good sample of nonvolatile elements. By a judicious combination of meteorite and solar data an attempt is made to reconstruct the chemical composition of the primordial solar system. But we cannot be sure that any rock we can lay our hands on is an honest sample of the solar system.

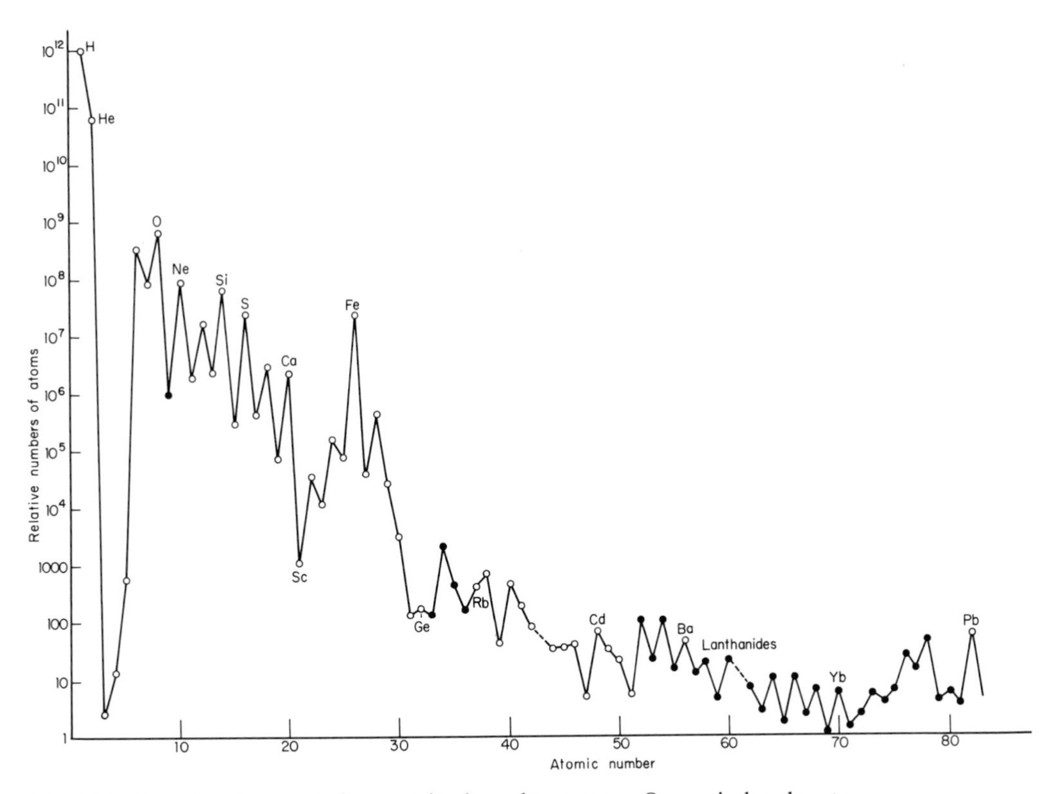

Fig. 104. The abundance of elements in the solar system. Open circles denote values taken from solar data as compiled by L. H. Aller in *Transactions of the Astronomical Society of Australia 1* (1968), 133. Filled circles denote values obtained from chondritic meteorites. Data for silicon and iron have been improved recently.

The moon rocks appear to exhibit effects of a complicated history of crushing, melting, and exposure to solar radiation. Possibly the best sample would be a chunk of a "new" comet or of a satellite of an outer planet of the solar system, both of which would require sophisticated space probes. It is unlikely that the abundance picture will be changed much from that given in Fig. 104.

The following characteristics are to be noted:

(1) Hydrogen is more abundant than all other elements put together; helium is runner-up, with a ratio of about one helium atom for every ten or a dozen hydrogen atoms.

(2) There is a deep minimum corresponding to lithium, beryllium, and boron.

(3) This deep minimum is then followed by a peak corresponding to carbon, nitrogen, oxygen, and neon.

(4) Following the oxygen peak there is an irregular decline until scandium is reached.

(5) A prominent abundance peak occurs at iron, following which the abundances drop off irregularly once more until we reach an atomic weight of about 100 (atomic number $Z = 45$).

(6) Beyond $Z = 45$, the decline is very gentle; there are minor peaks at barium and lead. After lead the curve drops off rapidly; bismuth is the last stable element.

(7) There is a pronounced difference between the abundances of elements with even atomic number and those with odd atomic number; the even-Z elements tend to be more abundant. Thus, carbon, oxygen, and neon are more abundant than nitrogen, fluorine, or sodium, while silicon and sulfur are more abundant than phosphorus. This is the rule of Oddo and Harkins.

This distribution of elements is not the result of mere chance but contains the essential clues to the mechanisms whereby the elements were created. A number of theories have been proposed, all of which are variants of two basic ideas: (i) the elements were created in an early stage of the evolution of the universe; (ii) hydrogen was the original element and all other elements were created from it by processes occurring in stars or in stellar envelopes. The first hypothesis is commonly associated with the "big-bang" theory of the origin of the universe. According to this view, all the matter of the universe was once compressed down into a remarkable substance called ylem, of such incredibly high density and temperature that the matter consisted essentially only of neutrons. When the universe started to expand, the neutrons broke down into protons and electrons. Protons captured neutrons to form heavy hydrogen, which in turn was soon built into helium. The difficulty with this hypothesis, which was proposed by Gamow, is that element building tends to stop at helium, since an alpha particle cannot capture a neutron. Particles of mass 5 (and also of mass 8) do not exist in nature. If some way could be contrived for getting around these difficulties, the Gamow theory would provide for the building of many familiar isotopes and atomic nuclei. One possible solution is for simultaneous collisions of three (rather than two) nuclei to occur, but the required conditions of density and temperature would appear to be difficult to achieve.

The arguments for element formation in stars are the following:

(1) Stars of markedly different chemical composition exist, the dif-

ferences being fundamentally of two types: (*a*) the metal-to-hydrogen ratio is abnormally low, as in the "subdwarfs"; (*b*) the atmospheres have abnormal abundance ratios involving heavier elements, such as carbon-to-oxygen and zirconium-to-titanium, and excessive amounts of manganese, mercury, gallium, or even more unusual elements.

(2) Stars of low metal-to-hydrogen ratio are invariably very old stars, although normal metal-to-hydrogen ratios may be found in stars of all ages.

(3) The peculiarities of the abundance curve are such that no one process or set of conditions seems capable of explaining them.

If we assume that *all* elements save hydrogen were manufactured in stars, we must assume that no first-generation stars exist; even the oldest known stars were seeded by material from yet older objects. On the other hand, if we postulate that the elements were manufactured at the dawn of creation, it is very hard to understand why certain extremely old stars such as HD 140283 should not have the same metal-to-hydrogen ratio as the sun. If we adopt the idea of a creation with a big bang, 10, 15, or 20 thousand million years ago, we might suppose that some of the heavier elements were produced at that time. There is some evidence for a "background" radiation throughout all the observable universe, corresponding to a temperature of 3°K. If this background radiation is interpreted as what is left of the radiation of the great fireball at the dawn of creation (big bang) and if one makes the best guess one can about the density of matter in the observable universe, one can estimate the conditions available for element building in the first few hours of the exploding universe. Calculations based on models of the big-bang universe produce relative abundances of hydrogen, deuterium, He^3, He^4, and possibly also Li^7 that are not far from the observed values. Heavier elements such as sulfur cannot be formed in this way in sufficient quantity. Additional heavy-atom nuclei might be produced in the cores of stars, particularly in massive ones. Some portion of this material must be returned to the interstellar medium out of which new stars are formed. The last stages of the evolution of a massive star may produce some of the most striking features of elemental abundances.

In fact, the Burbidges, Fowler and Hoyle, A. G. W. Cameron, and their associates have identified no fewer than eight different types of processes required to produce the observed elemental abundances, in addition to any possible manufacture of elements in an original big-bang explosion of the universe. We sketch the mechanisms briefly.

Hydrogen is converted into helium in the course of normal stellar energy production. Eventually, in the cores of some stars, helium may be converted into carbon by the "triple-alpha" process (suggested by Öpik and worked out in more detail by Salpeter) wherein three alpha particles are jammed together almost simultaneously. The finite, although brief (10^{-14} second), lifetime of the Be^8 nucleus formed when two alpha particles collide permits another alpha particle to be captured if the density and temperature are high enough:

$$He^4 + He^4 + He^4 \rightarrow C^{12}.$$

Determination of the rate of this process depends on difficult and sophisticated experiments, and some uncertainty remains. Once carbon is produced, collisions between C^{12} nuclei and alpha particles produce O^{16} in accordance with the reaction

$$C^{12} + \alpha \rightarrow O^{16} + \gamma.$$

Under conditions existing in stellar interiors, the reaction rates are such that carbon and oxygen nuclei appear with similar abundances, although in most stars oxygen is more abundant than carbon. In highly evolved stars of spectral classes R and N, carbon exceeds oxygen in abundance (see Chapter 4).

At temperatures in the neighborhood of 1,000,000,000°K, which can occur in the cores of highly evolved massive stars, even some carbon nuclei have energies sufficient to overcome the strong electrical repulsion between one nucleus and another. Then "carbon burning" can occur in reactions such as

$$\begin{aligned} C^{12} + C^{12} &\rightarrow Ne^{20} + \alpha, \\ \text{or} &\rightarrow Na^{23} + p, \\ \text{or} &\rightarrow Mg^{23} + n, \\ \text{or} &\rightarrow Mg^{24} + \gamma. \end{aligned}$$

Naturally, the protons and neutrons are absorbed very quickly and the alpha particles almost as fast. At these temperatures, many other reactions would occur, including spallation reactions, that is, the shattering of a nucleus by collision with another particle. An example is

$$O^{16} + He^4 \rightarrow 5He^4.$$

Detailed calculations by David Arnett and James Truran show that the most abundant nuclei, Ne^{20}, Na^{23}, and Mg^{24}, can be produced in just about the right ratios for the solar system.

If the star is so massive that further compression and temperature rise can occur, "oxygen burning" may take place with reactions like

$$O^{16} + O^{16} \rightarrow Si^{28} + He^{4}.$$

What happens as the temperature and density are raised to higher and higher values? Reactions will occur at a faster and faster rate; some nuclei will be built up, others will be shattered. Bodansky, Fowler, and Clayton concluded that at temperatures above 3,000,000,000°K an equilibrium is established involving reactions like

$$Si^{28} + He^{4} \rightleftarrows S^{32} + \gamma,$$

that is, silicon captures an alpha particle to form S^{32} with emission of a gamma ray; conversely, a S^{32} nucleus absorbs a gamma ray and undergoes disintegration to $Si^{28} + He^{4}$. They were able to reproduce the solar abundances by invoking an equilibrium situation in which silicon is "burned."

In such an equilibrium situation, the most stable nuclei, that is, the nuclei that hold their constituent particles most tightly, will be favored over those that hold them less tightly. In the interval between calcium ($A = 40$) and nickel ($A = 60$), iron will be the most favored element, followed by chromium, nickel, and other metals of the iron group.

Thus we can easily understand the high abundances of elements like neon, magnesium, silicon, and sulfur, which are presumably produced by carbon burning, oxygen burning, and so forth in the dense cores of highly evolved stars. Some of this material must have escaped to the interstellar medium from which a later generation of stars is born. The "iron peak" requires very high temperatures and densities, which can occur only in the cores of extremely massive stars. The problem of getting this material dispersed into the interstellar medium without modifying it is an extremely difficult one. It is generally supposed that such material is actually supplied by the explosions of supernovae.

Although the abundances decline rapidly at first, beyond iron the distribution flattens out and beyond germanium and tin the numbers fall off irregularly. These elements cannot be created by simply raising the density and temperature in the nuclear furnace. In fact, elements such as mercury, gold, or bismuth should not appear at all. On the contrary, raising the temperature and density would cause the shattering of iron nuclei into alpha particles, with considerable absorption of energy from the surroundings. Ultimately, if the temperature and density were increased indefinitely, there would be nothing left but neutrons.

The clue to the origin of these heavier elements was contained in

Gamow's hypothesis that all nuclei were built up by successive neutron captures. The problem was, if such elements were built in stars, where did the neutrons come from? A possible answer to this question was given by A. G. W. Cameron and by the Burbidges, Fowler, and Hoyle, who pointed out that the actual number of neutrons required, compared to the number of atoms of carbon, oxygen, and neon, for example, was very small, and that neutrons could be produced by reactions like $Ne^{21} + \alpha \rightarrow Mg^{24} + n$ or $C^{13} + \alpha \rightarrow O^{16} + n$. Once formed, these neutrons could be captured by nuclei of iron and other elements to build up successively heavier nuclei. A neutron carries no charge; hence there is no electrical repulsion and it can penetrate a nucleus with ease. It is believed that in violent events such as supernova explosions large numbers of neutrons may be produced suddenly. In the more orderly evolution of massive stars, neutrons can be produced more slowly.

Different types of nuclei are built up depending on whether the atoms are subjected to a high or a low density of neutrons. Suppose a nucleus of atomic weight A and charge (atomic number) Z captures a neutron. It becomes a nuclide of atomic weight $A + 1$, still with a charge Z. The ratio of neutrons to protons in the nucleus is shifted in favor of the neutrons. The nucleus may eject an electron and become a nuclide of atomic weight $A+1$ and atomic number $Z+1$. Thus we write the reaction as

$$(A,Z) + n \rightarrow (A+1, Z); \ (A+1, Z) \rightarrow (A+1, Z+1) + \beta^-,$$

where β^- stands for a beta particle, that is, an electron ejected from the nucleus. Speaking very generally, nuclei tend to maintain roughly comparable numbers of protons and neutrons. Among heavier stable nuclei, the number of neutrons exceeds the numbers of protons; for example, in Fe^{56}, $Z = 26$ and $A - Z$ = number of neutrons = 30. One cannot go on adding neutrons indefinitely without allowing the nuclei to emit beta particles in an effort to restore some balance.

If the density of neutrons is low, the time interval between the capture of one neutron and then another by the same nucleus is long enough for the ejection of a beta particle to occur. Suppose the density of neutrons is so very high that a nucleus captures a second neutron before it has had a chance to eject a beta particle, that is, $(A,Z) + 2n \rightarrow (A+2,Z)$ or even $(A,Z) + 3n \rightarrow (A+3,Z)$. Such nuclei will eventually decay by ejection of beta particles, but note that the resultant nuclei (decay products) will consist of combinations (A',Z') that cannot be produced by slow, leisurely addition of neutrons. In fact, nearly all the nuclei of heavier elements are produced by one (sometimes both) of these processes. A few exceptions are the so-called proton-rich nuclei, which appear to have been produced by violent processes in supernova explosions.

To summarize, heavy nuclei can be produced either in violent events connected with supernova explosions, wherein large quantities of neutrons are produced, or in the cores of massive stars, where neutrons are produced less copiously.

We can now find a plausible explanation for the heavy-metal Class S stars and the carbon stars. The S stars show an enhanced supply of metals of the zirconium group, including technetium ($Z=43$), which has been produced artificially but does not exist otherwise on the earth. Some of these stars show also great quantities of barium and other heavy elements. The carbon stars, with their excess of carbon over oxygen, and sometimes great enhancements of C^{13}, show evidences of nuclear reactions. Just how this material, which is created deep in the interiors of stars, is conveyed to their surfaces without blowing them to bits remains one of the great mysteries.

Three elements, lithium, beryllium, and boron, cannot be manufactured in the stars, for they are quickly broken down to alpha particles and protons. They appear in cosmic rays, however, and the general belief is that these nuclei are produced by breaking down nuclei of heavier elements such as oxygen or neon by impacts with very high-energy particles, events that may have occurred more often in the early history of the solar system. The low abundance of these elements is in harmony with the suggestion of a special mechanism for their formation. Some may have been formed in the initial big bang.

Although some progress appears to have been made in understanding the origin of the chemical elements, many mysteries remain. Consider the freak-composition stars of spectral classes A and B. Some of these objects are spectrum variables showing periodic changes in the intensities of lines of chromium and the lanthanide (rare-earth) element europium, as though different elements were concentrated in different areas on the star's surface. Some have magnetic fields that vary periodically in intensity; the magnetic fields of others vary irregularly with time. Some stars, such as 53 Tauri, show excess amounts of manganese, coupled with enhanced abundances of gallium and mercury. Others, like 3 Centauri and HR 8349, show excessive amounts of phosphorus. Scandium is enhanced in ϕ Herculis, while HR 6870 shows excessive amounts of chlorine, iron, titanium, and especially strontium. Even more puzzling, perhaps, is a star discovered by Przybylski in which iron is absent and the rare earth holmium is prominent.

It is generally believed that these anomalous chemical compositions are confined to the surface layers of stars. Accordingly, attempts have been made to explain them by postulating that protons, alpha particles, and other charged particles are accelerated to high energies by rapidly changing magnetic fields in huge areas on stars corresponding to sunspots. These particles then collide with the nuclei of iron, barium, oxygen, and other elements, causing the ejection of protons, neutrons, and other particles and the consequent formation of nuclei of manganese, gallium, phosphorus, and other elements that are not formed abundantly in the usual processes.

All such theories encounter severe difficulties. The anomalous elements must be concentrated in a thin layer. Great numbers of high-energy particles are required, and the amounts of oxygen, iron, and similar elements may not suffice for the numbers of anomalous atoms required. The presence of short-lived promethium in HR 465 strongly favors this theory, however.

It would appear, furthermore, that the substance from which the solar system was formed has had a complicated history. Much of this material was of primeval origin but some must have been processed through two or perhaps more generations of stars. Particularly significant clues appear to be supplied by the isotopes of heavy elements such as the rare gas xenon and the naturally occurring radioactive elements uranium and thorium. These data suggest that the material destined to be collected into the solar system may have been "seeded" by radioactive and other nuclei produced in the explosion of a supernova. In particular, the interval between the creation of some of these elements and the formation of the solar system was small compared with the 5 or 10 thousand million years that elapsed between the formation of the oldest stars and the birth of the sun. If we accept the picture of a repeated cycling of material through stars and the interstellar medium, coupled with the processes of element building, we would expect a correlation between the metal content of a star and its age. Young stars should be metal rich; ancient stars should be metal poor. Yet, although no young metal-poor stars are known, there occur stars of very great age with the same metal-to-hydrogen ratio as the sun or the stars of the Orion association. Possibly, element building proceeded vigorously during the early history of the Galaxy and at different rates in different places. Consequently the metal content of a star depended not only on when it was formed but on where it was formed—in a globular cluster like M 92 or in an open cluster such as M 67.

Another mystery is the seeming uniformity of the chemical composition of the accessible portions of the universe. The differences in the chemical make-up between stars in our Galaxy appear to exceed by far the established differences between our Galaxy and the Magellanic Clouds or the Triangulum spiral M 33. The detailed histories of these systems may all differ substantially, yet the same processes are at work everywhere.

Still another problem is posed by the helium abundance. If this element was produced in the big bang, the helium-to-hydrogen ratio would be about 0.1 by numbers of atoms. Indeed, calculations of evolutionary tracks of the ancient stars in globular clusters, carried out by John Faulkner and Icko Iben, suggest just this. On the other hand, the atmospheres of blue, highly evolved stars in globular clusters appear to have a very low helium abundance (Sargent, Searle, Greenstein, and Munch). Perhaps these stars resemble the freak-composition stars which appear also to be deficient in helium. Further, the solar helium abundance as inferred from the solar wind and solar cosmic rays seems to be abnormally small. Yet the helium-to-hydrogen ratio (by numbers of atoms) seems to be about 0.10 to 0.12 in our own and other galaxies.

10 Pulsating Stars

Although most stars in the Galaxy remain pretty much the same from day to day, from month to month, or even from millenium to millenium, some depart radically from such placid behavior. Although relatively few in number, these stars merit close attention, for such apparently abnormal objects may well supply vital clues to the theories of stellar constitution and evolution.

In this chapter we devote our attention to stars that change continuously in brightness, rising and falling in magnitude, although not necessarily in a strictly regular fashion. The *variable stars* are so named because they fluctuate in apparent brightness, some periodically and others irregularly, by amounts ranging from a few percent to a factor of several hundred. We concern ourselves here primarily with periodic variables, such as the RR Lyrae stars, and related variables found in globular clusters, the classical Cepheids, δ Scuti stars or dwarf Cepheids, the long-period variables, and the β Canis Majoris stars. All of these stars are believed to have one feature in common: their light variation appears to be produced at least partially by body pulsations of the star.

Figure 105 shows the location of different types of variables in a Hertzsprung-Russell diagram.

The Cepheid Variable Stars

Before we examine the question of what factors make a star variable, let us summarize some known facts about Cepheid variables. Named after the prototype, δ Cephei, discovered by Goodricke in 1794, these stars fall into two rather well-defined subgroups. One group comprises the cluster-type and Population II Cepheids that often are found in large numbers in globular star clusters. Those with periods ranging from about 90 minutes to 0.8 day and with amplitudes of about 1 magnitude are called RR Lyrae stars. Their distinctive light curves make them easy to recognize. Population II Cepheids of longer period include another distinctive group, called W Virginis stars. The second group comprises the classical or Population I Cepheids. All of these stars have periods greater than 1 day and display regularities that are missing from Population II Cepheids.

The first examples of these cluster-type variables were discovered in 1895 by S. I. Bailey in certain globular clusters. The prototype, RR Lyrae, however, was found by Mrs. Williamina P. Fleming in 1900 far from any globular cluster. It has a visual amplitude of about 0.79 magnitude, a period of 0.5668 day, and a range in spectral class from approximately B9 to F2. Spectroscopically these stars present some engaging problems. G. Preston noted that spectral classes estimated from hydrogen and the Ca II *K* lines could differ by as much as one whole spectral class, depending on the metal-to-hydrogen ratios in these stars. This difference, δS, is correlated with period and space velocity. RR Lyrae stars in our neighborhood of the Galaxy appear to be less metal deficient than those found at great distances in many globular clusters. Struve found the prototype RR Lyrae to show bright lines at certain phases; in this respect it resembles other Population II Cepheids. The shape of the light curves of these variables shows a progressive change with increasing period. An interesting characteristic of RR Lyrae stars, noted many years ago by M. Schwarzschild, is that in the color-magnitude diagram of the globular cluster M 3 all of these objects fall in a small domain. Furthermore, every star that falls within this domain is variable.

Great numbers of RR Lyrae stars occur in the central bulge of our Galaxy, and in fact the total number in the Galaxy may exceed 100,000. They also appear in certain dwarf galaxies. Determination of their absolute magnitudes is difficult. From the fortunate circumstance that RR Lyrae and several high-velocity subdwarfs constituted a moving cluster converging to a fixed point in the sky, Eggen found the mean absolute magnitude of RR Lyrae to be +0.65; several other RR Lyrae stars had absolute

Fig. 105. Variable stars and the Hertzsprung-Russell diagram. The main sequence is denoted by the solid line: positions of giants, subgiants, supergiants, and white dwarfs are indicated, and the domains of different kinds of variables are shown. The RV Tauri stars and semiregular red variables probably owe their variations to pulsations. The T Tauri stars are probably stars in the process of formation; the red flare stars are probably very young main-sequence stars. The positions of several representative type stars are indicated. The dwarf-Cepheid sequence actually may not cross the main sequence. Stars of Populations I and II are built on different models, so the overlapping of two different kinds of variables is not surprising. The internal structures of the stars differ from one type of star to another, but within each group they are probably similar.

magnitudes between +0.2 and +0.8. Thus they appear to be 40 or 50 times as bright as the sun.

Classical Cepheids show conspicuous changes in brightness and spectral class. Figure 106, due to Arthur Code, shows the spectral variation in the 7.18-day classical Cepheid η Aquilae as a function of phase in the light cycle. For comparison, spectra of certain supergiant stars are also reproduced.

In Fig. 107 the light-variation curve of η Aquilae is compared with its radial-velocity curve and the corresponding change in radius. Notice the steep rise to maximum light. The star also becomes bluer and the spectral class becomes "earlier." After maximum, the star declines more slowly in brightness; there is a hump on the descending branch for this particular star. Spectroscopic observations show that the spectral lines also undergo displacements, with the same period as the light variation. The radial-velocity curve is here plotted "upside down" in the sense that velocities of approach (negative velocities) increase upward numerically. The corresponding changes in the radius of the star are shown in the bottom curve. Notice that maximum light occurs before the star has reached maximum radius. That is, as conventionally plotted, the velocity curves of many Cepheid variables are almost exact mirror images of their corresponding light curves, with a maximum light occurring simultaneously with, or slightly earlier than, the greatest velocity of approach.

Mean spectral classes of classical Cepheids are correlated with their periods, stars of longer periods tending to have later spectral classes. The most important characteristic of classical Cepheids is the correlation between their periods and intrinsic luminosities; the brightest stars have the longest periods. In 1910 Miss Henrietta S. Leavitt at Harvard discovered this relation from her studies of the classical Cepheids in the Small Magellanic Cloud, one of the two nearest external galaxies. With

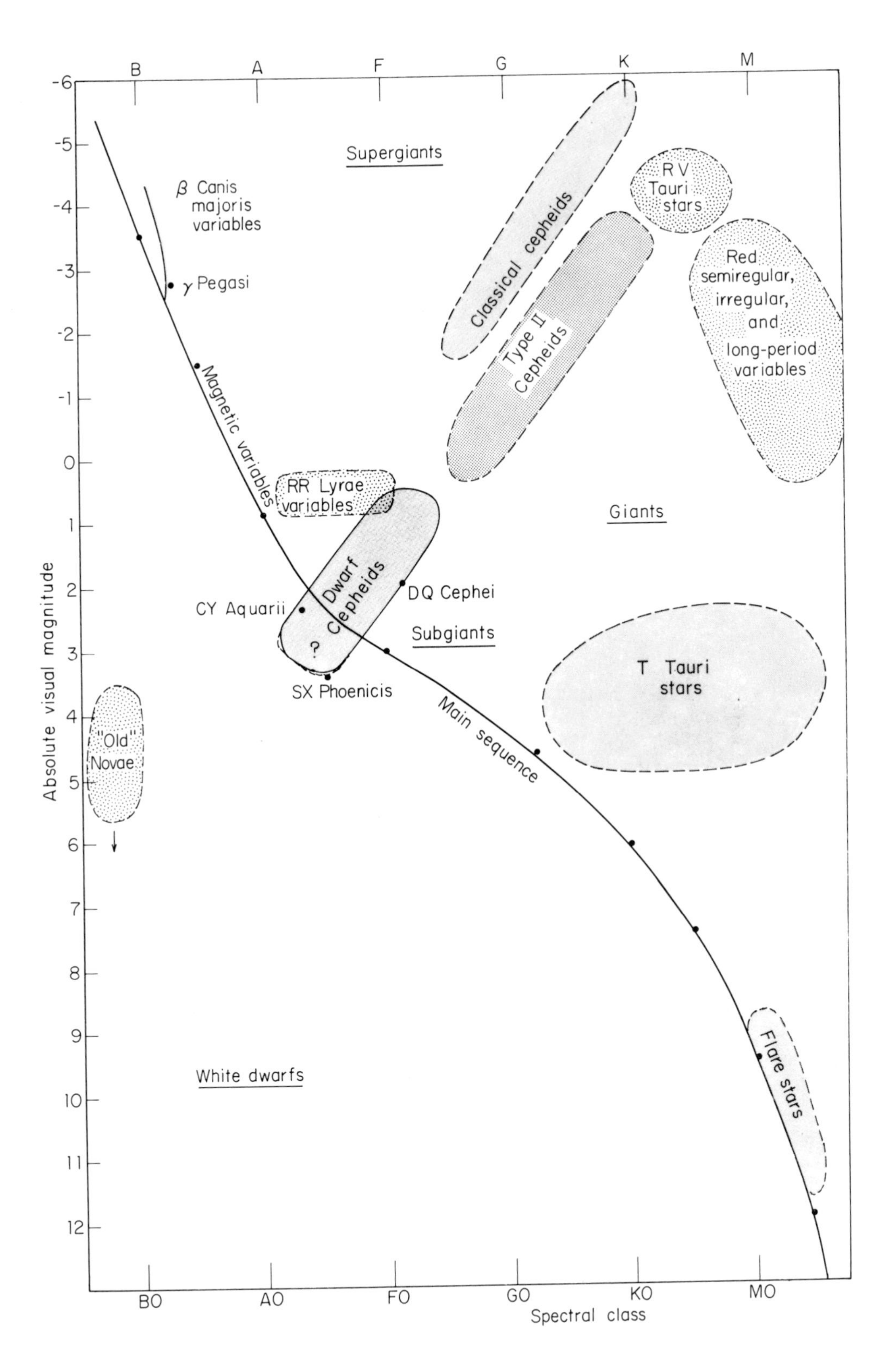
B
A
F
G
K
M
-6
-5
-4
-3
-2
-1
0
1
2
3
4
5
6
7
8
9
10
11
12
Supergiants
β Canis majoris variables
γ Pegasi
Classical cepheids
RV Tauri stars
Type II Cepheids
Red semiregular, irregular, and long-period variables
Magnetic variables
RR Lyrae variables
Giants
Dwarf Cepheids
DQ Cephei
CY Aquarii
?
Subgiants
SX Phoenicis
T Tauri stars
"Old" Novae
Main sequence
Flare stars
White dwarfs
Absolute visual magnitude
BO
AO
FO
GO
KO
MO
Spectral class

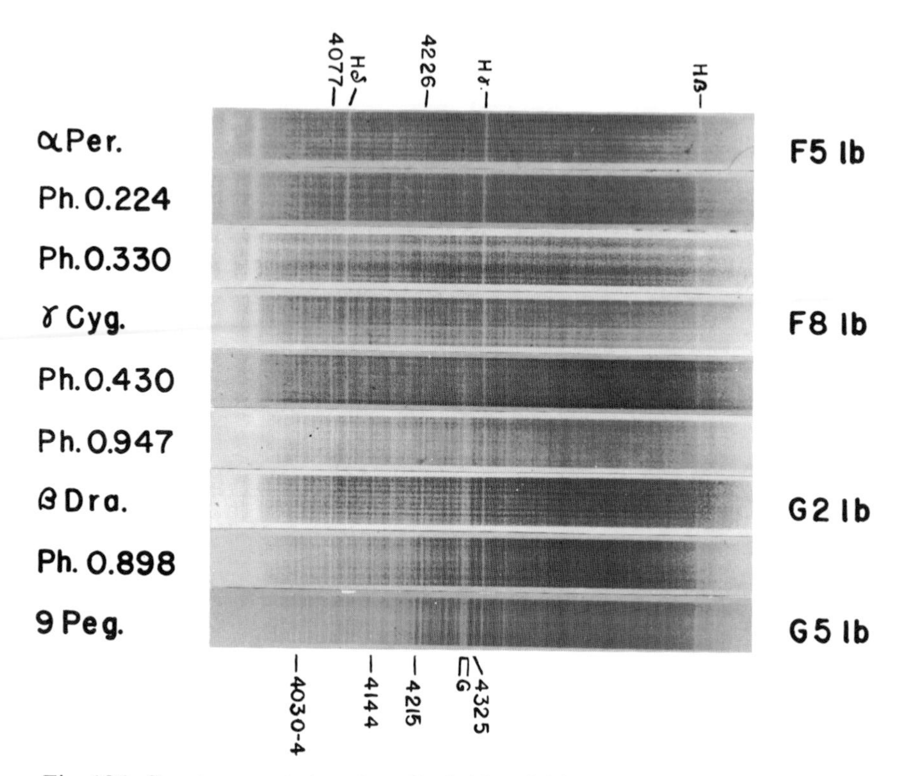

Fig. 106. Spectrum variations in a Cepheid variable. The spectrum of η Aquilae is shown at phases 0.224, 0.330, 0.430, 0.947, and 0.898. The spectra of several supergiant stars, α Persei, γ Cygni, β Draconis, and 9 Pegasi, are shown for comparison. (Courtesy Arthur Code.)

sufficient accuracy, all stars in this cloud can be regarded as being at the same distance from us. Thus, the relation between period and luminosity observed by Miss Leavitt represents an intrinsic property of these stars. If the Cepheids in the Magellanic Clouds are typical of all Population I Cepheids, we perceive that this period-luminosity correlation makes it possible to estimate the absolute magnitude and, therefore, the distance of any classical Cepheid whose period is known. This empirical relation has had important consequences for the exploration of the universe (see Bok and Bok, *The Milky Way,* p. 105; Shapley, *Galaxies,* p. 48), but in order to use it we must know the intrinsic luminosity of at least one Cepheid—or, in technical jargon, we must establish the zero-point of the period-luminosity law.

Classical Cepheids are luminous supergiants, that is, they are highly evolved stars, and none are close enough to us that reliable parallaxes

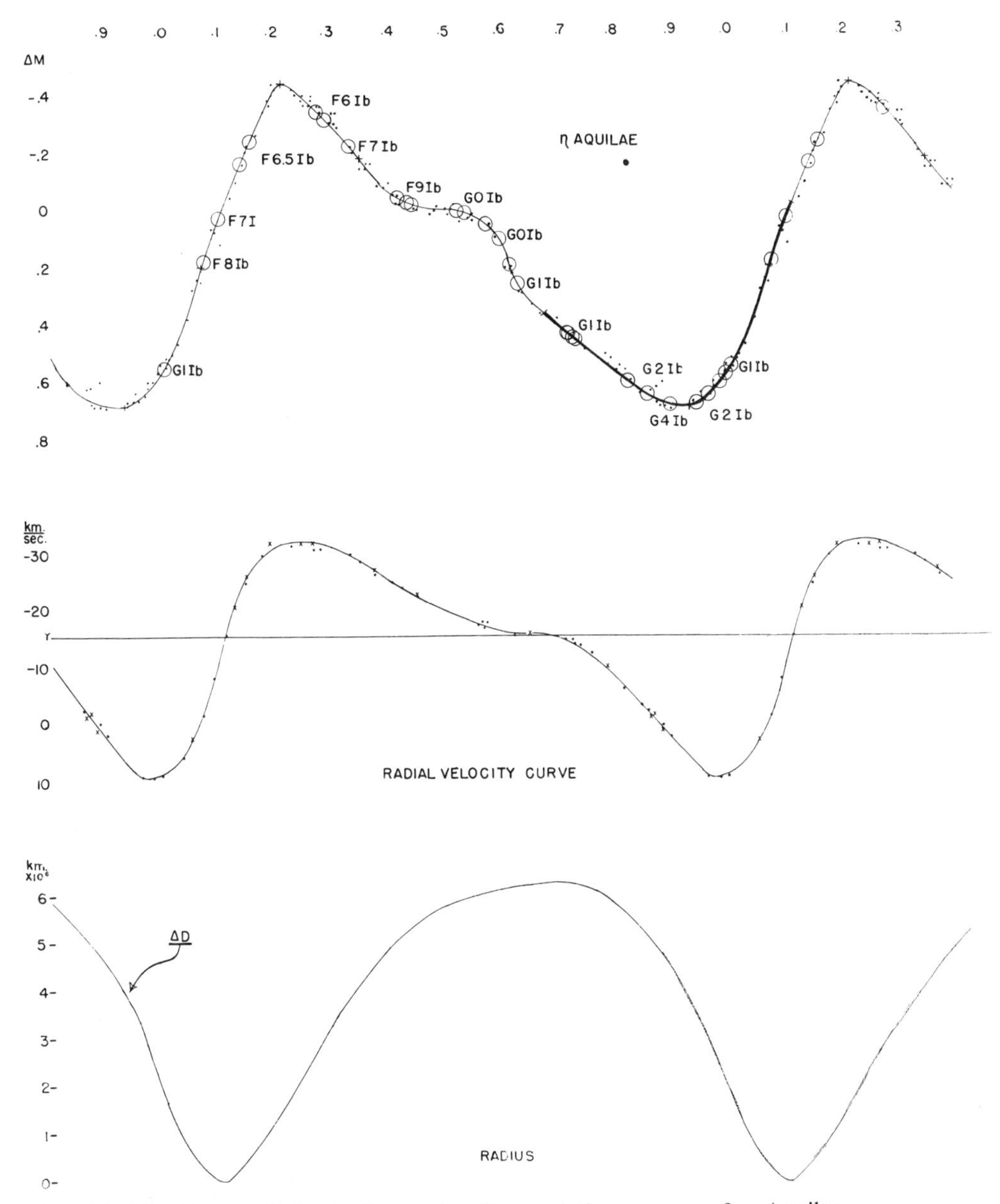

Fig. 107. The light, radial-velocity, and radius-variation curves of η Aquilae. All quantities are plotted against phase or fraction of period (1 unit = 7.177 days). The top curve shows the light variations δM in magnitudes; the spectral class corresponding to each phase is indicated. Notice that the spectral classes differ at points corresponding to the same brightness on the rising and falling branches of the light curve. The earliest spectral class comes *after* maximum light. Notice the bump on the descending branch of the curve, which is also mirrored in the curve of radial velocity (in kilometers per second). The lowest curve shows the change in radius (in millions of kilometers) as a function of time. Notice that maximum radius comes just before minimum light and minimum radius occurs on the rising branch of the curve. (Courtesy Arthur Code.)

can be found by the trigonometric method. Instead, reliance must be placed on less direct statistical methods and on the occasional involvement of classical Cepheids with open clusters (as first noted by John Irwin). In spite of great effort, the zero point was not adequately established until the mid 1950's, when it was found that previous determinations had given an underestimate of about a factor of 4 in luminosity. A classical Cepheid with a period of about 10 days has an absolute photographic magnitude of about −3.5.

Figure 108*a*, due to Shapley, shows the empirical period-luminosity curve as deduced for classical Cepheids in the Small Magellanic Cloud, and Fig. 108*b* that determined by Kraft for galactic Cepheids. Some of the scatter of the points along the curves is due to the finite thickness of the cloud and some to observational error, but some certainly is intrinsic. Two Cepheids of the same period may differ in luminosity by a factor of 2.

Evolutionary tracks of massive stars through the Cepheid region of the Hertzsprung-Russell diagram, as calculated by Baker and Kippenhahn and others, show that a given star may cross the region twice, and that in a given crossing the luminosity may change only slightly while the period changes significantly. Despite these complications, classical Cepheids continue to supply important clues for distances of remote stellar systems.

Studies by H. C. Arp and others demonstrate that Population II Cepheids scatter below the curve defined by the classical Cepheids. These Population II Cepheids, with periods greater than 1 day, are useful in deciding whether the stellar population in a remote system resembles a globular-cluster population or a spiral-arm (solar neighborhood) population, but they are inaccurate as distance indicators.

The Pulsation Theory

The light and spectral variations in Cepheids cannot be explained by eclipses. Furthermore, if the radial-velocity curves were attributed to the motions of the components of a spectroscopic binary, the orbits would have to be smaller than the stellar dimensions. Impressive observational and theoretical evidence supports the pulsation theory, proposed by Shapley in 1914. According to this view, Cepheid variables are continually in the process of swelling and contracting like so many huge balloons.

Just what are the conditions inside a star that would enable it to pulsate? Why do some stars pulsate and not others? We cannot look into the interior of a star, but any conjectures that we make should be guided by

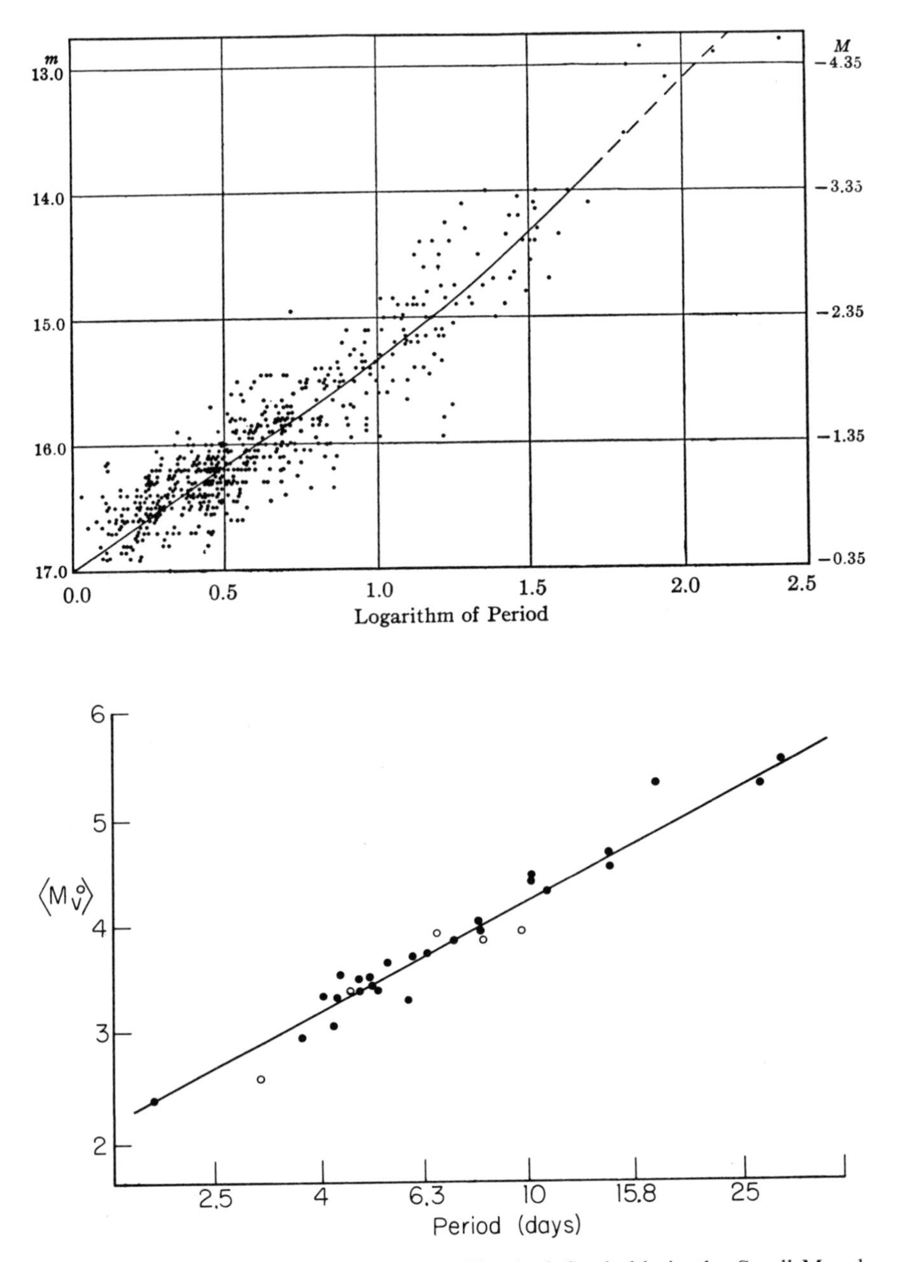

Fig. 108. Period-luminosity curves. (*a*) Classical Cepheids in the Small Magellanic Cloud. Both apparent (*m*) and absolute (photographic, *M*) magnitudes are shown. The scattering of the points about the curve is of much significance. (From Harlow Shapley, *Galaxies*, p. 51.) (*b*) Galactic Cepheids. (After R. Kraft, *Astrophysical Journal 134* (1961) 630; copyright University of Chicago Press.)

the known laws of physics. As we have seen, the reason why a star does not collapse under its own gravitation is that its interior is so hot. Although the average density of the sun exceeds that of water, we have good reasons for believing that its interior consists solely of hot, highly compressed gases. Gases at high temperatures exert considerable pressure as a consequence of the impact of high-speed atoms on one another. Radiation can also exert a pressure, although it is less important than gas pressure. At any point inside a normal star, radiation pressure plus gas pressure just suffices to support the weight of the overlying stellar layers.

Let us now suppose that the very neat balance between gravity and pressure is upset, perhaps by the gravitational pull of a passing star, or even by the slight adjustments the star has to make in its internal structure as it converts hydrogen into helium. When the pressure exerted upon a gas is lessened, the gas expands, just as bubbles rising from the depths of a pond grow larger as they approach the surface. As the heated stellar gases expand, they force the surface layers upward, but only for a time, because expanding gas cools and therefore exerts less and less pressure as the speeds of its atoms slacken. When the pressure becomes low enough, gravity reasserts control and expanded layers fall back. Does the star then regain its stability? No, for the momentum of the downward-moving gas is usually enough to cause it to overshoot the stable position. Once more the gases become compressed and heated enough to overpower gravity and the sequence is repeated.

The situation is not unlike that which prevails when a weight is suspended from a coiled spring. The spring stretches, and the tension increases until it just compensates for the weight. If we disturb the equilibrium by pulling slightly on the weight and then releasing it, the increased tension in the spring overpowers the force of gravity and the weight shoots up beyond its original position, until the force of gravity pulls it back. The oscillation that is set up continues until it is gradually damped out by friction.

The explanation of the Cepheid variation as the result of a contest inside a star between predominantly gas pressure on the one hand and the force of gravity on the other appears to fit the observed features of Cepheid variation. As long ago as 1879 Ritter demonstrated by sound mathematical and physical arguments that it was possible for a star to pulsate. But astronomers then knew little about the interpretation of the light curves and nothing about the variable radial velocities of the Cepheids. It was not until 1914 that Shapley showed that the pulsation hypothesis gave the most reasonable explanation of the variability of these stars. Finally, Eddington, a few years later, worked out a detailed mathematical theory of Cepheids. He proved that large Cepheids should pulsate more slowly than smaller, denser ones. In mathematical terms, he found that the product of the period and the square root of the average density should be nearly constant, a prediction that seemed to be in harmony with the observations. The period is measured directly but the density must be found from the radius of the star and its mass. The radius follows from the absolute magnitude and surface temperature, since the surface temperature determines the rate of radiation per unit area and the absolute magnitude gives the total luminosity, which is the surface area times the emission per unit area. The mass of the star is found by determining the point on the main sequence from which the star evolved and using the mass appropriate to that luminosity.

The Eddington theory, however, immediately ran into certain difficulties. First and foremost, the calculations showed that the energy of oscillation should gradually decay, just as the oscillation of a weight attached to a spring dies out, and that a normal Cepheid should stop pulsating in a mere few thousand years. Evidently, the mechanical energy dissipated must in some way be replenished from within the star.

Second, the theory predicted that the star should be brightest when it was smallest because the rise in temperature more than offsets the decrease of the surface area. Correspondingly, it should be faintest when largest and at both maximum and minimum the radial velocity should be zero. However, reference to Fig. 107 shows that at maximum and at minimum light the velocity is considerably different from zero.

Thanks to efforts of M. Schwarzschild, S. A. Zhevakin, J. Cox, A. Cox, R. F. Christy, and their associates, these difficulties have been resolved and the pulsation theory has yielded a wealth of dividends. What is required, however, is an exact knowledge of the properties of matter at the temperatures and pressures encountered in stellar interiors and the use of an exact physical theory and exact equations for handling them. Many features of a variable star's behavior depend on near cancellation of opposing effects, so precise calculations are necessary. The older approach, which relied on approximate and intuitive arguments, cannot be expected to yield exact agreement between theory and observation, but it did give Eddington's period-density relation.

Consider first the nature of energy flow in a pulsating star. The predominant flow is a steady stream outward; only in the outer 15–25 percent of the radius will there be a significant ripple arising from the alternate damming up and release of energy in pulsations. When a volume of gas is compressed, its temperature is raised. Ordinarily, the opacity of the material will be decreased so that energy will escape more readily. Under these circumstances the pulsations will tend to damp out. In certain outer regions of the star, where ionized helium prevails, compression may actually increase the opacity of the material. More energy is trapped during the compression phase and when it is released during the expansion phase it gives an extra boost to the outer layers, thus tending to build up the pulsations. There are also effects arising from the influence of changes in the molecular weight and from the geometrical property that contraction always tends to compress the gases. Energy tends to be dissipated in deeper layers, and the amplitude of the pulsation adjusts itself so there is no long-term storage or depletion of energy in the pulsating layers.

Whether or not a star will pulsate depends on the details of its structure and evolutionary history. John and Arthur Cox showed that, if one calculated a static model for a star in the "unstable" region of the Hertzsprung-Russell diagram, it would start to oscillate and the amplitude would build up until storage and dissipation balanced. As evolution carries the star out of the unstable region, the oscillations die away and the star ceases to pulsate. Pulsations affect the outer layers of the star; the core, where energy is generated by thermonuclear reactions, is left undisturbed.

Detailed calculations with precise models, taking into consideration the large amplitudes and the storing and dissipation of energy in the outer layers, account for the observed behavior of the velocity and light curves. In agreement with observations, the time of mean rising light is predicted to come near minimum radius. Christy was, in fact, able to explain exactly the light curves of many classical and cluster-type Cepheids. In particular, he accounted for the bumps on many curves as follows. A large acceleration of material in the outer layer where helium is becoming ionized sends a pressure pulse inward which bounces off the core and is reflected back to the surface. In one model, the secondary bump is an echo of a primary bump 1.4 periods earlier. Besides such echoes, Christy found resonance effects in that ingoing and outgoing waves may mutually interfere; sometimes they reinforce each other and sometimes they cancel. Hence one can understand the complex light and velocity curves observed. In all models, a change in radius of only a few percent suffices to account for the observed light variations.

The efforts of the Coxes were devoted mainly to an interpretation of classical Cepheids, whereas Christy's analysis gives us an understanding of RR Lyrae stars and Population II Cepheids. Further refinements are required, especially to take into account atmospheric phenomena and show how they are related to the internal structures of the stars. At the University of Michigan, many years ago, Curtiss and Rufus concluded that the displacements of the lines of neutral and of ionized atoms give different velocity curves. Since the ionized atoms may be presumed to exist at high levels in the atmosphere, where the density is low, and the neutral atoms at lower levels, the work of Curtiss and Rufus suggested that the different layers of the atmosphere are not oscillating in unison. More recently it has been observed in Population II Cepheids that at certain phases two sets of absorption lines, indicating different velocities, are observed. The separate segments of the velocity curves appear to represent successive shells of material ejected from the stellar surface. A mass of material is thrown off, slows down, and appears to fall back into the star. In the meantime, a second cloud of material is thrown off, the descending layer disappears, and the next cycle takes over. Abt found from a study of W Virginis that the two sets of absorption lines were formed in strata of different densities.

Eddington pointed out, too, that stars might pulsate in overtones, as well as in their fundamental frequencies, just as a musical string emits notes one, two, or three octaves apart, depending on how it is plucked. One example is the so-called "dwarf Cepheid" star δ Scuti, observed by E. A. Fath, which may pulsate not only in its fundamental period, but also in overtones.

Some of the best examples of overtone pulsations are found among RR Lyrae stars. Christy's theoretical studies indicate that stars of high luminosity-to-mass ratio tend to pulsate in the fundamental mode while those of low luminosity-to-mass ratio prefer the first overtone. At the dividing line between the two types of behavior, the state of pulsation depends on the past history of the system.

Another type of variable which may be related to the Cepheid and exhibit various phenomena of overtones is the RV Tauri stars. Here the observational data are more fragmentary. Atmospheres are more extensive and the interrelation between the atmosphere and the interior plays a more significant role in establishing the light curve.

The success of theory in interpreting pulsating variable stars constitutes one of the most impressive advances in modern astrophysics. Miss E. Hofmeister obtained an excellent agreement between the theoretical period—luminosity relation and the empirical one determined by Robert Kraft (see Fig. 108*b*). Christy's calculations indicate that it may be possible to determine mass, luminosity, radius, and helium content for these stars from observational data on light and velocity curves alone. The luminosities so determined appear to agree with data from other sources. For example, for β Doradus, whose period P is 9.84 days, Christy finds for the luminosity $L = 3.7 \times 10^3\ L_\odot$, or the bolometric magnitude $M_b = -4.2$, which fits the period-luminosity relation. Application of the theory (in this instance the transition period at which RR Lyrae stars switched from fundamental to overtone pulsations) gave $L = 46\ L_\odot$ for these stars in ω Centauri and $L = 37\ L_\odot$ for those in M 3. On the other hand, the masses seem to be lower than those found from evolutionary arguments both for RR Lyrae stars and for classical Cepheids. Do these stars lose mass during late stages of their evolution?

A curious feature of RR Lyrae stars is that the prototype RR Lyrae has a magnetic field that varied between −1580 and +1170 gauss with the secondary period of 41 days. What role do these magnetic fields play in pulsations?

The Long-Period Variable Stars

Another important group of variable stars, those of long period, present even more engaging and puzzling problems than do the Cepheids. The stars in this group are all cool red giants and supergiants, of spectral classes M, R, N, and S, radiating, at maximum brightness, about 100 times as much energy as the sun. Their periods range, in general, from 100 to 500 days and their brightnesses, as observed visually, fluctuate about a hundredfold, or five magnitudes.

Figure 109 shows the light curve of the brightest and most famous of the red variables, *o* Ceti, otherwise known as Mira ("The Wonderful"). Notice that, unlike the Cepheid variables, Mira does not always return to the same brightness at maximum, but is nevertheless then easily visible to the naked eye. At minimum, the star fades to about the ninth magnitude, and is visible only in the telescope. Like the Cepheids, the long-period variables seem to exhibit rough relation between period and spectrum, the stars of longer period being of later spectral type. Clearly, the long-period variables do not fall on the period-luminosity curve for the Cepheids, since a hypothetical Cepheid of such long period would be much brighter than any known red variable.

Most of the energy radiated by cool stars is concentrated in the form of invisible heat waves in the infrared region of the spectrum. This radiation can be measured by means of a vacuum thermocouple. A continuous electric circuit is made by joining two dissimilar metals, such as

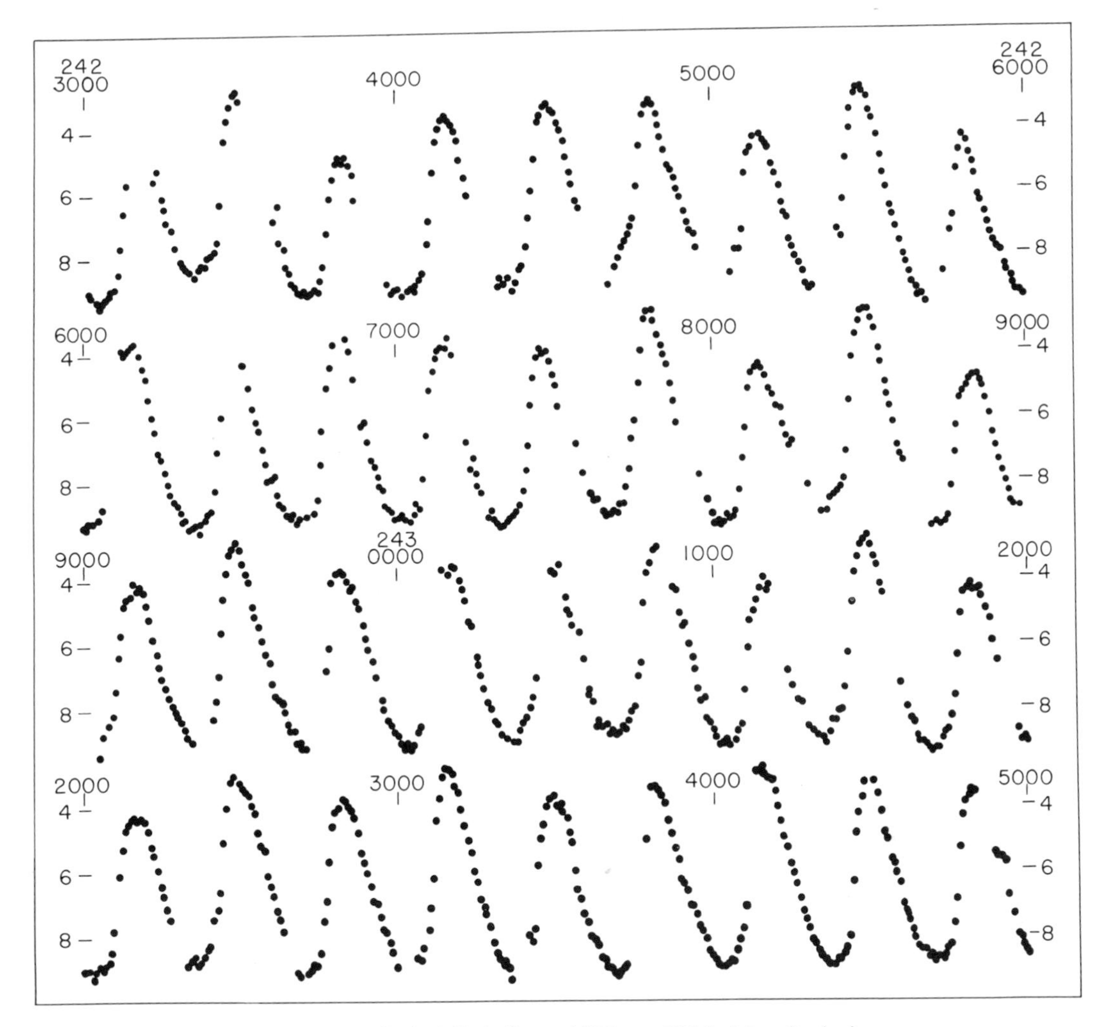

Fig. 109. The light curve of o Ceti (Mira) from 1922 to 1954. Magnitude is plotted against the Julian day. Most of the observations have been secured by members of the American Association of Variable Star Observers. (Courtesy Margaret W. Mayall.)

copper and iron. If one of the two junctions is heated by radiation falling upon it, while the other is shielded, an electric current is generated that is proportional to the amount of the incident radiation. This device responds to radiation of all wavelengths, but, since its operation depends on the actual heating of the junction by starlight, it is rather insensitive. With the aid of filters, one can measure the relative proportions of radiation in different parts of the infrared and hence determine the temperature. Long ago, Pettit and Nicholson at Mount Wilson used such a device to measure the radiation and the temperature of long-period variables. They found that at maximum light the temperature of Mira (spectral class M 6) is 2600°K and at minimum 1900°K. Similarly, the star R Leonis (M 8) varies from 2200 to 1800°K, whereas χ Cygni ranges from 2200 to 1600°K. In Chapter 4 we saw that the amount of energy radiated by a star is proportional to the fourth power of its temperature. Since the temperature of Mira changes only by a factor of 1.37, the energy emitted should vary from maximum to minimum by $(1.37)^4$ or 3.5. Yet the amount of variation shown by the light curve in Fig. 109 is over five magnitudes, or more than a factor of 100. This apparent anomaly may be partly explained as a consequence both of the difference in shape between energy curves at different temperatures and of the fact that the light curve of Fig. 109 represents the variation in visual light, whereas most of the emitted radiation is invisible to the eye. Figure 110 compares the computed energy curves for Mira at maximum and minimum light. The total radiation, represented by the area under each curve, has increased 3.5 times from minimum to maximum, but the visible radiation, of

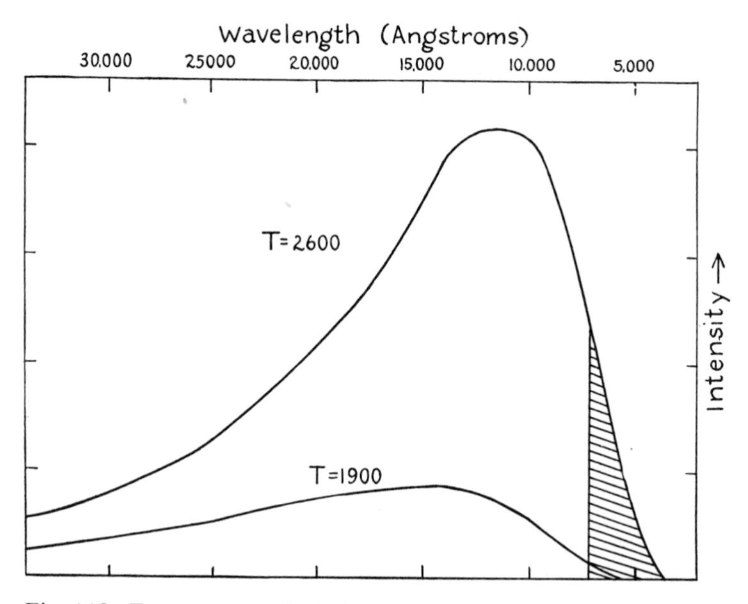

Fig. 110. Energy curves for Mira at maximum (2600°K) and minimum (1900°K). The shaded area shows the relative amounts of energy in the region visible to the eye. Notice how the change in visible light is very much greater than the bolometric change (that is, the change in total radiation). Curves such as these were first constructed for long-period variables by Joy.

wavelength less than about 8000Å, has gone up by a much larger factor. Only part of the discrepancy may be accounted for in this way, however. The change in visible light is still about 15 times as large as we would expect. Why? Because cool stars do not radiate like black bodies.

The Spectra of Long-Period Variables

The spectra of long-period variables are dominated by molecular bands, which, with high dispersion, may be resolved into individual lines. In variable stars of Class M, only the violet region is free from the obscuring bands of titanium oxide, which produce great gaps in the spectrum from 4600 to 6400Å. In cool carbon stars the blue and violet regions are largely blotted out by bands of carbon, and in the S stars these regions are obliterated by absorption by zirconium oxide. The absorption is strongest at the band heads and then gradually lessens as the separation of band components increases. We have previously mentioned that some carbon bands in the R and N stars are produced by molecules containing the carbon isotope of mass 13. As noted by McKellar and subsequently by Wyller, Climenhaga, and others, this isotope is often more abundant in these stars than on earth, where it constitutes 0.7 percent of all carbon atoms.

The spectra of long-period variables (Fig. 111) are also rich in dark atomic lines characteristic of low temperatures, except when these lines happen to fall in regions of band absorption. Thus the *D* lines of sodium, although quite strong in S stars, are smothered by the bands of titanium oxide in stars of Class M. As the variable star fades and its temperature falls, the dark lines change as we would expect them to do on the theory of ionization (Chapter 4). The *H* and *K* lines of ionized calcium fade, and easily excited lines of neutral atoms, such as that of calcium at 4227 Å, strengthen. Also, at the lower temperatures, additional compounds form and all the bands become intensified. Proper studies of the spectra of cool stars require high dispersion (see, for example, the studies by Y. Fujita and his associates) and much useful information is obtained in the infrared. Recently a number of very cool ("dark-brown") stars have been found by infrared detectors. Some of these are also variable stars.

The most perplexing problem in connection with the spectra of red variables is the appearance, before maximum light, of strong, bright lines, especially of hydrogen (Fig. 112). The lines appear generally in red stars that vary and the range of their intensity must be far greater than the range of light variation. They reach maximum intensity about one-sixth of the

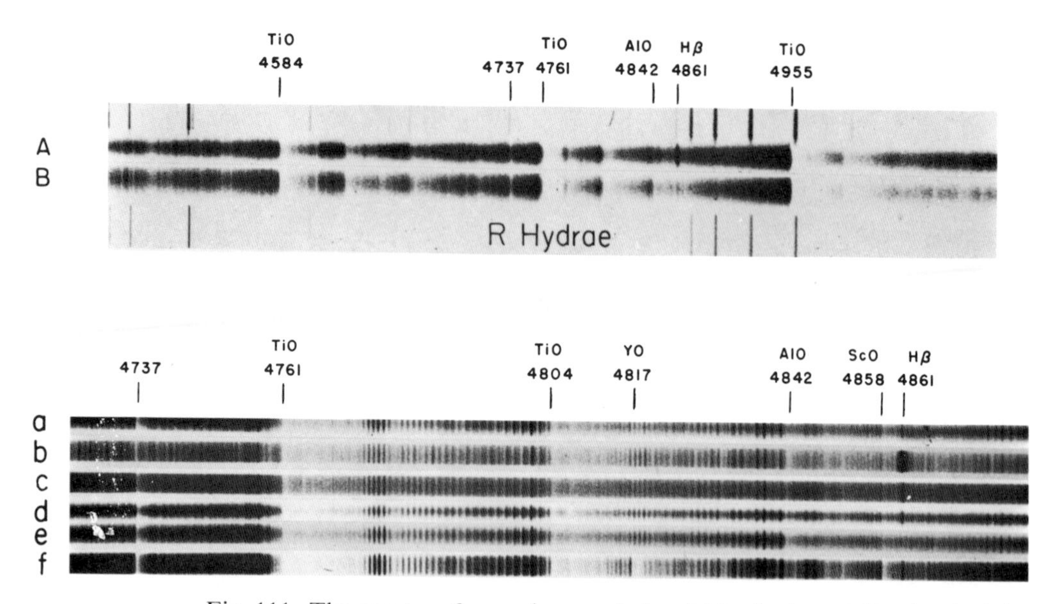

Fig. 111. The spectra of some long-period variables in the neighborhood of the blue hydrogen line Hβ (4861 Å). (*Top*) Two spectra of R Hydrae: (*A*) 1924; (*B*) 1940. The 4842-Å AlO band was very strong in 1950. (*Bottom, a, b, c*), o Ceti (Mira) at different phases; (*d*) R Hydrae; (*e*) R Leonis; (*f*) χ Cygni. Notice bands of AlO, ScO, TiO, and YO. (Courtesy A. J. Deutsch, P. C. Keenan, and P. W. Merrill, *Astrophysical Journal* (copyright University of Chicago Press), *136* (1962), 21, Fig. 5.)

period after maximum light. The phenomenon of bright lines in stellar spectra is not unusual, although the great majority of stars do not show them. Actually, bright lines are most likely to occur when a star is very hot and possesses an enormously distended atmosphere. The red variables, although supergiants with huge atmospheres, are cool objects. Furthermore, the bright hydrogen lines are not radiated by the outermost portions of the atmosphere, but at levels *below* those in which the molecules are absorbing. The evidence for this remarkable behavior comes from a close examination of the intensities of the bright hydrogen lines, in Mira and in other variables of Class M. In laboratory sources, in the solar chromosphere, and in the nebulae, the intensities of the bright lines of the Balmer series diminish regularly from Hα toward shorter wavelengths. But in the M stars, whenever a hydrogen line falls within a band of titanium oxide, it is greatly weakened. Thus the first two lines, Hα and Hβ, are much fainter than Hγ, which in turn is not as strong as Hδ. Similarly, the line Hε is weakened by its close proximity to the *H*

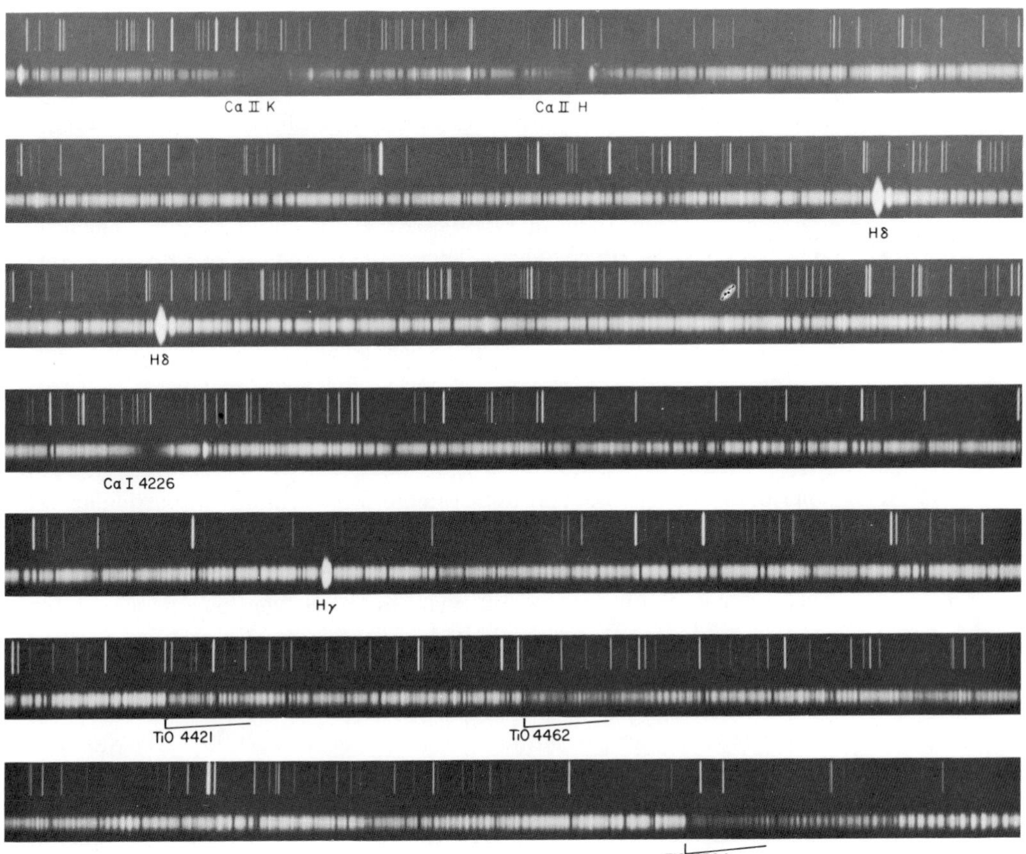

Fig. 112. A portion of the spectrum of *o* Ceti (Mira), as photographed by George Herbig with a dispersion of 2 Å/mm with the coudé spectrograph of the 120-inch reflector at the Lick Observatory, University of California. Notice the strong Hγ and Hδ emission lines of hydrogen and the weakness of the hydrogen Hϵ (3970-Å) line near the Ca II *H* line. The fine structure of the TiO bands and the rich metallic spectrum are well exhibited in this spectrogram.

line of ionized calcium. We can only conclude that the bright-line radiations are depleted by the absorption of overlying layers of titanium oxide and ionized calcium.

Although these red variable stars have been presumed to expand and contract somewhat like Cepheid variables, their spectroscopic behavior makes a detailed interpretation on the basis of any pulsation theory extremely difficult. Joy's measurements of the Doppler shifts of the dark lines and bands in Mira, and Merrill's similar studies of other long-period variables, show that the velocity variations are much more complex than in the spectra of classical Cepheids. Changes in the velocity of the reversing layer are small and impossible to reconcile with the type of atmospheric motion found in classical Cepheids. Joy found that the differences between various cycles of Mira are greater than the average change of velocity observed in a single cycle; the velocities at the bright maxima are appreciably greater than those observed at the fainter maxima. The bright lines always show an outward displacement, suggesting a hot layer rushing upward from some point of origin far below the layers where the titanium oxide bands are formed. The exact nature of this "hot front" remains obscure; it may have some of the properties of a shock wave. Merrill suggested that an advancing wave of heated gas and radiation may produce many of the observed features of the long-period variables. One fact that must be kept in mind is the vast extent of the atmospheres of supergiant stars in general and long-period variables in particular. The spectrum we observe is contributed by regions of widely varying temperatures and densities.

Another question that is still unanswered is whether changes in temperature and surface area as a red variable goes through its cycle are adequate to account for the observed variability of the light. The change in visible light for Mira is about 15 times as great as we would expect from a temperature variation between 1900 and 2600°. Allowance for changes in the surface area aggravates the discrepancy, for Pettit and Nicholson found that the radii of red variables are on the average 18 percent greater at minimum than at maximum. Two possible explanations have been offered. One is that when the temperature of the star is lower, near minimum light, the heavy absorption bands in the visual region become even heavier and may screen off considerably more photospheric light than at maximum. The other possibility is the actual veiling of the star by the formation of small solid particles. Merrill and others have pointed out that, since the atmospheric temperatures at certain phases are often below the boiling points of refractory substances like carbon or certain metallic oxides, clouds of liquid droplets or solid grains might condense from the gases of the star's upper atmosphere. The resulting veiling process would be a periodic phenomenon because the condensations depend on the temperature, but it would not be identical during the different cycles. Thus the observed irregularities in the amplitude of the visible-light variation could be understood.

All red supergiant stars seem to be variable in light to some extent. Many are quite irregular with a small amplitude, as is Betelgeuse; others are semiregular.

Differences in Chemical Composition in Red Variable Stars

All long-period variables are stars in an advanced stage of evolution; their spectra display a diversity that can be interpreted only in terms of differences in chemical composition. The Class N stars appear not only to have an excess of carbon over oxygen, but also other composition differences that set them apart from the M stars, namely, a greater proportion of heavy metals of the zirconium group. The S stars such as R Cygni, R Geminorum, and R Andromedae appear to have much more zirconium than titanium. Other variables such as χ Cygni fall between the M stars like R Leonis and R Hydrae and Mira, which appear to have about the same composition as the sun, and the "heavy-metal" S stars. In the S stars, not only is zirconium strengthened with respect to titanium, but the neighboring metals, strontium, yttrium, niobium, and molybdenum, are strengthened with respect to calcium, scandium, vanadium, and chromium as compared with a normal star. These composition differences all have important evolutionary implications.

The β Canis Majoris Stars

Not all stars that appear to owe their variations to pulsations are Cepheids or late-type stars. The β Canis Majoris stars have a small light variation. Their most remarkable property is the nature of their velocity changes. The line profiles vary in appearance, the lines being usually broadest when the star reaches its average (γ) velocity on the descending velocity curve. Double absorption lines are seen in some objects near the maxima and minima of the velocity curve.

The prototype, β Canis Majoris, was intensively studied by W. F. Meyer and by O. Struve and his coworkers. The velocity curve of this star consists of two interfering harmonics:

$$P_1 = 0\overset{d}{.}25002246\ (6^h0^m), \qquad \text{amplitude} = 5.8\ \text{km/sec};$$

$$P_2 = 0\overset{d}{.}2513003\ (6^h2^m), \qquad \text{amplitude} = \begin{cases} 4.2\ \text{km/sec}\ (1909\text{–}1931) \\ 3.0\ \text{km/sec}\ (1931\text{–}1938) \end{cases};$$

$$\text{beat period} = 49.1236\ \text{days}.$$

The line changes are correlated only with the period P_2. The light curve probably also has similar harmonics. Sometimes, as in σ Scorpii, only a single period, P_2, is observed, but the amplitude changes, and there are pronounced variations in the line widths. Variation of the β Canis Majoris type is restricted to a small spectral range (B1–B3, near the main sequence) but the stars show no evidence of any temperature change during the oscillation. McNamara finds that the β Canis Majoris stars fall above the main sequence and that they exhibit a period-spectrum relation, the earlier spectral types having the longer period. Ledoux has suggested that the light variations might be produced by nonradial oscillations, that is, deformations of the star from a spherical shape, but the problem is far from being solved.

The β Canis Majoris stars represent only one of several types of variables that have more than one period of light variation and exhibit the phenomenon of beats. We have already mentioned δ Scuti. RR Lyrae has two periods—a fundamental around 1/2 day and a harmonic or overtone of about 1/4 day. These two periods interfere to produce a harmonic with a period near 41 days. SX Phoenicis, which has the shortest primary period known, 79 minutes, has a beat period of about 280 minutes. AI Velorum is an even more remarkable star. In the course of a single day, its light curve runs through nine cycles of different shape, which Walraven found to be the consequence of six different superposed sinusoidal oscillations.

Faint Short-Period Variables

Stellar pulsations appear not to be confined to stars that are very much brighter than the sun. Intensive investigations by T. Walraven, O. J. Eggen, Harlan J. Smith, O. Struve, and others have substantiated the existence of plusating stars that are not much more luminous than the sun.

SX Phoenicis is a pulsating subdwarf star (spectral class A5) for which Smith finds an absolute magnitude of +3.9 from its trigonometric parallax. CY Aquarii, with a period of 90 minutes, is a similar star, of absolute magnitude +2.5, as perhaps are several other stars, including δ Scuti, VZ Cancri, DQ Cephei, and AI Velorum. Smith proposes to designate such objects as dwarf Cepheids. Their periods are less than one-fifth day, they are about two magnitudes fainter than the RR Lyrae stars, and they show a definite period-luminosity relation. Some have light curves that indicate a single fundamental period of vibration; others, such as δ Scuti, have two or more periods with resulting beats. (The existence of pronounced period-luminosity relations among classical Cepheids, dwarf Cepheids, and β Canis Majoris stars implies that within each group the period multiplied by the square root of the density is a constant, $P\sqrt{\rho} = K$, but that K changes from one group to another.)

In spite of the brilliant theoretical successes of J. Christy, E. Hofmeister, A. Cox, S. Zhevakin, J. Cox, and others, and the extensive observational programs of Joy, Deutsch, Keenan, Preston, Kraft, J. Smak, and many others, numerous fundamental problems remain. Certainly one of the most basic is the question of absolute magnitudes. It is in this context that the variables in the Magellanic Clouds are so valuable, since their intrinsic luminosities are known. Thus, investigations such as S. Gaposchkin's survey of light curves of variables in the Magellanic Clouds and the Galaxy should pay rich dividends as large telescopes become available in the Southern Hemisphere.

11 Exploding Stars

The Novae

On the evening of June 8, 1918, there suddenly appeared in the constellation Aquila a first-magnitude star, which was noted independently by dozens of people. Older photographic records at some observatories showed that on June 5 the star was of the eleventh magnitude, as it had been for the previous 30 years. Forty-eight hours later it was of the sixth magnitude—a hundredfold increase in brightness. Still increasing on June 9, its magnitude was −0.5, so that it outshone all the stars in the sky but Sirius and Canopus. After attaining the peak of its splendor, the star began to fade, rapidly and erratically at first and then more slowly. In 18 days it had declined to the third magnitude, and it faded from naked-eye view about 200 days later. Its present brightness is close to its original value before the outburst, but it is still really five or six times as luminous as the sun.

Bailey estimated that 25 such novae brighter than the ninth magnitude at maximum flare up in our Galaxy every year, although only eight or nine have been conspicuous to the naked eye since 1900. Many novae are found near the central bulge of the Galaxy; because of their high luminosity at maximum, they can be seen to great distances. H. C. Arp found a similar spatial distribution to hold in the Andromeda galaxy M 31, with a frequency of 26 per year.

Before the advent of large telescopes and photography, these flashing apparitions were regarded as actually new stars and even now the designation *nova* remains in common usage, although photographs show that bright novae such as Nova Aquilae 1918 existed as faint stars before their outbursts. All available evidence indicates that a nova is the result of a stupendous stellar explosion wherein the outermost layers of a star are violently wrenched away. The causes of these outbursts are unknown, but they appear to be confined to small blue stars (with surface temperatures of the order of 50,000°K) which are members of close binary systems.

Such violent explosions of a star's surface must be the result of some deep-seated disturbance. Actually, at any point in the interior of a normal star the weight of the overlying material is just balanced by the outward pressure of gas and radiation. As the temperature of a gas rises, the pressure also rises, because the gas tries to expand. Let us suppose now that, owing to some sudden liberation of energy, a stratum of gas somewhere inside the star becomes overheated. If the star does not have a safety valve by means of which some of the excess energy may escape, the mounting pressure of the expanding gas may so overbalance the weight of the material above it that the entire overlying surface layer of the star will be violently ejected into space.

The conclusion that novae are the results of explosions in the surface layers of stars has been deduced from the remarkable spectral changes that accompany such "outbursts." Sometimes, as with Nova Delphini (Figs. 113 and 114), the changes are extremely complex. In a more typical nova, such as Nova Cygni 1920 or Nova Vulpeculae, the changes are less involved and may be described as follows. During the early stages of a nova outburst, the ejected surface layer swells up like a balloon. As it grows larger, it becomes brighter in total light, although the apparent photospheric temperature actually drops. At maximum light, the spectrum is continuous and crossed by dark lines, as it was before and during the rise in brightness. Evidently, the photosphere, although enormously expanded, is still intact. The dark lines of hydrogen and of ionized metals are produced in that portion of the atmosphere which is between the surface of the main expanding cloud and the observer. While the dense expanding cloud is expanding rapidly, the atoms in the line of sight are in rapid motion toward the observer (Fig. 115). Consequently, the absorption lines are shifted toward the violet from their normal positions by the Doppler effect; from the magnitude of the shift we are able to

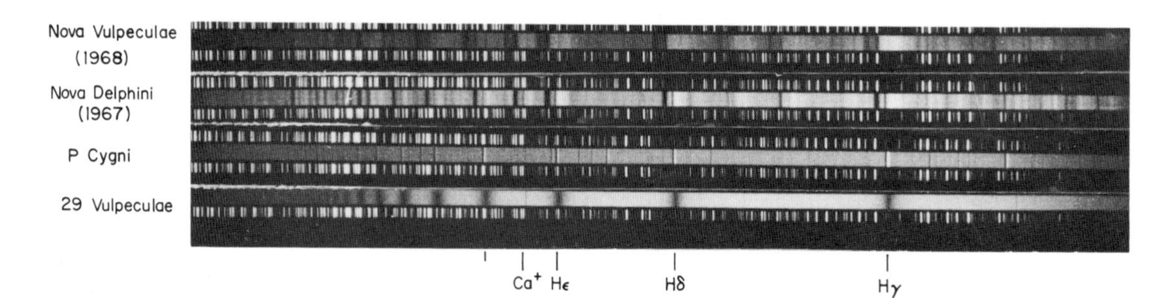

Fig. 113. Spectra of novae following maximum light: the spectra of two novae, Nova Vulpeculae (1968-I) and Nova Delphini (1967) are compared with the spectra of the emission-line star P Cygni and the normal (spectral class A0V) star 29 Vulpeculae. Notice the broad emission lines with absorption on their violet edges in the spectra of the two novae (see Fig. 116). Nova Delphini showed extremely complex changes in its spectrum during the many months it lingered near maximum light. Bright hydrogen lines, displaying sharp absorption on their violet edges, are the most prominent features of the spectrum of P Cygni, a star which rose from obscurity in 1600 in novalike fashion, faded a bit, rose again, and finally settled down to a fifth-magnitude star. Notice that the emission lines are much narrower and less prominent than in the spectrum of the novae. (Ojai Observing Station, University of California, Los Angeles.)

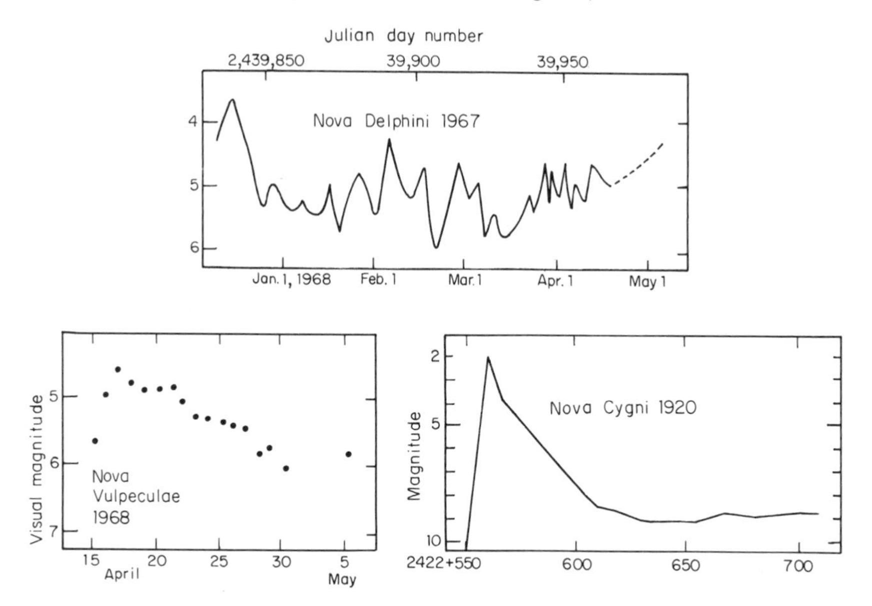

Fig. 114. Light curves of three novae; the visual magnitude is plotted against date or Julian day. Nova Cygni (1920) and Nova Vulpeculae (1968-I) showed normal behavior, that is, a rapid decline following maximum light. By 8 May 1968 Nova Vulpeculae had faded to magnitude 6.5; thereafter it declined steadily. By contrast, Nova Delphini (1967) remained near maximum light for many months. Its light curve near maximum light, due to G. Alcock and L. J. Robinson, showed complex changes; later it faded normally. Alcock discovered both Nova Delphini and Nova Vulpeculae. The light curve for Nova Cygni (1920) is due to Leon Campbell; those for Nova Delphini and Nova Vulpeculae are from *Sky and Telescope* (June 1968), pp. 397, 399.

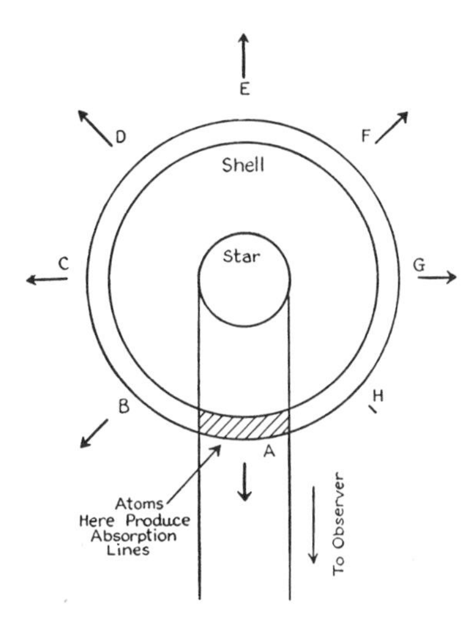

Fig. 115. Schematic diagram of an expanding shell about a nova. The arrows indicate the directions in which the various parts of the shell are moving. Only the atoms in the shaded part of the shell are between the star and the observer and hence produce absorption lines.

deduce the rate of expansion. Expansion speeds of 1000 kilometers per second are not uncommon.

When the ejected layer has expanded yet further, its gases become so attenuated as to be virtually transparent. Under these circumstances we see through to the far side, getting a view of the whole expanding shell. The continuous spectrum fades and the bright lines emerge, much widened by the Doppler effect (Fig. 113). The star itself continues to emit a continuous spectrum. But the light from all parts of the expanding shell that do not lie between the star and the observer will consist of bright lines. The arrows in Fig. 115 mark the direction of motion of atoms in different regions of the expanding envelope. The atoms at points *B* and *H* are moving toward the observer; their lines will, consequently, be shifted toward the violet. At points *C* and *G* the motion is at right angles to the line of sight and the Doppler shift is zero. At *D* and *F* the atoms are receding and the bright lines from these regions of the shell will be displaced toward the red. Summing up the contributions from different parts of the shell we find that the net effect of the expansion is to broaden out the emission lines (Fig. 113). At the same time the material at *A* absorbs light from the continuous spectrum of the star, thus producing dark lines. Since the atoms at *A* are approaching the observer, the dark lines will appear displaced from the corresponding emission lines toward the violet.

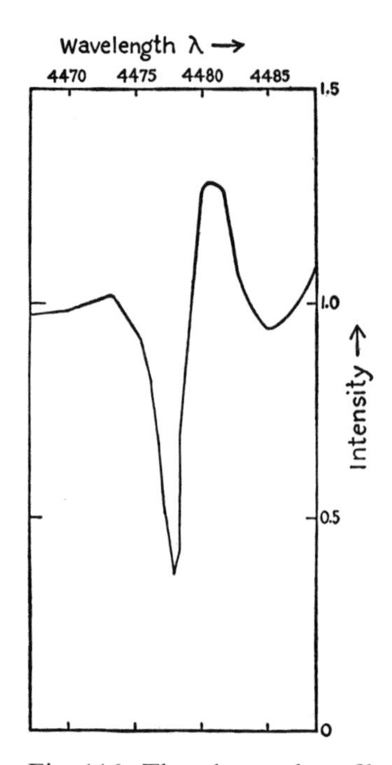

Fig. 116. The observed profile of the 4481-Å line of Mg II in DQ Herculis. Intensity is plotted against wavelength, the intensity of the continuous spectrum being taken as 1.0. The absorption component (to the left) is produced by the gases in front of the star (area *A* in Fig. 116). The relative strength of the emission component due to gases at points *B* to *H* in Fig. 116 increases as the shell becomes more and more attenuated. (From a plate taken December 15, 1934 at Lick Observatory, University of California.)

In Fig. 116 is shown the observed profile of the 4481 Å line of ionized magnesium in Nova DQ Herculis. Notice the strong absorption component on the short-wavelength (left) side of the emission component.

After the original shell has become fairly attenuated, we observe diffuse high-excitation lines of permanent gases that originate in a continuously ejected cloud close to the star, while the principal shell is expanding far from the star. Also, the weakening of the continuum means that the gases are becoming increasingly transparent. The level to which we can see in the ejected material becomes deeper and deeper as the amount of material ejected becomes less and less.

As the gas expands further, the bright lines of the metals steadily weaken, while those of the light gases like hydrogen, nitrogen, and oxygen remain prominent. This phenomenon is easily explained. The metal atoms have low ionization potentials, that is, the outer electrons of these atoms are not very tightly bound and are easily torn away. When the density is high, the separation between atom and electron is short lived, but in the tenuous shell of a nova, an electron once lost is not easily replaced. The metal atoms tend to lose not one but many electrons, and multiply ionized atoms absorb and emit light almost exclusively in the inaccessible far-ultraviolet region of the spectrum. The permanent gases are much more difficult to ionize. Thus their radiations persist long after those of the metals have disappeared.

At a yet later stage of the nova development the most prominent lines are the so-called "forbidden" radiations characteristic of planetary and diffuse nebulae. At this stage the gases have a density of the order of a million million times less than that of the air we breathe. Meanwhile, the continuous spectrum of the star has declined almost to the same feeble intensity it had before the outburst, while the bright lines of the nebular shell fade more slowly. The core of the nova may also show strong emission lines. Eventually, after the disturbance has subsided, the gaseous shell dissipates into space, to contribute to the debris of the interstellar medium, and the nebular spectrum disappears. The star returns to its normal, preoutburst state, perhaps to erupt again after a period of many thousands or even millions of years (or for certain recurrent novae after 10 to 40 years).

The spectroscopic evidence that novae do indeed cast off shells of gas has been confirmed by direct observations of Nova DQ Herculis (1934), Nova GK Persei (1901), and Nova V603 Aquilae (1918). Six months after the Nova Aquilae outburst, a faint greenish envelope or shell could be seen in the telescope; since then the shell has been expanding uniformly at the rate of about 2 seconds of arc per year (Fig. 117). Although the shell ejected by Nova Aquilae appeared roughly spherical or at least symmetrical, that cast off from Nova Persei is peculiarly unsymmetrical, as though most of the material came from a single hemisphere.

The existence of expanding shells aids in the determination of the distances of novae. By Doppler's principle, we obtain the actual rate of expansion in kilometers per second from measurement of the spectral-line displacements. If we now observe the cross motion, the apparent rate of expansion, in seconds of arc per year, from the direct photographs, we may compute the distance. The values obtained in this way for Nova Herculis, Nova Aquilae, and Nova Persei are about 730, 1200, and 2000 light-years, respectively.

In Nova Persei yet another type of phenomenon appeared. After the outburst, there were seen diffuse, irregular clouds of rapidly changing

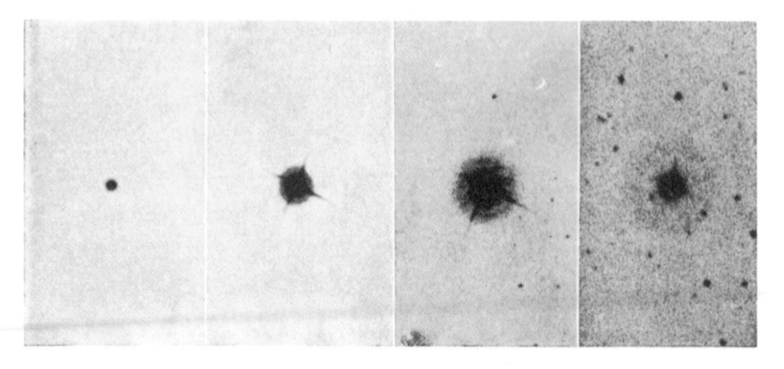

Fig. 117. Development of the shell about Nova Aquilae taken (*left to right*) in 1922, 1927, 1933, and 1940. The last picture was taken in the light of Hα (the red hydrogen line), since the nebulosity was too faint to be seen in photographic light. (Mount Wilson Observatory.)

Fig. 118. Nebulosity about Nova Persei (1901), photographed by Baade, December 13, 1939. (Mount Wilson Observatory.)

form not far from the star. The interpretation of this feature is that the light of the nova illuminated surrounding masses of interstellar gas and dust. The rate of illumination of these clouds gave a distance estimate of 400 light-years, in contrast to the determination of a distance of 2000 light-years by the method of the expanding shell. It has been shown, particularly by Oort, that the latter method is the more reliable.

Most novae appear to behave in a remarkably similar fashion, the chief difference being one of speed of expansion and light change. Thus 100 days elapsed before Nova Herculis fell three magnitudes below maximum brightness. Nova Aquilae (1918) and Nova Cygni (1920) diminished in brightness by three magnitudes after only 8 and 16 days, respectively. From a careful study of the seven bright novae that have appeared thus far in the 20th century, D. B. McLaughlin, of Michigan, found that those novae for which the light faded more slowly had lower initial expansion velocities, and showed longer time intervals between the various spectroscopic phases that we have described above. Furthermore, the stars that declined very rapidly had reached greater maximum brightness than those that declined more slowly, an effect that Arp finds duplicated in the Andromeda galaxy. McLaughlin emphasized that novae are remarkably similar regardless of the rate of development:

"Speculating a little more freely, the fact that we can express the rate of development in terms of a single parameter—the velocity—suggests that the phenomenon of a nova is due to a single sudden release of energy within the star. The great explosion represented by the rise to maximum light determines completely the train of events which follows during months or years afterward, just as the trajectories of the fragments of a bursting bomb are determined at the instant of the explosion. Whatever differences we observe between the spectra of novae having equal rates of development would then depend upon chance irregularities in the density of the ejected matter, on the composition and degree of turbulence of the gases, and on the orientation of the main erupted masses with respect to the line of sight."

The old novae, long after outburst, are found to be small, intensely hot stars. The equality of post- and prenova luminosities seems to indicate that they were probably similar objects before and after outbursts. These small, hot stars appear, from their luminosities, radii, and densities, to be much closer to white-dwarf stars than to main-sequence objects.

The U Geminorum Stars

Closely related to conventional novae are the remarkable variables of the SS Cygni or U Geminorum type, which are sometimes called "dwarf" novae. They remain nearly constant for intervals ranging from 3 weeks to 6 months, suddenly brighten twenty- or a hundredfold, and then decline. Some of them flicker irregularly at minimum light with a time scale of 1–10 minutes and a range less than 0.1 magnitude. Similar variations are found in old novae.

At minimum these stars show bright, wide, hydrogen lines and bright lines of Ca II. The color corresponds to that of a yellow star of spectral class G or K. At maximum the stars become blue; their spectra have broad, shallow absorption lines, that is, wide lines with high central intensities. These properties seemed strange until it was recognized that we were dealing actually with composite spectra, due to binary stars, a clue that emerged as of utmost importance.

Novae as Binary Stars

After it declined to minimum, DQ Herculis developed rapid light variations with a period of about 1 minute. While studying these fluctuations in 1954, Merle Walker found DQ Herculis to be an eclipsing binary with the shortest known period of any star, 4 hours 39 minutes. A preliminary analysis of the light curve indicated that the radii of the stars are about one-tenth that of the sun, and the masses are very small indeed. The secondary star (whose spectrum was never observed) might be a late M dwarf, but there is no escaping the fact that the mass of the nova component is much less than that of the sun.

The eclipse itself is not an ordinary stellar eclipse but rather one of a semitransparent ring or disk around the hot star. This ring is made fluorescent by the ultraviolet quanta emitted by the hot star. The absolute visual magnitude of the blue star is about +8.5. The 73-second period of variability, interpreted by Walker as the "pulsation" time of the blue star, has been used by Kraft to estimate its mass from the period-density relation $P\sqrt{\rho}$ = constant appropriate for a pulsating star, together with the exact relation between mass and radius that white dwarfs obey. Kraft found its mass to be 0.12 that of the sun; the star is truly a white dwarf. Several other old novae have been identified as members of binary systems; for example, T Aurigae 1891 is an eclipsing binary with a period, according to Walker, of 4 hours 54 minutes.

Table 10, due to Kraft, gives data for three old novae for which mass estimates have been made. Note that the masses range from about 0.1 to nearly 3 solar masses. Kraft concludes that novae require certain special types of binary systems; probably most of them belong to the old Population I or disk population rather than to Population II.

The composite spectra of the U Geminorum stars are readily understood; the stars are all binaries. For example, Joy found SS Cygni to be a spectroscopic binary with a period of 6 hours 38 minutes and spectral classes dG5 and Be. Kraft was able to identify the ancestors of the

Table 10. Masses of three novae. (Courtesy R. Kraft.)

Object	Spectral class[a]	M_V (red)	Period	Mass (Sun = 1) Red	Blue
T Coronae Borealis	Be, *g*M3+	+0.2	227^{d}.6	>3.7	>2.6
GK Persei (1901)	Be, K, +	+4.5	1^{d}.904	⩾0.56	⩾1.29
DQ Herculis (1934)	Be, −	>9	4^{h}39^{m}	0.20	0.12

[a] The notation B*e* indicates a class B star with emission lines; *g*M3 means that the spectrum resembles that of a giant star.

U Geminorum stars as close, virtually contact, binaries of the W Ursae Majoris type. These systems have typical periods of 0.37 day, masses typically 0.8 and 1.5 that of the sun for the two components, and $M_V \sim 4.5$. U Geminorum stars have nearly the same periods, separations, and masses. The blue stars are white dwarfs, the red are underluminous.

Origins of Novae

It seems firmly established that novae can originate only in certain very special double-star systems, while the W Ursae Majoris stars evolve into U Geminorum stars.

Evolution in close binary systems produces some spectacular effects. The more massive star of the pair will evolve more rapidly and, as it swells up, the outer envelope becomes entangled with the gravitational field of the other star. Most of the expanding envelope escapes into space but some becomes entrapped by the other star. Gradually, the initially more massive star loses its outer strata and the core settles down as a white dwarf. Meanwhile the secondary is continuing its evolution and eventually it, too, begins to lose its outer envelope into space. Once again it is likely that some of the escaping material will be captured by the companion, but now the companion is a white dwarf and the addition of new material to such a star may produce instabilities. It is not difficult to understand why novae are repeaters, but the exact details of the mechanisms are obscure.

Spectroscopic observations of novae at minimum are extremely difficult. We observe the spectra not just of the two components but often of only one of them and a more or less extensive ring of gas. The white-dwarf component may not be observed at all.

The Supernovae

In August 1885, a new star suddenly appeared near the center of the Andromeda nebula, attaining a brilliance about one-tenth the luminosity of that whole galaxy. It reached the sixth magnitude, declined ten-thousandfold in brilliance in 6 months, and then disappeared. Its position in the nebula and its spectrum, which was not like that of ordinary novae, indicated that this was not a foreground star. Similar objects, attaining luminosities comparable with that of the nebula in which they appear, have been detected in other and more distant spirals.

In 1917, Ritchey found two novae on photographs of the Andromeda nebula but these were thousands of times fainter than the 1885 object. Subsequent studies by Hubble and especially by Arp at Mount Wilson Observatory showed that every year there appear about 25 or 30 novae in the Andromeda nebula, which at maximum never become brighter than the fifteenth magnitude. Even though the absolute magnitude of the brightest of these novae is -8.5, they are mere pygmies beside the spectacular object of 1885. Such stars, which reach luminosities comparable with those of the galaxies that contain them, are generally called supernovae. In recent years many of these objects have been discovered in external galaxies.

Zwicky has found that supernovae appear with very different frequencies in different types of galaxies, so average figures may be misleading. In some galaxies such as Sa spirals none have been observed. Shapley found they appeared at a rate of four or five a century in Sc systems, that is, spirals similar to the Triangulum system. One galaxy is known in which three such outbursts have occurred, two in 1921 and one in 1937. Several similar instances have been reported by Zwicky.

There exist at least two types of supernovae (Zwicky has suggested as many as five); they should not be confused with population types. To make matters really confusing, type I supernovae may be associated with the globular cluster or so-called Baade Population II stars, whereas the type II supernovae are probably connected with the Population I (spiral arm) stars. Type II supernovae are never found in elliptical galaxies but appear to be associated with the arms of spirals.

Zwicky's type III supernovae may be only variants of type II. These objects appear to hover near maximum for weeks rather than only a few days as do types I and II. The featureless spectrum of the best example (in NGC 4303) indicates that large masses of gas are ejected at high temperature. Zwicky placed a supernova (in NGC 3003) with a unique

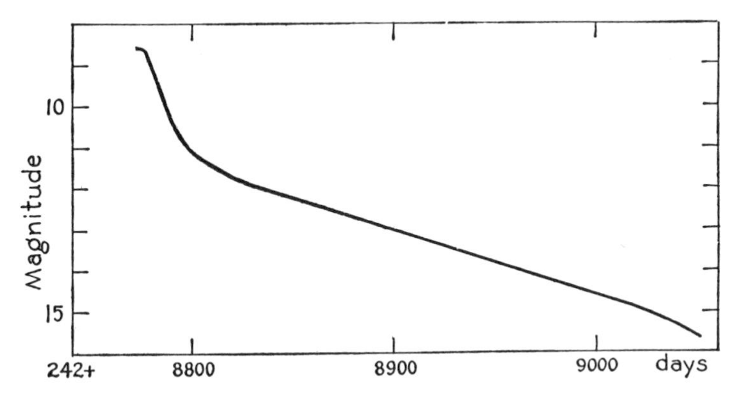

Fig. 119. The light curve of the supernova in IC 4182. (After D. Hoffleit.)

light curve in a separate class, type IV. His type V appeared to be represented by η Carinae, which is described in Chapter 12.

A typical light curve of a type I supernova is shown in Fig. 119. At maximum these stars are normally several hundred or even more than 1000 million times as bright as the sun. Their brightness rises very rapidly, then passes through a maximum phase that may last for several days. When it has fallen to about a tenth of its maximum value it begins to decline steadily at a rate of about 1 magnitude every 2 months. The total fall in brightness is probably at least a millionfold. At maximum these stars show little ultraviolet light, so that their surface temperatures are probably relatively low. To radiate as much light as they do, supernovae must swell up to gigantic proportions, probably many times larger than the orbit of the earth. It seems probable that, as in a nova, a supernova results from a star or some massive object blowing itself almost completely to pieces. The operation thus occurs on a truly prodigious scale. Whereas in the usual nova the velocity of ejection amounts to a few hundred kilometers per second, the material from a type I supernova may be thrown out at a rate of 30,000 kilometers per second.

Since supernovae are observed in distant galaxies, studies of their spectra require large telescopes. R. Minkowski, who made the first thorough study of supernova spectra, found that they consist entirely of a few ill-defined broad bright bands (Fig. 120), with no trace of sharp lines, except for two of the so-called forbidden lines of doubly ionized oxygen (see Chapter 7) at 6300 and 6363 Å, which appear at a late phase. The bands are broader and the excitation is evidently higher in type I than in type II. The positions and relative intensities of these broad features change with time, perhaps because the atomic radiations change in

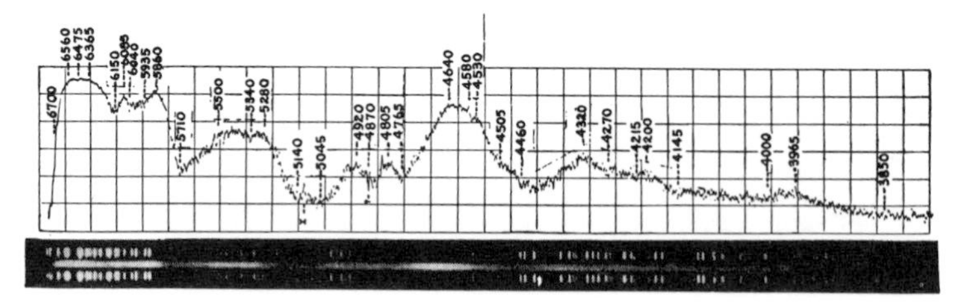

Fig. 120. Spectrum of the supernova in IC 4182, photographed by Minkowski on September 10, 1937. A microphotometer tracing, which registers the degree of blackening on the original negative, is shown above the spectrum. Notice that there are no sharp lines in the spectrum of the star. (Mount Wilson Observatory.)

importance as ionization and excitation vary. Their great breadth is to be understood in terms of the huge velocities of ejection. They never develop a novalike spectrum.

Type II supernovae reach luminosities about 1.2 magnitudes lower than those of type I. Their light curves are not similar to those of type I, nor even similar to one another. The velocities of ejection of the shell seem to be less than for type I supernovae. The spectra of type II supernovae have, however, been interpreted by F. L. Whipple and Mrs. Cecilia Payne-Gaposhkin at Harvard, who suggested that they are very similar to those of ordinary novae except that the enormous velocity of expansion in supernovae so widens the individual bright lines that they overlap with each other to form an effective continuum. To strengthen the argument, Whipple and Mrs. Gaposhkin constructed sets of so-called "synthetic" spectra, calculating how the spectrum of a supernova would appear if its atmosphere contained various mixtures of elements such as helium, carbon, oxygen, nitrogen, and iron in various stages of ionization and excitation, and were in a state of rapid expansion. If one assumes that hydrogen is relatively less abundant than in ordinary stars, although helium and iron are very prominent, then these synthetic spectra fairly closely resemble the spectra of type II supernovae, although the spectra of novae of type I are not so well explained.

Supernovae in Our Galaxy

Five, probably six, more or less well-observed supernovae appear to have been noted in our own Galaxy. In 1572 Tycho Brahe observed a star that outshone Venus and was easily visible in broad daylight. It has never been identified optically, although if it had been an ordinary nova it would now have an apparent magnitude of about +11, lie at a distance of 150 parsecs, and be easily detectable. No such star has been found, but a strong radio source has been observed near the position indicated by

Tycho. It is undoubtedly the supernova remnant and consists of a shell with a diameter of about 7.5 minutes of arc, which corresponds to about 11 parsecs at the probable distance of 5000 parsecs. With correction for space absorption, the absolute magnitude of the supernova at maximum must have been −19.6, that is, its luminosity was equal to 4000 million suns! Minkowski, who observed a few very faint remnants of the shell, regards Tycho's nova as the prototype of supernova type I.

Another probable type I supernova was the star observed in 1604 by Kepler; a remnant of the nebulosity was discovered by Baade. It is also a strong radio source. The supernova of 1006 observed by the Chinese, reached $m = -8$, but available data are very poor, partly because of its unfavorable position ($\delta = -42°$). An even more unfavorably placed object, the supernova of A.D. 185 at $\delta = -62.5°$, was also observed by the Chinese. The most famous of supernovae, the Crab nebula and the remarkable object η Carinae, will be discussed in the next chapter.

The positions of defunct supernovae are often revealed by rapidly moving or expanding clouds of gas, which are sometimes detected as emitters of strong radio-frequency radiation. The Network or Veil nebula in Cygnus shows motions that can be interpreted as a simple radial expansion. The individual elements of the "net" consist of filaments about 1 parsec long and 0.01 parsec wide, while the entire shell has a mass about that of the sun. It appears to be an old supernova shell which is being slowed down by interaction with the interstellar medium. Another example is IC 443. Both of these nebulae are abnormally strong radio sources. Cassiopeia A, which is one of the strongest radio sources in the sky at low frequencies, appears to be the relic of a supernova. It lies in a heavily veiled region of the Galaxy. Minkowski's measurements suggest a distance of 3400 parsecs, a radius of 2 parsecs, and a mass twice that of the sun for the nebulosity, which consists of small condensations. If the fast-moving condensations of this nebulosity have not been slowed down, the outburst occurred about A.D. 1700.

Three possible supernova remnants have also been identified in the Large Magellanic Cloud.

Origin of Supernovae

During the course of its outburst, the total light emitted by a type I supernova is about 3.6×10^{49} ergs, or equivalent to that emitted by the sun in 300 million years. We do not know how to estimate the total power output, which may be much larger since great quantities of energy may be emitted in the x-ray and radio-frequency regions.

In any event, large amounts of energy must be released very suddenly. A favorite hypothesis is the collapse of a massive star during the last stages of its evolution. For example, it is conjectured that, with increasing density and temperature, the predominantly iron core of such a massive star would convert itself to helium with the absorption of large amounts of energy. The star would then implode, surface hydrogen would become mixed with the superheated interior, and the entire mixture would explode like a nuclear bomb.

Another suggestion is a collapse in which the core of a star suddenly attains a density greatly in excess of even that of a white dwarf. Electrons are jammed into protons to form neutrons yielding stars of very small dimensions. Cameron and also Ambartsumian and Saakian envisaged even more exotic stars, with densities of 10^{19} grams per cubic centimeter, wherein the unstable particles of high-energy physics became the stable building blocks of matter. It is possible that pulsars are the neutron-star cores of former supernovae (see Chapter 12). To date we have not been able to identify any supernova with a previously existing normal highly evolved star. Stars, even massive ones, normally appear to go out—not with a bang but a whimper.

In Chapter 9 we discussed the formation of stars from condensations of large masses of material in the interstellar medium. Suppose a nonrotating mass of 200–300 solar masses started to condense into a star; it could not form a stable body because the required temperature would be too high. The mass would pull itself together and then explode as the nuclear fuel burned too quickly. That such a process may occur is suggested by runaway stars which may have been formed in the following way. Suppose a binary system is formed in which one component has a mass of 300 suns and the other a mass of 20 suns. The first component never evolves into a stable star—it explodes. The companion, which had been traveling at a high speed in its orbit, is suddenly freed, as a whirling stone flies away when the string breaks. Several examples of young, massive, high-velocity stars are known. Unfortunately their ages are so great as to preclude any hope of finding supernova remnants.

Such supernova remnants as are known all pose engaging problems (see Chapter 12). Inwardly they seem to be involved in magnetic fields and to generate high-energy particles, electrons, protons, or even cosmic rays.

12 High-Energy Astronomy

Insofar as its gross properties are concerned, a star is one of the simplest objects in nature. It shines because a high internal temperature is necessary to provide the pressure that supports the weight of the overlying layers. At these temperatures, protons move with such high speeds that they may penetrate the nuclei of heavier atoms, permitting the release of large amounts of energy. The nuclear furnace maintains the high temperature, sustaining the energy flow to the surface. Likewise, the radiation from a normal gaseous nebula such as that in Orion can be understood. In simple terms, ultraviolet energy received from illuminating stars is degraded by seemingly well-understood atomic processes. The astrophysicist refers to the radiation from the bright photosphere of the sun as thermal radiation, that is, radiation emitted by hot matter simply because it is hot.

In recent years, however, attention has increasingly been focused on a host of phenomena that can be loosely called "nonthermal." More particularly, they pertain to situations in which relatively small numbers of particles are endowed with very large amounts of energy. The manifestations of these effects are felt throughout the entire electromagnetic spectrum, as we shall see.

Specifically, in our Galaxy, there is generated a flux of energy equivalent to the output of 10 million suns that is invested in an attenuated stream of atomic nuclei whose average energy is 1000 million electron volts. These are the cosmic rays, which have long been recognized as high-energy particles impinging on the earth from outer space, but whose possible connection with identifiable stars and nebulae has only recently been recognized. X-ray emission from celestial sources and copious amounts of radio-frequency emission from certain gaseous nebulae often appear to be associated with high-energy particles. Supernovae remnants include some of the best examples of these nonthermal sources; let us examine one of them in more detail.

The Crab Nebula

The Crab Nebula, so named by Lord Rosse, who made a remarkable drawing of it a century ago, is one of the most interesting of all gaseous nebulae. Direct photographs with appropriate filters show that it consists of a rather diffuse or amorphous mass upon which is superposed a network of intricate filaments that is slowly expanding much like the shell around a nova (Figs. 121 and 122). Measurements by J. C. Duncan suggested that the mass had expanded from a single origin, probably an exploding star, and that the outburst occurred approximately 900 years ago. This time interval is important because in 1054 Japanese and Chinese astronomers independently noticed a bright, temporary star in the same part of the sky. It became brighter than Jupiter, reaching a visual magnitude of −3.5, dimmed gradually, and disappeared from view after about 650 days.

Spectrographic observations have yielded valuable information on the expansion of the Crab Nebula. The filaments give a bright-line spectrum resembling that of a planetary nebula; the amorphous mass, which contributes about 80 percent of the light from the nebula, yields a continuous spectrum. An early spectrogram obtained by N. U. Mayall at the Lick Observatory is shown in Fig. 123. The slit of the spectrograph is placed along a diameter of the nebula. Notice that the strong 3727-Å line is bow shaped. This effect is just what we would expect in an expanding shell of gas. At the center of the nebular "disk," the gas on the side toward us is rushing in our direction and the line is shifted toward the violet, while the material on the opposite side is moving away and the line is shifted toward the red. Hence the spectral line appears split in the central regions of the nebula. At the edges of the nebula, the material is moving across the line of sight, and the line-of-sight speed of the ex-

Fig. 121. The filaments of the Crab Nebula, photographed by Guido Münch on a 103a-E emulsion at the prime focus of the 200-inch telescope on September 22, 1966. He used an interference filter with a 50-Å pass band and an exposure of 70 minutes to obtain the filaments in the light of Hα and forbidden radiation of ionized nitrogen.

panding shell is zero; hence the shift of the spectral line is zero here. Now the maximum separation of the two components of the line is proportional to twice the speed of expansion of the shell. On the other hand, measures on direct photographs give the apparent (angular) rate of expansion. Clearly, if we know the actual speed of expansion and the angular speed of expansion, we can calculate the parallax. There is a slight complication in that the nebula is elliptical in shape; it appears to be an ellipsoid

Fig. 122. The "amorphous mass" of the Crab Nebula, photographed by Guido Münch on a yellow-sensitive (103a-D) emulsion at the prime focus of the 200-inch telescope on October 22, 1963. He used a yellow (Corning 3484) filter and Polaroid filter so oriented as to give minimum transmission in the north-south direction. The photograph registers essentially the continuum radiation; only the strongest filaments can be seen faintly in the light of the forbidden radiation of neutral oxygen. North is at the top; east is to the right.

whose longer (major) axis expands at a different rate than the shorter (minor) axis. Allowing for these effects, the velocity of expansion along the major axis is 1720 kilometers per second, which corresponds to an angular rate of expansion of 0.2 second of arc per year. Now, at a distance of 1 parsec, a transverse (proper) motion of 1 second/year would

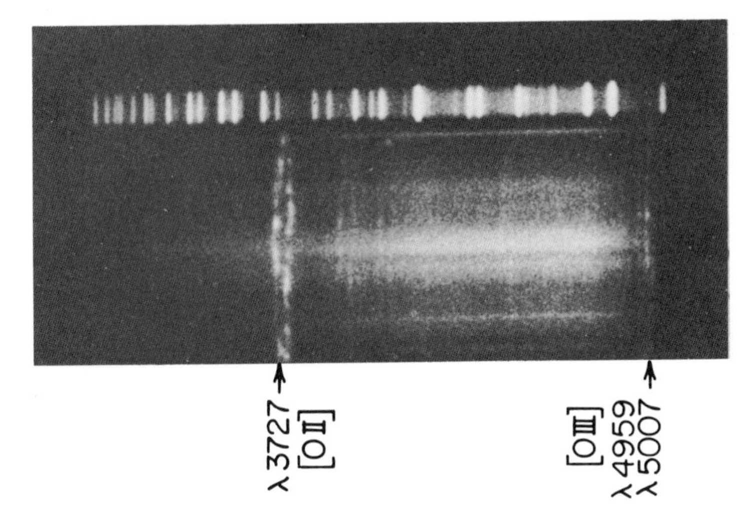

Fig. 123. An early spectrogram of the Crab Nebula, obtained in the thirties by N. U. Mayall with the nebular spectrograph on the Crossley Reflector at Lick Observatory; notice the strong continuous spectrum and bright bow-shaped forbidden lines of ionized oxygen [O II] at 3727 Å. More recent work by Virginia Trimble shows that in addition to the expansion there are complex internal motions.

correspond to 1 astronomical unit/year = $(1.49 \times 10^8$ km/yr) $(3.156 \times 10^7$ sec/yr) = 4.74 km/sec. Hence a linear velocity of V km/sec corresponds to an angular transverse motion of μ sec/yr at a distance of d parsecs, in accordance with the relation $V/4.74\mu = d$(parsecs). Putting in the numerical values for V and μ, we get $1720/(4.74 \times 0.20) = 1800$ parsecs for the distance. Then, if we allow for space absorption, the absolute visual magnitude of the star at maximum must have been about −18.2. In any event, the star must have been many times as bright as an ordinary nova, and hence it probably was a supernova.

It was probably not a type I supernova like Tycho's star or the object observed in IC 4182. The expansion velocity of 1700 kilometers per second is much smaller than the 20,000 kilometers per second typical of such supernovae. Furthermore, its rate of decline was smaller.

The system of filaments, which probably has a mass about equal to that of the sun, seems, according to Woltjer, to have an over abundance of helium. This system extends throughout the volume occupied by the diffuse mass. This, the so-called amorphous mass, which actually exhibits a distinctive structure, shows brightness variations near its center. There occur "ripples" which appear to travel with a speed of about 47,000 kilometers per second. The outer portion appears to expand at the same rate as the system of filaments.

The Crab Nebula is an intense source of radio-frequency radiation, and in fact was one of the first radio sources to be identified with an optical object. The extent and detailed structure of the diffuse mass appears to be the same in the radio-frequency range as in the optical range. The brightness distribution and polarization effects differ in the two ranges; the nebula is more uniformly bright in the radio-frequency range than in the optical range, but the radio-frequency polarization is more strongly concentrated toward the center than is the optical. Radio waves, like light and other electromagnetic waves, are emitted by any hot source; the intensity of this "thermal noise," as it is called, depends on the temperature of the source and on the quantity of emitting material along the line of sight. The radio-frequency radiation from the Crab Nebula is definitely not of thermal origin. It is much too intense for the known surface brightness of the object at optical wavelengths and the intensity does not have the right dependence on wavelength. Furthermore, both optical and radio emission from a thermal source are always unpolarized.

Thus the significant clue was provided by Dombrovsky's discovery in 1954 that the light of the Crab Nebula was polarized. Photographs secured by Baade, who took a series of plates through a Polaroid filter set successively at different position angles, show striking changes. Polarizations up to 60 percent are readily recognized; some regions may be 100-percent polarized.

Shklovsky suggested that the process of emission of energy by the Crab Nebula is similar to that of a synchrotron, a device for producing high-speed electrons for investigations in nuclear physics. The emission pattern of an accelerated electron whose velocity approaches that of light is quite different from that of a more slowly moving one. In a synchrotron the electrons are constrained to move in a circular path by the action of a magnetic field perpendicular to the plane of their motion. Because the electrons move in a curved path, they are constantly being accelerated and hence radiating energy. The light is emitted, not primarily in a direction perpendicular to the electron's motion, but rather in a narrow cone in the direction of motion; furthermore it is polarized. We may visualize the electron as a toy engine running on a circular track; the beam from its headlight (across which a piece of Polaroid is placed) represents the emitted light. A given electron will appear to emit a momentary pulse in the direction of the observer once each revolution, but there are hordes of electrons so that any flicker is wiped out. The radiation is emitted at the expense of the electron's kinetic energy, which decays away continuously. The energy distribution of the emitted radiation differs very markedly from that of a black body; there is a sharp high-frequency cutoff that depends on the orbital period of the fastest electrons.

Synchrotron radiation appears to explain both radio-frequency and optical emission from the Crab Nebula and the nonthermal radiation of other old supernova remnants. Probably the strength of the magnetic field is about 10^{-3} gauss in the Crab Nebula, roughly 1500 times smaller than the earth's magnetic field. The Crab Nebula is also a strong emitter of x-rays. Is it possible that the excitation is derived from an incredibly small, fantastically hot star that radiates mostly in the x-ray region? Now the moon occasionally passes in front of the Crab Nebula; one such occultation occurred in July 1964. If the x-ray source is stellar it would disappear instantly when it was occulted by the moon; if it is an extended source, it would fade away gradually. Bowyer, Byram, Chubb, and Friedman of the Naval Research Laboratory observed this occultation with detectors flown in a rocket above the earth's atmosphere. The results clearly demonstrated that the x-ray source was not a point source but an extended one with an angular width of about 1 minute of arc. It has been argued that this x-ray emission is also synchrotron radiation, but this question cannot be settled until the polarization of these x-rays is measured.

The spectrum of the Crab Nebula has been observed from 30-meter radio waves to the x-ray region (0.2 Å), with large gaps in the far infrared and ultraviolet (Fig. 124). The total emitted power is about 10^{38} ergs per second, or 25,000 times that of the sun. Nearly all of this energy is radiated as synchrotron radiation. Hence there must exist a mechanism that supplies energy to accelerate electrons to very high energies at this rate; presumably it also accelerates heavy particles as well. The ultimate "powerhouse" appears to be a rapidly flickering source of radio-frequency (and in this instance also optical) radiation, known as a "pulsar" (see below). Pulsars are currently interpreted as incredibly dense, rapidly spinning "stars" consisting of neutrons.

Eta Carinae

Another puzzling object is the remarkable star (?) η Carinae, observed by Halley in 1677 as a star of the fourth magnitude. In 1751 it was about as bright as Polaris and thereafter showed fluctuations. By 1843 it had become more luminous than Canopus; then, unlike a nova or supernova, it

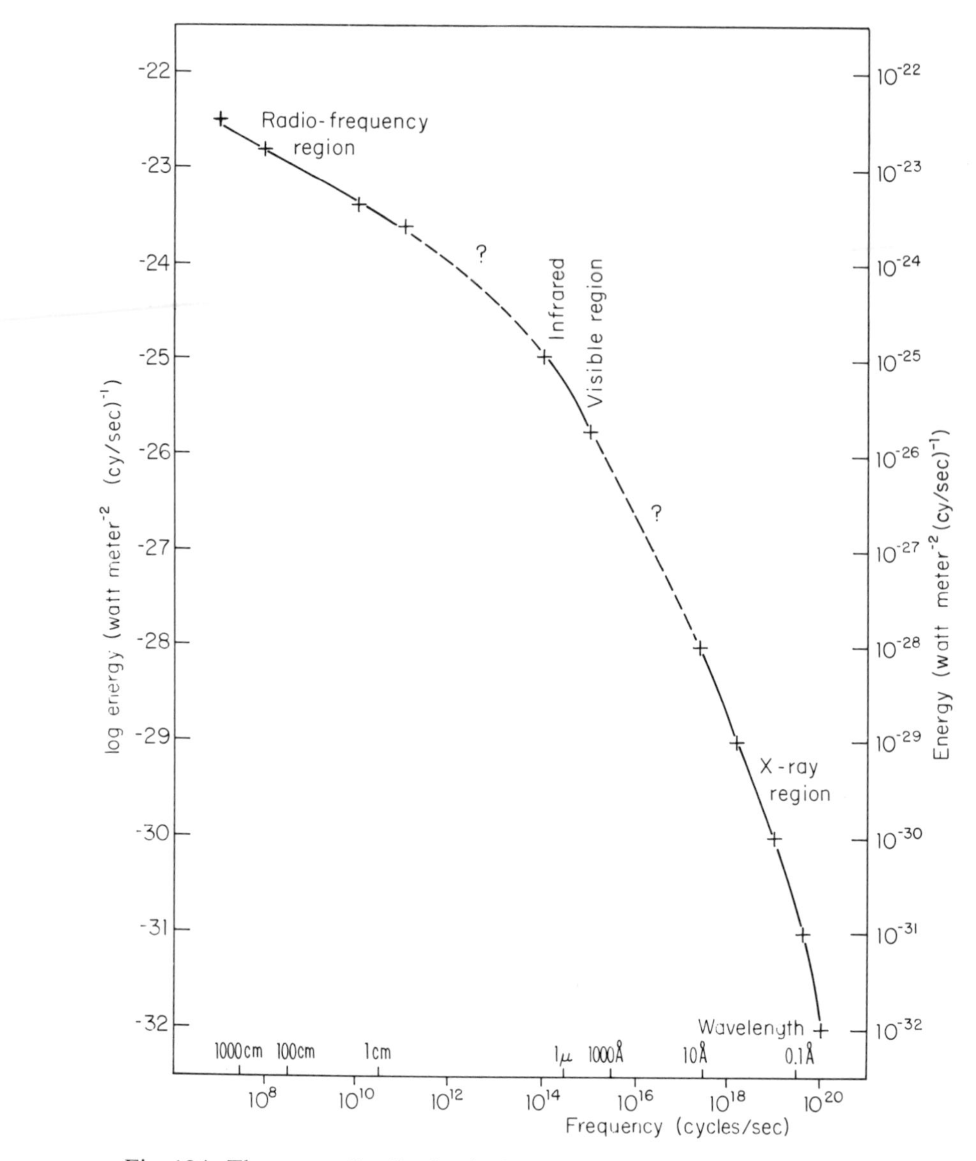

Fig. 124. The energy distribution in the continuous spectrum of the Crab Nebula. Each scale division corresponds to a tenfold change in intensity or frequency. Data from the radio-frequency, visual, and x-ray regions are combined in this figure. The solid segments of the curve represent observed data; the dashed segments are interpolated over unobserved regions.

remained bright for several years. By 1856 it had started to decline, fading to the second magnitude by 1858 and to the sixth by 1867. Since then it has fluctuated irregularly, with a minor outburst in 1893.

In the telescope it now shows an intensely red nonstellar nucleus with a uniform core of diameter about 1.5 seconds of arc. Surrounding this is an irregular, elongated nebula about 12 seconds of arc in diameter. Both the brightness of the nucleus and the structure of the nebula are variable.

The spectrum of the nucleus is extremely complex (Figs. 125, 126, and 127). The stronger lines show broad emission profiles, often with sharp, narrow emission spikes. There is also violet-displaced absorption. For example, in 1961 the Hγ profile showed dips corresponding to shells moving outward with velocities of 48, 120, and 480 kilometers per second.

In the small nebula, Gaviola found small outward-moving bright re-

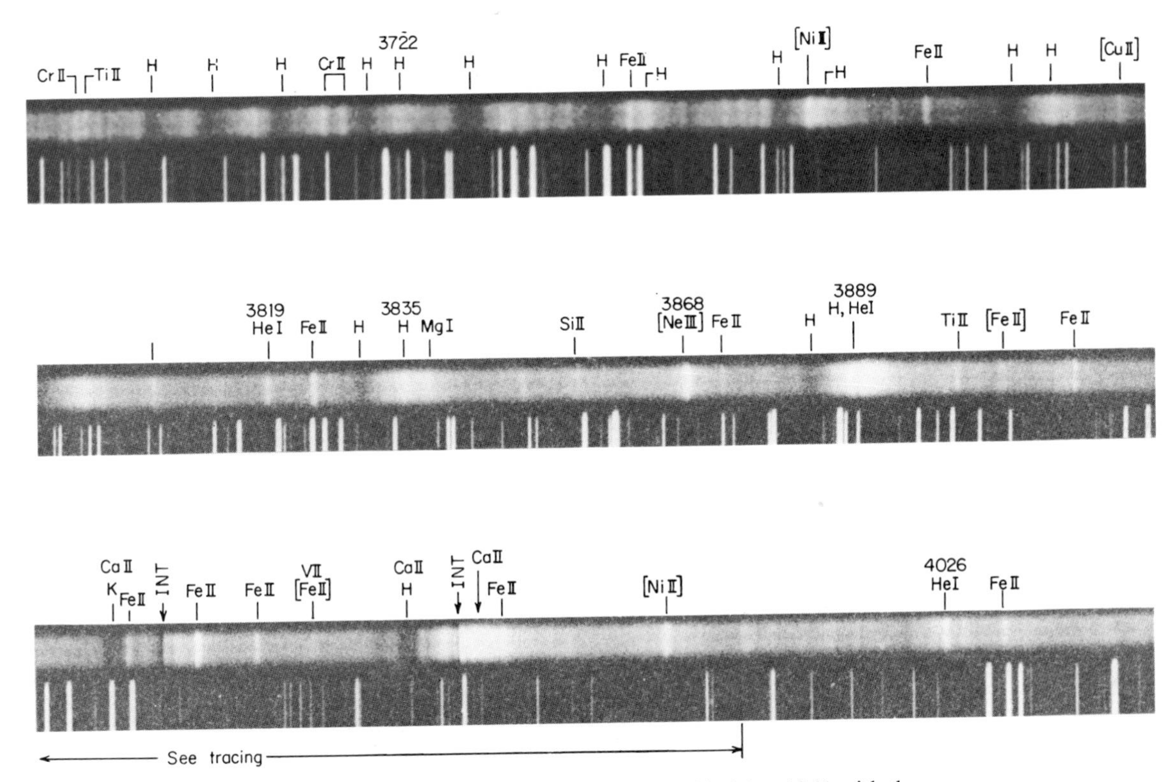

Fig. 125. A portion of the spectrum of η Carinae, obtained in May 1961 with the coudé spectrograph of the 74-inch reflector at Mount Stromlo, showing the complex structure of some of the lines. The dark lines marked "Int" indicate interstellar absorption. The portion marked by the arrow corresponds to Fig. 127.

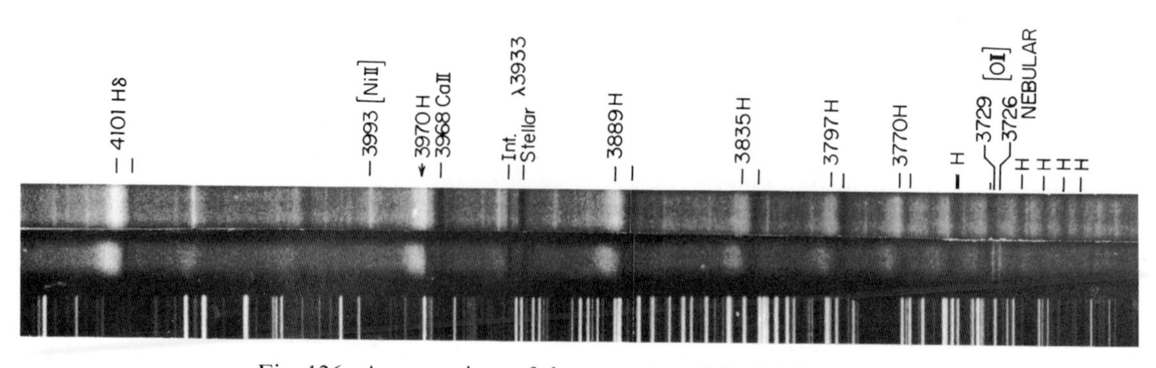

Fig. 126. A comparison of the spectrum of the bright core of η Carinae (*upper*) with that of the surrounding shell (*lower*); the diameter of the core is 1–2 seconds of arc. Notice that many of the sharp lines in the bright core assume a fuzzy appearance in the shell and that absorption lines are displaced. These effects are probably produced by electron scattering by the gas of the expanding shell. (Cerro Tololo Interamerican Observatory.)

gions or "condensations" which were probably ejected in the 1843 outburst. If one correlates the rate of movement of these condensations with the widths of the broad emission features in the spectrum (which correspond to an expansion rate of about 500 kilometers per second), one derives a distance of 1200 parsecs and a maximum luminosity $M_v = -12$ (or even as much as -15 if there is heavy space absorption). Evidently η Carinae was not a type I or a type II supernova. (Zwicky assigns it to a special class V.)

Strong polarization effects have been found in the small nebula by Thackeray and by Visvanathan, who finds that not only the continuum but also the lines are polarized. He concludes that the effects are produced by electron scattering in a nonspherically symmetric shell—much as polarization is produced by scattering by electrons in the corona of the sun.

Synchrotron radiation cannot be excluded. Indeed, Rodgers and Searle measured the intensity of the background continuous spectrum and found it to resemble that of intense nonthermal sources. McCray assumed that the optical continuous spectrum of the core is synchrotron radiation. He developed a model of the nebula with a magnetic field of about 10^{-3} gauss and a low-frequency cut-off that acts to suppress radiation in the radio-frequency range, to explain the fact that η Carinae is not a radio source. He predicts a measurable flux of high-energy $\dot{\gamma}$-rays, that is, very hard x-rays, so far not confirmed by observation.

The physical conditions in the radiating volume in η Carinae are very complex. Oxygen seems to be deficient. Searle and Rodgers suggest that

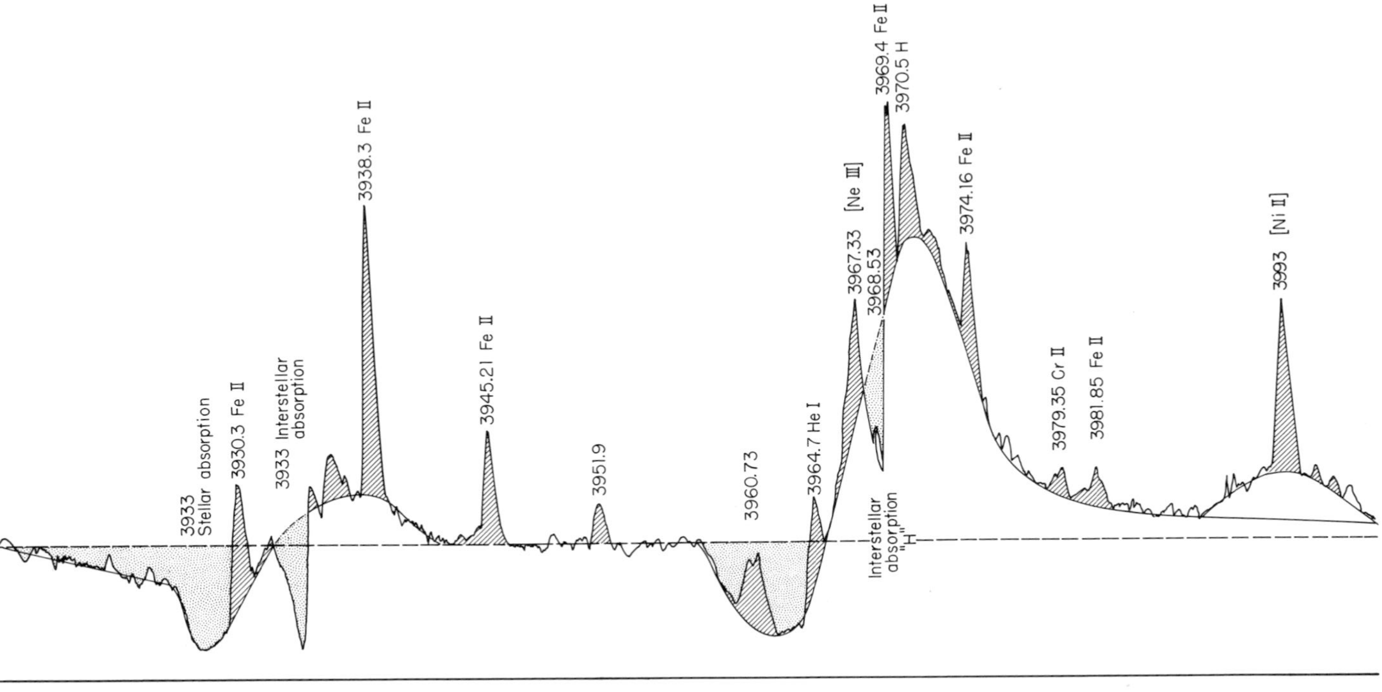

Fig. 127. An intensity tracing of a portion of the spectrum of η Carinae. The dashed straight line indicates the assumed position of the background continuous spectrum. Notice the broad dome-shaped profiles of the strong lines of ionized calcium (3933 Å), ionized calcium (3968 Å), superimposed on the 3970-Å line of hydrogen and, curiously, the 3993-Å forbidden line of ionized nickel. The sharp interstellar-absorption lines of ionized calcium (3933 and 3968 Å) are to be compared with the broad absorption lines of ionized calcium in the star η Carinae itself. The lines of ionized iron (Fe II), chromium (Cr II), and helium (He I) and the forbidden lines of ionized nickel [Ni II] and neon [Ne III] all show sharp spikes.

no star is observed at all; what we see are emissions of discrete blobs of gas in an extended volume. The original star may have literally blown itself to bits.

High-Energy Particles and Cosmic Rays

Supernova remnants such as the Crab Nebula seem to be objects in which electrons are accelerated to high speeds. It is likely that heavier particles are accelerated there as well, for these objects may be sources of high-energy particles or cosmic rays (see the section on pulsars).

More than 60 years ago it was noted that an isolated charged electroscope would gradually lose its charge. This was attributed at first to ionization produced by natural radioactivity in the earth, but a balloon experiment by Hess in 1912 showed that the radiation came from above and could truly be called cosmic radiation. High-energy particles impinge on the earth's atmosphere or on solid structures on its surface, shatter atomic nuclei, and produce hosts of secondary or shower particles (see Fig. 93). It is these fragments that are customarily observed. To observe the primary particles one must fly detectors in balloons at heights above about 25 kilometers.

Although some cosmic rays are produced in violent flare activity on the sun and some energetic particles are accelerated in the solar system, most of these particles—indeed all of the high-energy ones—come from outside the solar system. In fact, magnetic fields of the order of 10^{-4} to 10^{-5} gauss with irregularities of a size comparable with the earth-moon system tend to deflect out those with energies less than 100 million electron volts. Hence we have accurate data only for cosmic-ray particles of greater energy. The observed variations are all due to the action of magnetic fields in the solar system.

The energy of the average cosmic-ray particle is about 1000 million electron volts. The density of the particles is about one per 1000 cubic meters, but, since they travel with nearly the velocity of light, every second about six of them would pass through an area of 1 square centimeter in space. Cosmic rays of all energies appear to come from all directions of space with equal probability; there is no preferred direction. Their actual trajectories are modified, of course, by the earth's magnetic field. They drop off very rapidly in numbers with increasing energy. That is to say, if we define $N(E)$ as the number with energy greater than E, their distribution very closely obeys the law $N(E) \sim E^{-\alpha}$, where $1.5 < \alpha < 2.1$ for energies per particle between 10^{10} and 10^{19} electron volts.

How do solar and galactic cosmic-ray particle types compare? Biswas and Fichtel showed that energetic particles from the sun have the same relative abundances of helium and heavier elements as the solar photosphere. Other cosmic rays show a larger percentage of heavy particles and also substantial numbers of the cosmically rare nuclei of lithium, beryllium, and boron. The latter are produced by the shattering of heavier nuclei struck by cosmic rays and of heavy cosmic-ray particles by encounters with interstellar atoms. The fractional abundance of these vulnerable nuclides thus gives an idea of the amount of matter penetrated by high-energy particles; it is equivalent to a sheet of steel approximately 1 centimeter thick. Since the density of normal atoms in the galactic plane is known, we can estimate from this figure the length of time these high-energy particles have spent in the galactic plane. It turns out to be a few million years. An important quantity involved is the interaction cross-section, or target area, for collision. This is appropriate to the sizes of nuclei, about 10^{-26} square centimeter, rather than the sizes of atoms, 10^{-14} to 10^{-18} square centimeter. Electrons are also present among cosmic rays, most of them accelerated by the same mechanisms that operate on the heavy particles. They act as tracers of cosmic-ray generators. Thus the high-energy electrons in the Crab Nebula strongly indicate that cosmic-ray particles are produced there.

All our data are consistent with the idea that cosmic rays fill the disk of the Galaxy in a more or less steady-state fashion. Probably they fill a volume greater than that actually occupied by the galactic disk. The cosmic-ray particles of highest energy, that is, those with energies greater than 10^{20} electron volts, may be of extragalactic origin, but they possess only about 0.001 percent of the total energy. The behavior of cosmic rays is certainly strongly influenced by the galactic magnetic field, but we do not know enough about this field to make detailed calculations. It is of interest that the cosmic-ray data suggest an average particle density in the galactic plane, about 5 atoms per cubic centimeter, greater than previously assumed, but these results are consistent with some calculations made on the basis of stellar motions. The magnitude of the magnetic field required to retain the cosmic rays in the Galaxy is about 5×10^{-6} gauss.

It is an amusing fact that on scales greater than 10 million kilometers we can regard the cosmic-ray particles as constituting a gas, but one with some peculiar features; in particular, the distribution of particle velocities is very different from that of a normal gas. The cosmic-ray gas is exceedingly hot and tenuous, but its pressure is actually greater than that of the particles of the interstellar medium, being about one-millionth atmosphere; the individual particles are separated on the average by 10 meters. The velocity of "sound" in this gas varies between 10 and 100 kilometers per second. Since the cross-sections for interaction are about 10^{-26} square centimeter, that is, some 10^{-8} to 10^{-10} times smaller than those for collisions between atoms in a gas, the cosmic rays interact very little with other components of the interstellar medium. There would be no interaction at all if it were not for the magnetic field.

The Origin of Cosmic Rays

Cosmic rays induce radioactivity in meteorites, which can be measured by very sensitive techniques and indicates that over the last 10 to 100 million years cosmic-ray activity has been very nearly constant. One possible reason is that there occur repeated violent outbursts in the Galaxy, either in individual supernovae or in the central regions, and that the effects of these outbursts get smoothed out.

If cosmic rays originate in our Galaxy or local group of galaxies, the rate of generation of cosmic-ray energy is about 10 million times the rate of output of solar energy. Unless their orbits are so arranged that the particles spend much of their time outside the galactic plane, their lifetimes do not much exceed a million years. Only the most energetic particles originate outside the Galaxy.

Charged particles with high energies appear to be generated whenever ionized gases in magnetic fields experience violent disturbances. Some mechanism converts a large fraction of the energy of the ionized gas, or plasma, into kinetic energy of a few fast particles—hence x-rays, gamma rays, and cosmic rays must all be related. Examples of these processes are seen in laboratory plasmas, in solar flares, and in the interaction of the solar wind with the magnetized shell around the earth. Other examples appear to be supernova ejections, galactic-core explosions, quasars, and radio galaxies.

Only supernovae and violent explosions in the cores of galaxies appear to be adequate to supply the energies required. Stars like the sun can supply only negligible amounts of cosmic rays.

What physical mechanisms are capable of accelerating particles to such high energies? Several schemes have been proposed, though no final answer can be given at the present time. One possibility is that particles are accelerated by a mechanism akin to that employed in terrestrial devices, that is, by a resonance phenomenon of some sort. A number of years ago, Fermi suggested that a few of the faster particles in an ionized gas might be reflected from ionized, moving gas clouds that contained magnetic fields. He showed that, if fast particles were present to begin with, the mechanism would operate to speed them up, but very long times would be required.

A continuous supply of high-energy cosmic rays may be produced by pulsars (see below). Sudden outbursts of cosmic-ray activity, however, may be associated with the detonations of supernovae. Colgate and his associates noted that the front of a shock wave running outward from an explosion of a supernova into a gas of steadily decreasing density will be accelerated until all the material near the wave front moves with nearly the velocity of light. Thus a whole cloud of particles may be accelerated to very high energies.

X-Ray Astronomy

We have already referred to the Crab Nebula as a source of x-rays. Actually, x-ray emission from the solar corona had been predicted by Edlén; it was observed in 1949 by Tousey and his associates of the Naval Research Laboratory, who fired rockets above the earth's atmosphere. The sun is so feeble an x-ray source, however, that if its example were emulated by other stars there would be no point in looking for x-rays outside the solar system.

In 1960, however, Giaconni, Gursky, Paoline, and Rossi discovered a celestial source of x-rays in the constellation Scorpius. Subsequent observations by Giaconni's group, by Friedmann, Byram, Bowyer, and Chubb, by Clark, Kraushaar, and associates at the Massachusetts Institute of Technology, by a Lockheed group, by McCracken and the Fenton brothers and their associates at the Universities of Adelaide and Tasmania, respectively, and by many others have rapidly expanded our knowledge of these remarkable objects. About 23 discrete sources are now known; many tend to cluster near the galactic plane. X-ray emitters fall into four categories: (1) a diffuse, roughly isotropic background, (2) starlike objects such as Scorpius X-1 and Cygnus X-2, (3) supernova remnants such as the Crab Nebula and Cassiopeia X-1, and (4) extragalactic nebulae such as the great elliptical galaxy M 87 in the Virgo cluster. The x-ray spectrum appears to be continuous; no x-ray lines have ever been observed from cosmic sources.

The strongest x-ray source in the sky is Scorpius X-1, which was identified with a star by Oda, Osawa, and Jugaku in Japan. Its optical spec-

Fig. 128. The nonthermal jet in the elliptical galaxy M 87, photographed with the 120-inch reflector at the Lick Observatory, University of California. This jet is a strong source of radio-frequency radiation.

trum resembles that of an old nova but we cannot be sure that the optical spectrum refers to the same thing that emits the x-ray spectrum. Scorpius X-1 is so strong that it actually affects the ionization of the earth's ionosphere. The total energy output (nearly all in x-rays) is about 10,000 times that of the sun. It also emits in the radio-frequency range, where it is a faint variable source. Cygnus X-2, which appears to be a binary, may be a somewhat similar source; both it and Scorpius X-1 show rapid variations in brightness. The Crab Nebula is the second strongest x-ray source and its size and position seem to coincide rather well with the optical source. The elliptical galaxy M 87 is one of the weakest sources identified, about 0.3 percent as bright as Scorpius X-1. Optical observations show it to be characterized by a strong jet that is polarized in the optical region (Fig. 128). It is a strong emitter of radio-frequency radiation.

What produces x-ray emission in a heated gas? The simplest mechanism involves encounters between free electrons and ions—the so-called free-free emission or bremsstrahlung. The electrons are decelerated and therefore radiate energy. This process produces the normal, so-called thermal radiation of hot gas in the radio-frequency range. Such emissions can occur in the x-ray region if the gas temperature is 1,000,000° or more.

Nonthermal radio-frequency radiation is produced by synchrotron

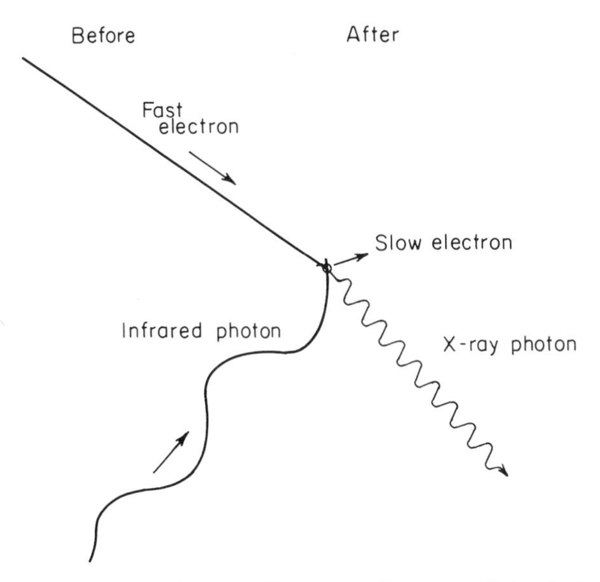

Fig. 129. The inverse Compton effect: a collision between a fast electron and a low-energy, infrared photon results in the production of a slow electron and an x-ray photon.

emission from high-speed electrons in a magnetic field. In order to produce x-rays by such a mechanism, electron energies of the order of 10^{13}–10^{14} electron volts would be required, if the magnetic field was 0.0001 gauss (a huge field for interstellar space).

A third mechanism, favored by Felten and Morrison, is the so-called inverse Compton effect. A high-energy electron may collide with a low-energy photon. The electron departs with greatly reduced energy and a high-frequency photon is created (Fig. 129). For example, an electron with 10^9 electron volts of energy may collide with a photon of energy 0.001 electron volt (corresponding to only 3°K) to produce an x-ray photon. A similar electron colliding with a photon of starlight (corresponding to a temperature of, say, 4000°K) would produce a gamma ray. In a magnetic field of 2 microgauss, this same electron would produce synchrotron emission at a wavelength of about 100 meters. The diffuse background radiation has a greater flux than that of all discrete sources. Its most likely origin, according to Hamilton and Francey, is the inverse Compton effect of high-speed electrons, trapped by magnetic fields in a halo or cloud around the galaxy, and interacting with photons of starlight and of the 3°K background radiation (see the next section). The x-rays from discrete sources may arise from bremsstrahlung and even from synchrotron mechanisms. Scorpius X-1 and Cygnus X-2 can be interpreted in terms of thermal radiation from sources with temperatures of 10^6–10^8°K. A different dependence of emissivity on energy is exhibited by the Crab Nebula and by M 87, where the emission is proportional to (energy)$^{-3/4}$.

X-ray observations promise to be increasingly important in the future as they enable us to probe the properties of hot clouds of gas at a temperature of 10,000,000°K, study residues of supernovae explosions, and secure better data on the density and chemical composition of the interstellar gas.

The Cosmic Microwave Background

At the other extreme from the isotropic x-ray background is the so-called 3.5°K microwave background. In Chapter 7 we remarked that interstellar absorption lines all corresponded to transitions from the very lowest level of the atom or molecule involved. Swings called attention to a CN line whose lower excitation potential was 0.0048 electron volt above the ground level. Absorption from this level can be explained if the atoms are bathed in black-body radiation corresponding to 3 or 4°K. Radio-frequency measurements at wavelengths of 3 and 7 centimeters indicate a background radiation corresponding to 3.5°K. Presumably this radiation is a residue of the "primeval fireball," that is, it is a relic from the very earliest stages of the universe, which cooled from an enormous initial temperature to 3°K as a result of the expansion.

Radio Galaxies

Although a supernova remnant may radiate as much energy in the radio-frequency range alone as the sun emits in all wavelengths, the total radio emission from a normal galaxy is such that these objects are difficult to observe. Over the observable radio-frequency range none emits more than 10^{32} watts; that is, the radio-frequency energy is never more than about 200,000 times the power output of the sun. Such emission from ordinary galaxies is rare.

There exists a class of objects, called radio galaxies, that emit, in the radio-frequency range alone, up to 2×10^{10} times the total energy output of the sun. Some of these, such as NGC 1218, appear to be giant elliptical galaxies, with masses perhaps 10 times that of our Galaxy. Sometimes the emission originates from a small, intense blob or filament immersed within a faint, extended halo structure. An example is provided by the galaxy M 87; as we have seen, this system contains a small, blue, outward-moving gas jet whose light is polarized (Fig. 128).

Many radio galaxies show a remarkable double structure, which has been studied particularly by Bolton, Matthews, Maltby, and Moffat. A single, optically observed galaxy is flanked by two large blobs that emit intense radio-frequency radiation. These blobs may have diameters ranging from 10,000 to 100,000 parsecs, and spacings up to 300,000 parsecs. A good example of a radio galaxy is Cygnus A; its two components are separated by 100 seconds of arc, which corresponds to 80,000 parsecs. Each component has a diameter of 20 parsecs and the two are connected by a faint bridge of luminous matter.

The typical picture, then, is of a galaxy that may show no particularly strange features optically, but on opposite sides of which and at distances of many tens of thousands of parsecs are two huge clouds emitting intense radio-frequency radiation. These clouds are invisible optically. Hence they are not thermal emitters but are radiating energy that has been stored in high-speed particles. The clouds must have originated from some kind of explosion in the central galaxy and have been hurled outward in opposite directions. From the separation of the blobs, one can get a lower limit to their ages, and from their rate of emission of energy, which is typically 10 million times that of a normal galaxy such as our own, we can estimate the total amount of energy that must have been stored.

The energy stored in high-energy electrons and the magnetic fields needed to constrain them amounts to about 10^{61} ergs. If all of the mass of the sun could be converted into energy, the total energy supplied would be (mass of sun) $\times$ (velocity of light)$^2 = 10^{54}$ ergs. Thus the energy stored in these radio-galaxy blobs must be equivalent to the total annihilation of 10 million stars like the sun—and much of this is invested in high-energy particles.

Since, in actual practice, thermonuclear reactions release at most only about 1 percent of the mass as energy, the actual masses involved in violent processes that produce an intense radio galaxy would involve 1000 million stars of solar mass. That is, the total amount of energy radiated by 1 percent of all the stars in our Galaxy during their entire lifetimes would be equivalent to the energy involved in one of these radio galaxies. How can such large amounts of energy become invested in processes involving such high-energy particles? We do not know, yet. Large-scale motions of vast quantities of gas have been observed in the nuclei of certain galaxies, such as M 82; these events have been interpreted as violent eruptions. No adequate theory has ever been proposed, largely as a consequence of the lack of adequate data.

Quasi-Stellar Objects

The most puzzling objects known in the astronomer's universe are the quasi-stellar radio sources, often called quasars. They embody a host of amazing, seemingly mutually inconsistent, properties. Whatever comes out to be the correct interpretation, we can safely bet that they will emerge as the most astonishing of nature's powerhouses.

No better illustration of the complementarity of radio and optical astronomy exists than the discovery of the quasars. Radio astronomers had detected a number of objects that seemed to be point sources and for which accurate positions could be obtained. Optical astronomers found objects resembling stars at these positions. Their spectra showed a few diffuse emission lines which could not be identified until it was realized that the spectra were subject to enormous red shifts.

More than 200 of these objects have been observed; they are blue in color and fall close to the position indicated for a radio-frequency source. Similar blue objects that are not radio "stars" are also known. Some are stars, others are similar to quasars except that they emit no abnormal radio-frequency radiation. Both types are often called quasi-stellar objects or QSO's. Down to the 18th magnitude there are probably about 3 QSO's per 10 square degrees, but the number of quasars, which are strong radio-frequency sources, is only about 1.4 per 1000 square degrees.

Most quasars are stellar. Measurements with radio-frequency interferometers indicate that their diameters are less than 0.002 second of arc. They appear to be uniformly spread over the sky, showing no association with known galaxies or even clusters of galaxies.

Their optical spectra (which have been studied by Maarten Schmidt, J. L. Greenstein, R. Lynds, the Burbidges, J. B. Oke, E. T. Wampler, and others) are sensational. Most of the radiation is due to a continuum whose intensity falls off toward the ultraviolet in a manner suggestive of synchrotron radiation rather than the combined contributions of many stars. Infrared observations show that QSO's differ from one another; some are bright in this region while others are faint. Upon this continuum are occasionally superposed bright lines that typically contribute about 20 percent of the total brightness. These bright lines are often 20–30 Å wide, but it is not clear whether the broadening is caused only by large-scale mass motions or by scattering of relatively narrow lines by fast-moving electrons. From the character of the bright-line spectrum, notably the presence of numerous forbidden lines but also the absence of lines of [O II] (which are characteristic of low-density nebulae), it is estimated that the gas has a density of 10^7 electrons and ions per cubic centimeter. Thus, the bright-line emission comes from a rarefied gas that more closely resembles a nova shell than a planetary nebula.

Most remarkable of all is the red shift, sometimes so huge that lines normally in the inaccessible ultraviolet now fall in ordinary spectral regions. Let z denote the ratio $\Delta\lambda/\lambda$ of the Doppler displacement of wavelength to the original wavelength; z values between 0.06 and 2.3 have been found. Thus, emission lines such as Lyman α (1216 Å) and lines of N IV (1488 Å) and C IV (1663 and 1550 Å) are displaced into the observable range; if $z = 3$, Lyman α falls near 3650 Å. If we interpret these displacements as arising from velocities in the line of sight, this means that QSO's can have speeds, *away* from the observer, that are substantial fractions of the velocity of light itself. Furthermore, the absorption lines sometimes do not give the same z-values as the emission lines, suggesting an expanding (or in one instance a contracting) shell of gas.

Of crucial importance in interpreting QSO's are the variability and the polarization in their optical and radio-frequency emission. Two factors are important in the variability: the scale or amplitude of the variation and the time intervals involved. If a source shows variations with a period of 1 month, its size cannot much exceed the distance a light signal will travel in 1 month, or 5000 astronomical units. Harlan Smith and Dorrit Hoffleit have traced back the variations of the brightest quasar, 3C 273, for 80 years on Harvard Observatory plates (3C refers to the third Cambridge catalogue of radio sources). This source shows roughly rhythmic variations with a period of 13 years, upon which are superposed sudden flashes. Other quasars are more spectacular. In a matter of months, 3C 446 increased twentyfold in brightness. Measurements by Oke and others indicate that some quasars have variations within 24 hours, which means that the diameter of the actually varying region is only 20–30 times the diameter of the solar system. The nucleus of a quasar may change in luminosity, but the surrounding gas cloud, which may occupy a thousand times the volume, does not. Kinman and his associates found curious variations in the optical polarization. During an outburst of 3C 446, the polarization angle rotated 90° in a month and then returned to its original value. H. D. Aller and F. T. Haddock found that several quasars show time variations in position angle and amount of polarization at a frequency of 8000 MHz, corresponding to a wavelength of 3.75 centimeters.

Interpretation of Quasi-Stellar Objects

How are these remarkable data to be understood? Crucial to the argument is the interpretation of the red shifts of the spectral lines. If we assume that red shifts are due to the expansion of the universe, enormous masses and luminosities packed into extremely small dimensions are required. An alternative hypothesis is that quasars are small objects hurled from the nucleus of our Galaxy, but then enormous kinetic energies are required. If we assume that they are ejected from other galaxies, some blue shifts should be observed; none have ever been found. Another proposal is that a small but extremely massive object produced the necessary red shift by gravitational action. It is difficult then to explain the relatively narrow forbidden emission lines, which are characteristic of relatively low densities. A low-density envelope is hard to reconcile with a huge gravitational field. Either there must be an enormous central mass (equivalent to a thousand galaxies) or else the object must consist of an attenuated gas inside a small, fantastically massive star cluster. Alternatively, it may be supposed that red shifts are caused by some as yet undiscovered law of physics.

Let us return to the assumption that these red shifts represent recession velocities in an expanding universe (see Shapley, *Galaxies,* chap. 7) and compare the luminosities of QSO's and normal objects. The bright quasar 3C 273 has a luminosity equivalent to about 1000 normal galaxies, or 40 of the brightest. Also, in 80 years, 3C 273 has shown no tendency to decline. Schmidt has concluded that the ages of quasars may lie between 1000 and 1,000,000 years. If the optical power output is of the order of 10^{46} ergs per second, then the total energy output would lie between 3×10^{56} and 3×10^{59} ergs. Such energies correspond to the complete annihilation of 1000 to 1,000,000 suns. The actual masses involved would have to be much greater, since only a small fraction of each gram can be converted to energy. If nuclear processes are involved, 10^6–10^9 solar masses are required, and a large fraction of this energy must be liberated within a small volume.

Perhaps significant clues are provided by the so-called Seyfert galaxies, which have small bright nuclei and inconspicuous outer arms. The nuclei produce broadened emission lines suggestive of large-scale turbulence and show variations with a period of about 1 week. Seyfert galaxies are bright in the infrared; their total bolometric luminosity is about 10^{36} kilowatts, not much lower than the total estimated luminosity of a QSO (in the expanding-universe interpretation). Such a galaxy, with very faint outer arms and placed at a great distance, might resemble a QSO. Other distant galaxies, called N galaxies because their nuclei are bright, and also Zwicky's compact galaxies recall QSO's. Measurements by Oke of the N-type radio galaxy 3C 371 showed that the energy output of its nucleus and the distribution of intensity with wavelength resembled those of a quasar. Furthermore, it showed brightness variations of 0.15 magnitude in a few days and variations of 1 magnitude in a 2-year interval. Hence, quasars may represent extreme limits of a type of galaxy that is not uncommonly rare. Another point is that some quasars appear as double radio sources, separated by about 30 seconds of arc, with an optical point source. Thus they resemble radio galaxies.

A crude model, which represents a blending of many suggestions, involves a central, massive, probably rotating object that supplies the energy output—and the power requirements are indeed staggering. Among the hypotheses proposed are: (*a*) star collisions or supernova explosions or both in a dense, small star cloud of very large mass (Gold, Burbidge, Field, and others), and (*b*) gravitational collapse of a large mass, typically 100,000 solar masses (Fowler, Hoyle).

The supernova and collision theories have some attractive features, but the detonation or destruction of stars in rapid succession and in such a way as to reproduce the observed light curves, polarization effects, and so on taxes the ingenuity of the theoretician. Gravitational collapse of a large mass presents some severe difficulties. Another theory, due to Sturrock, finds analogies between quasars and solar flares. Large quantities of energy are liberated when magnetic fields are destroyed. Magnetic fields must be present since there must be a surrounding region where fast electrons are accelerated in such fields to produce synchrotron radiation.

The forbidden emission lines are probably produced in a surrounding filamentary medium hundreds or thousands of parsecs in diameter. Intertwined with this material are more extensive, cooler clouds that produce the absorption-line spectrum. In some quasars, vast extended jets may exist; in 3C 273 and 3C 279, they may extend to distances of thousands of parsecs.

Perhaps QSO's represent a distinct stage in the history of a galaxy. There is much evidence to support the suggestion that our Galaxy was originally a spherical object, with the then-existing stars moving in highly elongated orbits tilted at all angles with respect to one another. These extreme Population II stars had metal-to-hydrogen ratios much less than the solar value. Then, perhaps rather abruptly, the Galaxy developed into the flattened system we now know; a huge amount of material experienced nuclear transformations so as to produce the present proportion of heavy elements that we find in the sun, but also in the very ancient open star cluster M 67. Stellar motions and chemical compositions are consistent with the idea that a catastrophic rearrangement took place when the Galaxy was young. This may have been the quasar event.

Pulsars

Early in 1968, Hewish and his associates in Cambridge, England, discovered the first of a type of pulsating radio source or "pulsar." More than 95 percent of the time a pulsar emits no radiation toward the observer. Then it radiates abrupt pulses (Fig. 130) with durations of the order of 0.002–0.01 second (typically 0.01 second). These pulses can be complex, showing oscillations of duration less than 0.001 second. In some instances the intensity can vary from zero to its full value in 0.0001 second. The time interval between pulses is typically about 1 second, but can range from 0.0331 second to 3.75 seconds with a surprising constancy of period. For CP 1919 (CP means Cambridge pulsar), the period is 1.3370113 seconds; for CP 0328, it is 0.714518603 second; that is, the pulsar behaves like a clock with a fantastically accurate movement. The heights of the pulses show variations over times of the order of minutes, hours, or even months. Sometimes a pulsar will be observed at one time and not at another, or an inferior record will be obtained. Secondary periods may possibly exist, too. Pulse widths tend to increase with period.

Later, it was discovered that pulsar periods were not immutable; several seemed to be growing longer, at rates of the order of 1 part in 10^{15}. Thus, the pulsars would require millions of years to decay. One object, the Vela pulsar, showed a remarkable discontinuity in an otherwise steady lengthening of period.

A. G. Lyne and F. G. Smith found pulsars to show linear polarization, but the most remarkable is the Vela source. V. Radhakrishnan, D. J. Cooke, M. Komesaroff, and D. Morris found that in each pulse the plane of polarization turns through nearly 90°, but at the same time the amplitude changes considerably.

Before we try to interpret these remarkable observations, let us mention ways in which pulsars can be used as tools for probing the interstellar medium. Hewish and his associates noted that pulsars emitted over a wide range in radio frequency, but the lower the frequency, the later the pulse is received. They concluded that this "dispersion," or dependence

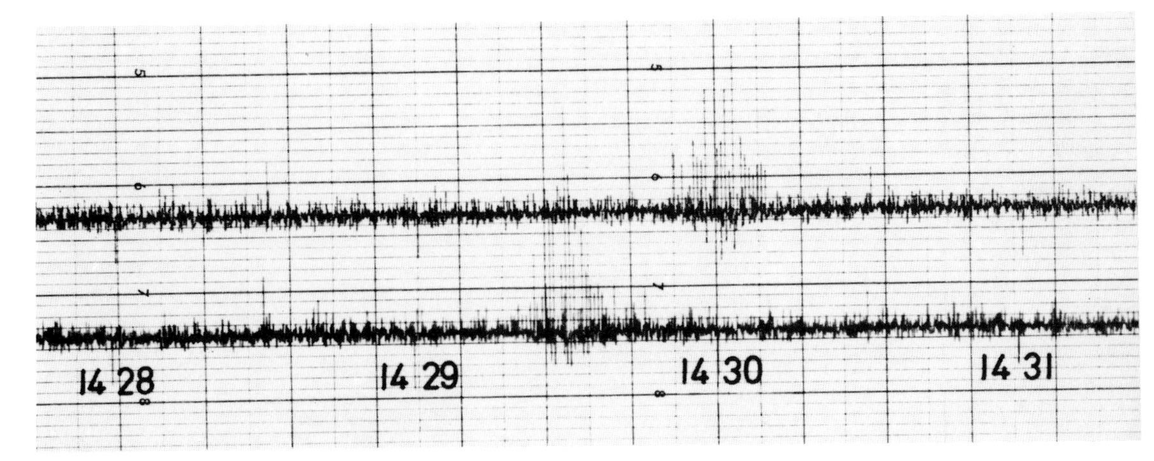

Fig. 130. A photograph of a chart-recorder trace showing the discovery of the pulsar MP 1426 in October 1968. The observation was made with the east-west arm of the Molonglo radio telescope. The lower trace records the signal from a beam directed slightly east of the meridian, the upper trace the signal from a beam slightly west of the meridian. Strong pulses can be seen on both channels. The difference in transit times on the beams is proportional to their separation, thus proving that the signals originate outside the earth. This pulsar has been observed on many subsequent occasions, but not as well as on the original record shown here. It lies close to the Southern Cross in the sky but has not been identified with any optical object. (Courtesy Michael Large, University of Sydney.)

of travel time on frequency, was caused by interstellar electrons. The amount of the delay depended on the total number of electrons in the line of sight. Thus, the time delays were much greater for pulsars near the galactic equator. This dispersion effect makes it difficult to observe distant pulsars, since a given receiver accepts radiation over a finite band width. Then each pulse tends to be confused with the following pulse if the pulsar period is short.

These dispersion effects can be used to estimate pulsar distances if we know the electron density along the line of sight. For this purpose, the interstellar medium may be approximated as cold, dense clouds intermixed with a hot, tenuous, ionized gas having a density of less than 0.1 electron per cubic centimeter. Observed dispersions show that typical pulsars lie at distances of less than 1000 parsecs. Some independent check is desirable. In some instances, one can observe pulsars behind cold, neutral hydrogen clouds that produce absorption at a wavelength of 21 centimeters. Since the distance of the cold hydrogen cloud can be found by other means, that of the pulsar can be estimated. Comparisons

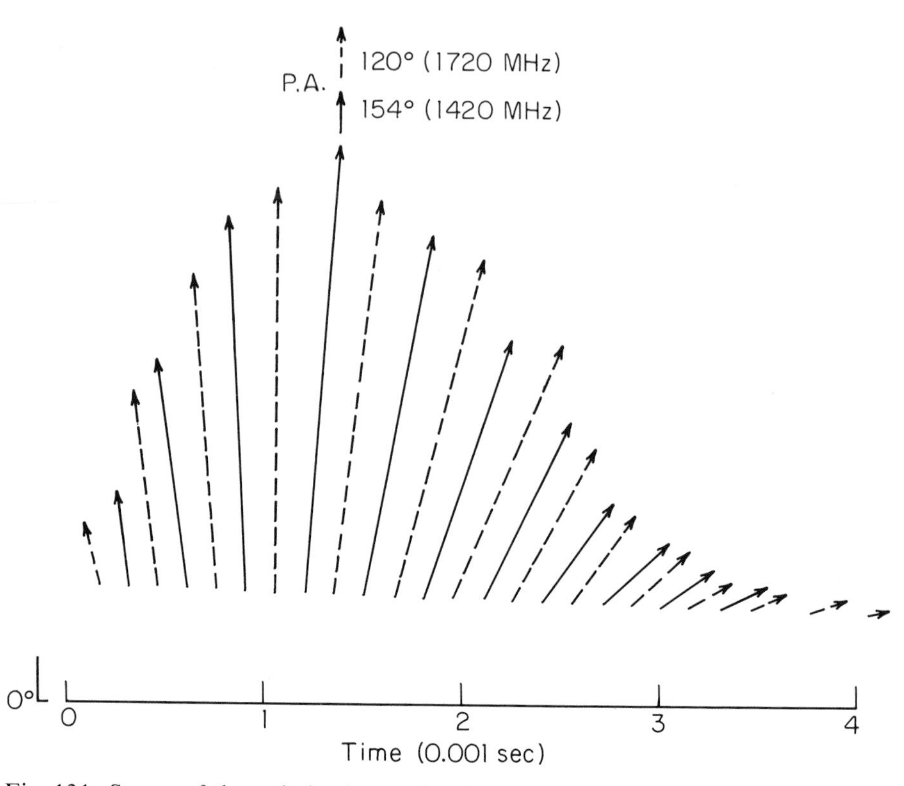

Fig. 131. Sweep of the polarization angle of the pulsar PSR 0833–45 during a pulse. (Courtesy V. Radhakrishnan and D. J. Cooke, Radiophysics Division, C.S.I.R.O., Australia.)

of the two techniques indicate densities of electrons or fewer per 100 cubic centimeters. The best means of getting distances, however, is by optical identification of the sources.

The plane-polarized radio-frequency radiation of a pulsar will suffer rotation of the plane of polarization as it passes through an ionized gas in a magnetic field. The amount of this Faraday rotation (see Chapter 7) depends on the square of the wavelength, the total number of electrons in the line of sight, and the strength of the magnetic field. The total number of electrons in the line of sight is known from the dispersion effect and we can measure the rotation of the plane of polarization for different wavelengths; hence we can determine the magnetic field. The Vela source, which gives a magnetic field of 0.8 microgauss, is possibly the best example, but observed values show a huge spread, due to a variety of magnetic-field direction cancellations. That is, if in one region along the line of sight the magnetic field is directed, say, north and in another region south, the Faraday rotation from the first region will be at least partially canceled by the contribution from the second. The net rotation would correspond to a much weaker uniform field than the true fields actually present.

The Nature of Pulsars

What, then, are the pulsars? No progress toward answering this question could be made until pulsars were identified with optically known objects. The first real breakthrough came when M. I. Large, A. E. Vaughan, and B. Y. Mills identified one pulsar in the middle of the Vela supernova remnant. It has a period of 0.089 second, the shortest known up to that time. Then D. C. Staelin and E. C. Reifenstein found that the Crab Nebula contained a pulsar with a period of 0.033 second. W. J. Cooke, M. J. Disney, and A. J. Tayler at the University of Arizona found optical pulses similar to the radio pulses. Then, by a television-type technique, J. S. Miller and E. J. Wampler at the Lick Observatory found that the pulsar was actually the star originally suggested by Baade, and that its maximum intensity was about 50 times its minimum intensity. The light was plane polarized and the period equaled the radio period. Similar pulses were found in the infrared by G. Neugebauer, R. B. Leighton, J. A. Westphal, and associates; x-ray pulses were found from rocket observations and very hard x-ray pulses from previously obtained balloon observations.

The large amplitude and short period mean that the source must be small, since otherwise radiation from different parts of the spherical globe would be blurred out by a factor of at least the ratio of the radius of the star to the velocity of light. If variations occur in 0.0001 second, the diameter of the source cannot much exceed 30 kilometers. The rate of emission of energy per unit area exceeds that of the sun in all wavelengths by many millions of times.

The Crab pulsar is slowing down. If we interpret it as a rotating body with a radius of 10–15 kilometers, the energy loss is about 10^{38} ergs per second, that is, about 20,000 times the energy output of the sun. This energy supply could account for the total energy output from the Crab Nebula if rotational energy is efficiently converted to particle energy. How can such energy conversion occur, and, above all, how can densities as great as 10^{12}–10^{15} grams per cubic centimeter be obtained? Presumably, also, the rotating star carries a magnetic field which may be as high as 10^{12} gauss.

The answer appears to be neutron stars. In Chapter 9 we described white-dwarf stars whose densities could be 100,000 times that of water, representing a situation where the pressure was so great that electrons were completely stripped from atomic nuclei. Then electrons and nuclei would move quite independently of one another and obey different gas laws. In a star like the sun, the electrons would be able to exert enough pressure to balance the heavy weight of the overlying layers. This would no longer be true for a very massive star and the density would continue to rise. Eventually, when the density was pushed up to and beyond a million million times that of water, the very nuclei would be forced into contact with one another, and the electrons would be pushed back into them. One by one the positive charges in the nuclei would be canceled and neutrons would be created. Ultimately, we would have a rapidly spinning object consisting of nothing but neutrons and presumably carrying a strong magnetic field.

Various models have been proposed to account for the way in which a spinning neutron star converts rotational energy to accelerate particles up to cosmic-ray energies. T. Gold imagined that the rotating object would have a magnetized region, or magnetosphere, which rotated as a solid body with it, out to the point where the rotational velocity equaled the speed of light, and that particles could be shed in the forward direction. P. Goldreich suggested that particles might escape along the polar lines of force. J. Gunn and J. Ostricker proposed a model in which the magnetic axis is inclined to the axis of rotation, as it is in the earth. Each model required magnetic fields of the order of 10^{12} gauss, but were able to explain the observed slowing down of pulsars. Difficulties remain, however. The Vela pulsar showed a remarkable discontinuity in its otherwise steady slowing down.

At last the energy source of the Crab Nebula and perhaps also the source of cosmic rays appears to have been found—and in an object that represents an entirely new, superdense state of matter. Most stars expire as white dwarfs; none has yet been observed to produce a supernova. Supernova remnants may be observable only for a few millenia; the pulsar may be optically observable only for a few decades or centuries, while persisting as a radio-frequency source for a million years.

It would be satisfying if we could close this account of stars and nebulae with some far-reaching conclusions that would tie the facts together in one nice, tidy package. Unfortunately, this is not possible. Today there are more puzzles than we dreamed of a generation ago. The problem of the origin and nature of the energy sources associated with intense non-thermal emitters such as quasars and radio galaxies is far from being solved. The origin of supernovae and their apparent evolution into neutron stars cannot be regarded as understood. Yet in other areas considerable progress appears to have been made.

We have seen how the chemical composition of a stellar atmosphere or a gaseous nebula may be deduced from its spectrum, and that the same chemical elements, in about the same proportions as in the sun, are found at the limits of the observable universe. We have sketched in broad outlines the history of a star from its formation to its death as a white dwarf. Almost by chance, as a consequence of the action of gas and radiation pressure and magnetic fields, blobs of dust and gas attain a sufficient mass and density for gravity to take hold. Then the star contracts quickly until the temperature rises high enough for nuclear reactions to occur. The star shines by converting hydrogen into helium—first as a main-sequence star and later as a giant or supergiant when the hydrogen in the core is gone. Then, as the nuclear fuels are exhausted, the outer parts of the star escape back into space and the core settles down as a white dwarf. The ejected material mixes with the interstellar gas and eventually from this material new stars are formed. The evolution of a star is not complicated, in spite of the fact that a high-speed computer is needed to track its changes through the most interesting phases of its evolution. The brightest jewel in the sky does not have the complexity of a single lotus flower.

The earth—the solid ground under our feet, the mountains that glisten in the sunset, the steel of our great industrial civilization, the salt of the oceans—and the rest of the solar system are but the ashes of long-dead stars that shone as the bright gems of heaven for 5 million millenia before the creation of the earth.

What then of the future? The sun will gradually brighten and the temperature of the earth will rise, until eventually the oceans boil away and the earth becomes but a burned cinder. The sun will experience only a relatively brief life as a giant before its outer envelope escapes into space as the core shrinks to a white dwarf with a density between a hundred thousand and a million times that of water.

Each electron and each atomic nucleus is assigned a place in the giant, incredibly dense crystal that had once been a star. No particle can move without another's taking its place. Everywhere there exist complete harmony and total organization; there is no disorder and there are no dissenters. No deviations are permitted anywhere, at any time. This is the ultimate death of matter, from which there is no resurrection, for material that gets locked up in such a state stays there until the end of time.

What happens to the material that escapes from the dying sun and drifts out into the clouds of interstellar smog and gas? We will never know, but perhaps the vision of St. John the Divine was as good as any other: "For I beheld a new heaven and a new earth, for the old heaven and the old earth had passed away."

Appendix A Designations of Stars and Nebulae

The brighter stars in each constellation are denoted by letters of the Greek alphabet (see below) and the name of the constellation (in the genitive case); for example Betelgeuse = α Orionis, Rigel = β Orionis. Usually the letters are assigned to the stars in order of brightness, but sometimes in order of position, as for the Big Dipper. For fainter naked-eye stars, Flamsteed numbers are sometimes used, for example, 53 Tauri. Most bright stars are listed in the *Bright Star Catalogue* of Yale University Observatory, where they are designated by their numbers in the *Harvard Revised Photometry* (HR numbers).

Stars between the naked-eye limit, near the sixth magnitude, and the ninth of tenth magnitude are listed in the *Bonner Durchmusterung,* or BD, catalogue, or in the Henry Draper, or HD, catalogue. In the BD catalogue, which has been extended to far southern skies in the *Cordoba Durchmusterung,* stars are listed in zones of declination. Thus BD +30° 3639 means star number 3639 in the zone at +30° declination (it would pass directly overhead for an observer at 30° N latitude). The HD catalogue lists magnitudes, positions, and spectral classes according to the equatorial coordinate system of 1900, in order of right ascension. Both catalogues include bright naked-eye stars and one star may be given several designations; for example ι Herculis = HD 160762.

Stars below the limits of faintness of these catalogues have their own particular designations, as for variable stars and special objects such as quasistellar objects or quasars, and pulsars. Variable stars are designated by letters preceding constellation names: RY Tauri, AX Persei, R Andromedae, and so on. Radio sources are often designated by their number in a catalogue of such sources; thus, 3C 273 is number 273 in the third Cambridge catalogue. Pulsars are indicated by notations such as CP 1919, which denotes a "Cambridge pulsar"; the number gives information on its position. Other objects are simply described by their positions with respect to well-known objects.

Bright nebulae and star clusters are listed in Messier's catalogue, but for most of these objects we depend on Dreyer's careful compilation in his *New General Catalogue,* NGC, and the two supplementary *Index Catalogues,* IC. These catalogues give a brief description of each object together with its position for 1865 (right ascension and north polar distance), plus precession constants so that the position may be quickly brought up to date. More recent catalogues of diffuse nebulae have been published by Hugh Johnson, W. W. Morgan, S. Sharpless, and S. Cederblad. The most complete listing of planetary nebulae is in L. Perek and L. Kohoutek's *Catalogue of Galactic Planetary Nebulae,* which lists these objects according to their galactic coordinates, but a finding list by 1950 positions is also given.

The Greek alphabet is as follows:

α	alpha	ι	iota	ρ	rho
β	beta	κ	kappa	σ	sigma
γ	gamma	λ	lambda	τ	tau
δ	delta	μ	mu	υ	upsilon
ϵ	epsilon	ν	nu	ϕ	phi
ζ	zeta	ξ	xi	χ	chi
η	eta	o	omicron	ψ	psi
θ	theta	π	pi	ω	omega

Appendix B Relations Between English and Metric Units; Very Small and Very Large Numbers

In this book we have generally preferred to use metric units—the *centimeter* as the unit of length, the *gram* as the unit of mass, and the *second* as the unit of time. These quantities constitute the c.g.s. system of units. We tabulate here the relations between the more frequently used metric and English units.

1 inch = 2.54 cm = 25.4 mm
1 meter = 39.37 in.
1 kilometer = 0.621 mile
1 mile = 1.609 km
1 liter = 1000 cm^3 = 1.06 qt
1 $in.^3$ = 16.387 cm^3
1 ounce = 28.35 gm
1 kilogram = 2.20 lb (avoird.)

We use the *Kelvin* or absolute temperature scale. Zero degrees Kelvin is −273° Centigrade; hence 273°K is 0° Centigrade or 32° Fahrenheit.

In expressing very large or very small numbers, we use powers of 10; for example, we write 30,000,000,000 as 3×10^{10} (3 followed by 10 zeros). For a very small number like 0.000,000,000,1 we write 1×10^{-10} ($1/10^{10}$); thus 1.71×10^{-16} means 1.71 divided by 10^{16}.

Appendix C Some Physical Quantities and Relations Useful in Astronomy

1. Definitions and Units of Force, Energy, and Power

In various chapters we have had occasion to refer to force, energy, and power; we summarize here the relations between these various physical entities.

Acceleration is the rate of change of velocity. If velocity is expressed in centimeters per second, acceleration is expressed as centimeters per second per second. The acceleration due to gravity at the surface of the earth is 980 (cm/sec)/sec, or 980 cm/sec^2, which means that the velocity of a freely falling body is increased by 980 cm/sec during every second of its fall.

Force is defined as mass times acceleration and the unit of force in the c.g.s. system is the *dyne*. A force of 1 dyne will give a mass of 1 gram an acceleration of 1 cm/sec^2. The force of gravity upon objects at the earth's surface is 980 dynes per gram of mass.

Pressure is force per unit area and is usually measured in dynes per square centimeter or in *atmospheres*. One atmosphere of pressure is 1,013,246 dynes/cm^2. It is equivalent to the pressure exerted by the weight of a column of mercury 760 mm high at 0°C.

A force of 1 dyne acting over a distance of 1 centimeter will do 1 *erg* of work. Since the erg is a very small unit, we often use the *joule;* 1 joule is equal to 10^7 ergs. The ability of a system to do work is called *energy,* which is measured by the work that is done. Energy appears in various forms, as mechanical, electrical, thermal, and others. The unit of heat energy is the *calorie,* which is the amount of heat needed to raise the temperature of 1 gram of water through 1 C degree. The *mechanical equivalent of heat* is the ratio of a quantity of work to the quantity of heat into which that work may be converted; if the work is measured in joules and the heat in calories, the mechanical equivalent of heat is 4.185 joules/calorie.

The rate of doing work is called *power* and is expressed in horsepower, watts, or kilowatts. One *watt* of power is equivalent to the deliverance of 1 joule, or 10^7 ergs, of work per second. A *kilowatt* is 1000 watts and amounts to a work rate of 10^3 joules/sec or 10^{10} ergs/sec. A *horsepower* is equal to 746 watts.

2. Some Relations Concerning Gases

The pressure of a gas, which is the force it exerts per unit area on the walls of its container, is related to the mass of the gas, the volume in which is is enclosed, and the temperature.

Standard conditions of temperature and pressure are 0°C or 273°K and 1 standard atmosphere or 760 mm-of-mercury pressure. Under standard conditions, 22.415 liters of a gas will weigh μ grams, where μ is the molecular weight of the gas—28.02 for molecular nitrogen N_2, 32.00 for O_2, and so on.

The *gas law* is

$$PV = RT,$$

where P (dynes/cm²) is the pressure of the gas, V (cm³) is the volume occupied by 1 mole, or μ gm, of the gas, $R = 8.314 \times 10^7$ ergs/deg mole is called the gas constant and T (°K) is the temperature of the gas. The gas law is frequently written in the form

$$p = nkT,$$

where p (dynes/cm²) is the pressure, n is the number of atoms or molecules per cubic centimeter, and k is *Boltzmann's constant,* 1.380×10^{-16} erg/deg.

Example. If the electron pressure in the atmosphere of the sun is 10 dynes/cm² and the temperature is taken as 5800°K, what is the number of electrons per cubic centimeter? We have

$$10 = n \times 1.380 \times 10^{-16} \times 5800,$$

whence

$$n = 1.25 \times 10^{13} \text{ electrons/cm}^3.$$

3. Definition of Electric Charge in the Electrostatic System

Similarly charged bodies repel, oppositely charged bodies attract each other. Suppose two small insulated bodies (pith balls are often used in such experiments) are similarly charged and placed a distance r apart in vacuum. They will repel each other with a force F given by *Coulomb's law:*

$$F = \frac{q_1 q_2}{r^2},$$

where q_1 and q_2 are the charges on the two bodies. If $F = 1$ dyne, $q_1 = q_2$, and $r = 1$ cm, the amount of charge q_1 or q_2 is 1 electrostatic unit (esu). Charge is also measured in *Coulombs;* 1 coul = 3×10^9 esu.

4. Table of Physical Constants

Velocity of light	$c = 2.99793 \times 10^{10}$ cm/sec
Constant of gravitation	$G = 6.673 \times 10^{-8}$ dyne cm²/gm²
Volume of 1 mole (0°C)	22.4136×10^3 cm³
Standard atmosphere (pressure)	1,013,246 dynes/cm²
Melting point of ice	273.16°K (absolute scale)
Mechanical equivalent of heat	4.185 joules/calorie
Acceleration due to gravity	$g_0 = 980.665$ cm/sec²
Density of oxygen gas (0°C)	1.429×10^{-3} gm/cm³
Avogadro's number (number of atoms or molecules per mole)	$N_0 = 6.02252 \times 10^{23}$/mole
Loschmidt's number (number of atoms or molecules per cubic centimeter at 0°C and 1 atmos)	$n_0 = 2.6873 \times 10^{19}$/cm³
Charge of electron	$\epsilon = 4.8029 \times 10^{-10}$ esu
Mass of electron	$m = 9.1091 \times 10^{-28}$ gm
Mass of proton	$M_p = 1.67252 \times 10^{-24}$ gm
Mass of hydrogen atom	$M_H = 1.67343 \times 10^{-24}$ gm
Radius of first Bohr orbit	$a_0 = 5.29167 \times 10^{-8}$ cm
Mass of proton/mass of electron	$M_p/m = 1836.12$
Gas constant per mole	$R_0 = 8.3143 \times 10^7$ erg/deg mole
Boltzmann constant	$k = R_0/N_0 = 1.38054 \times 10^{-16}$ erg/deg
Planck constant	$h = 6.6256 \times 10^{-27}$ erg sec
Rydberg constant for hydrogen	$R = 109677.58$ cm⁻¹

5. The Radiation Laws

In Chapter 4 we explained how the astronomer measures the temperature of a star by studying:

(*a*) The distribution of radiation intensity with respect to color or wavelength (application of Planck's law);

(*b*) The wavelength of the position of maximum intensity (application of Wien's law);

(*c*) The amount of energy radiated per unit area of the surface (application of Stefan's law).

These laws refer to the emission of energy by perfect radiators. In 1859, Kirchhoff showed that for any temperature *the ratio of the emissive power of a body to its absorptivity is a constant for all objects and equals the emissive power of a black body,* that is, one that absorbs all the radiation that falls upon it. It is a matter of familiar experience that dark-colored objects are much better absorbers of heat than light-colored ones, and they are also much better emitters of energy. No perfectly black surface has been produced but it is possible to realize experimentally the essential conditions of a black body, insofar as we wish to study radiation.

Planck's law gives the relation between the intensity in a frequency interval $\Delta\nu$, at frequency ν, for a temperature T, in unit solid angle (there are 4π unit solid angles in a whole sphere), as

$$I_\nu \Delta\nu = \frac{2h\nu^3}{c^2} \frac{1}{e^{h\nu/kT} - 1} \Delta\nu,$$

where h is Planck's constant, k is Boltzmann's constant, and c is the velocity of light. If we want the amount of radiation flowing over all directions we multiply this expression by 4π; to compute the energy density we multiply $I_\nu \Delta\nu$ by $4\pi/c$. If we wish Planck's law in wavelength units instead of in frequency units, we make use of the relations

$$\nu = \frac{c}{\lambda} \qquad \text{and} \qquad \Delta\nu = \frac{c}{\lambda^2}\Delta\lambda,$$

and obtain

$$I_\lambda \Delta\lambda = \frac{2hc^2}{\lambda^5} \frac{1}{e^{hc/\lambda kT} - 1} \Delta\lambda,$$

which is the form in which this radiation law is most often written.

Stefan's law gives the relation between the total amount of radiation emitted by a black body and the temperature of the body. It is

$$E = \sigma T^4,$$

where E (erg cm^{-2} sec^{-1}) is the rate of emission of energy, the Stefan-Boltzmann constant $\sigma = 5.6697 \times 10^{-5}$ erg cm^{-2} deg^{-4} sec^{-1} (or 5.67×10^{-8} watts m^{-2} deg^{-4}), and T (°K) is the temperature.

Example. What would be the amount of energy radiated per square centimeter per second by the surface of a star whose temperature is 5700°K?

$$E = 5.67 \times 10^{-5}(5700)^4 = 5.99 \times 10^{10} \text{ ergs/cm}^2 \text{ sec}$$
$$= 5.99 \text{ kilowatts/cm}^2.$$

Wien's law gives the wavelength at which the intensity of the radiated energy is a maximum. The relation is

$$\lambda_{max} T = 0.28978 \text{ cm deg.}$$

Example. At what wavelength in the spectrum is the intensity a maximum for a star whose temperature is 5000°?

$$\lambda_{max} = \frac{0.2897}{5000} \text{ cm} = 5.794 \times 10^{-5} \text{ cm} = 5794 \text{ Å},$$

since 1 Å = 10^{-8} cm. Hence the maximum intensity occurs in the yellow near 5800 Å.

Stefan's law and Wien's law may be derived from Planck's law. For a derivation of Planck's law and further discussion of radiation definitions and relations see L. H. Aller, *Astrophysics I: Atmospheres of the Sun and Stars* (Ronald Press, New York, ed. 2, 1963), chap. 4.

Appendix D Astronomical Constants

One astronomical unit	149,597,892 km = 92,955,635 mi (Venus radar)
One light-year	9.4605×10^{17} cm
One parsec	3.086×10^{18} cm = 3.26 light-years
Mass of the sun	1.989×10^{33} gm
Radius of the sun	6.960×10^{10} cm
Mean density of the sun	1.41 gm/cm^3
Surface gravity of the sun	2.740×10^4 cm/sec^2
Energy radiated by the sun	3.84×10^{33} erg/sec (Labs and Neckel)
Absolute bolometric magnitude of the sun	+4.76 (Kron and Stebbins)
Absolute photovisual magnitude of the sun	+4.84 (Kron and Stebbins)
Temperature T_{eff} of the sun	5776°K (Labs and Neckel)
Mass of the earth	5.977×10^{27} gm
Mean radius of the earth	6.3710×10^8 cm
Mean density of the earth	5.517 gm/cm^3
Surface gravity of the earth	980.665 cm/sec^2
Number of seconds in 1 sidereal year	3.1558×10^7

Appendix E Stellar Magnitudes and Colors

1. The Relation Between Apparent Magnitude, Absolute Magnitude, and Distance

Magnitude differences are related to ratios of brightness by the basic expression

$$0.4(m_2 - m_1) = \log l_1/l_2,$$

where m_1 and m_2 are the apparent magnitudes of two stars and l_1 and l_2 are their corresponding apparent brightnesses. One magnitude difference corresponds to a brightness difference of 4 decibels.

To obtain the absolute magnitude, which is defined as the magnitude the star would have if it were at a distance of 10 parsecs, we note that brightness varies inversely as the square of the distance. Hence if l is the apparent brightness of a star at the distance r and L is the brightness it would have at the distance R,

$$l/L = (R/r)^2.$$

Taking logarithms and setting $R = 10$ parsecs,

$$\log l/L = 2 \log R - 2 \log r = 2 - 2 \log r$$

since $\log 10 = 1$.

If M is the absolute magnitude of a star of apparent magnitude m, the relation between brightness and magnitude becomes

$$2.5 \log l/L = M - m,$$

or

$$M = m + 5 - 5 \log r.$$

If we use parallax p instead of distance, then, since r (parsecs) $= 1/p$ (seconds of arc),

$$M = m + 5 + 5 \log p.$$

Example. γ Geminorum, visual magnitude 1.93, has a parallax of 0".040 according to the Yale *Catalogue*. What are its distance, absolute magnitude, and absolute brightness, and how do they compare with those of the sun? The distance of the star is $r_s = 1/0.040 = 25$ parsecs, and $\log r_s = 1.40$. Hence the star's absolute magnitude is

$$M_s = 1.93 + 5 - 5 \times 1.40 = -0.07.$$

The star lies above the main sequence, from which it has evolved.

The absolute visual magnitude of the sun $M_\odot = +4.84$. Hence γ Geminorum is $4.84 - (-0.07) = 4.91$ magnitudes brighter than the sun.

The basic magnitude relation holds for absolute as well as apparent magnitudes. Hence

$$0.4\,(M_\odot - M_s) = \log L_s/L_\odot,$$

where L_s is the absolute brightness of the star and $L_\odot$ that of the sun. Hence, to compare γ Geminorum and the sun,

$$0.4 \times 4.91 = 1.964 = \log L_s/L_\odot,$$

whence $L_s/L_\odot = 92$.
Thus γ Geminorum is nearly 100 times as bright as the sun.

If there is absorption of light in space by obscuring matter (see Chapter 7), the equation relating apparent and absolute magnitudes must be modified. If the light of a distant star is dimmed P magnitudes by interstellar clouds, then

$$m_{\text{obs}} = m_{\text{true}} + P;$$

hence

$$M = m + 5 - 5 \log r - P.$$

The amount of interstellar absorption may be inferred from its effect on the colors of distant stars, since the interstellar cloud reddens as well as dims starlight.

We must now examine the question of stellar colors.

2. The Kinds of Magnitudes

The magnitude of a star depends on the color sensitivity of the detector employed. With a photocell and a violet filter, a red star may appear as a faint object, although it may yet be quite bright to the eye. Visually, a star such as χ Cygni may be of the tenth magnitude, yet a thermocouple, which measures the total energy that reaches it, may show that this star is emitting as much radiation as a normal star of the fifth magnitude. Therefore, in expressing the magnitude of a star, we must specify the color sensitivity of the detection system employed.

Suppose we observe a star with a photoelectric photometer equipped with a blue filter. The red, yellow, most of the green and violet, and the ultraviolet light will be cut out. Only blue light over a range of a few hundred angstrom units will affect the sensitive surface of the photocell.

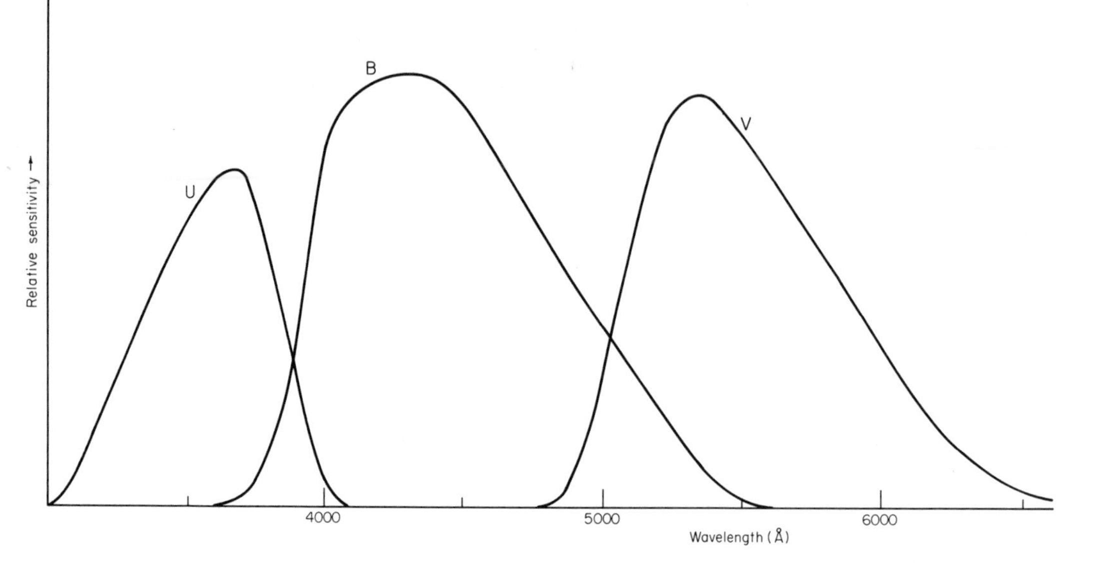

Fig. E–1. The wavelength dependence of the response of the *U, B, V* color system. These curves give the response for a *U, B, V* photometer as employed in a Cassegrain or Newtonian telescope, with reflection from two aluminized mirrors and atmospheric extinction for a star observed in the zenith at Mount Wilson. (After Johnson, Code, and Melbourne, as quoted by H. C. Arp, *Astrophysical Journal 133* (1961), 875.)

Experiments have shown that, although radiations covering a span of several hundred angstroms may fall on the photocell, the magnitudes determined will be just the same, to a fair degree of approximation, as though all the light were concentrated at one mean wavelength, which is called the *effective wavelength*, λ_{eff}.

The modern *U, B, V* system of photometry, due to Harold Johnson, employs filter-photocell combinations with the following characteristics (Fig. E–1):

Color	λ_{eff} (Å)	$(\Delta\lambda)_{1/2}$ (Å)	$F_\lambda\,(0,0)$ (10^{-7} erg cm^{-2}(100 Å)$^{-1}$ sec^{-1})
U	3650^{-100}_{+200}	530	4.35
B	4400^{-70}_{+100}	1000	7.20
V	5470^{-10}_{+30}	850	3.92

The effective wavelengths are given for a star of temperature 10,000°K.

The corrections to the effective wavelength of −100, −70, and −10 Å apply for a star of temperature 20,000°K, whose energy distribution is richer in ultraviolet light, and the corrections +200, +100, and +30 refer to a cool star of 4,000°K, whose radiation is mostly concentrated in the red; see C. W. Allen, *Astrophysical Quantities* (Oxford University Press, New York, 2nd ed., 1963), p. 195. The "half-band" width $(\Delta\lambda)_{1/2}$ is the width of the wavelength range over which the sensitivity exceeds half its maximum value. The last column gives the flux received (for a pass band of 100 Å) from a star of magnitude 0. The zero points are so adjusted that a main-sequence star of spectral class *A*0 has exactly the same magnitude in each system, that is, $U - B = B - V = 0$.

Other combinations of photocells and filters are also employed. Thus for work in the red and infrared Harold Johnson has introduced the systems R ($\lambda_{\text{eff}} \sim$ 7,000Å), I (9,000 Å), J (12,500 Å), K (22,000 Å), L (34,000 Å), M (50,000 Å), and N (102,000 Å = 10.2 μm). It is possible to employ narrower pass bands and to select wavelengths to emphasize certail spectral features, such as those depending on luminosity or metal abundance, for example.

3. Color Indices, Spectral Classes, and Bolometric Corrections

The difference between the magnitudes of a star as measured in two color systems is called a color index. The most commonly employed color indices are $B - V$ and $U - B$. Consider, first of all, stars that are unaffected by space absorption.

The $B - V$ color index then depends simply on the temperature. It is negative for very blue stars, such as the nuclei of planetary nebulae, because such objects are brighter in the blue than in the visual spectral regions. For cool stars, however, the $B - V$ index is positive and may become very large. Thus it serves as an index of temperature; in a Hertzsprung-Russell diagram V is plotted against $B - V$.

What happens if we plot $U - B$ against $B - V$? Consider first stars that are unaffected by space absorption, and suppose that stars radiate as black bodies. Then a graph of $U - B$ against $B - V$ would simply compare the slopes at two points in the Planckian curve and we would expect to get very nearly a straight line. The actual two-color graph for main-sequence stars resembles a "lazy *S*" (Fig. E–2.) Although $B - V$ changes smoothly with temperature, $U - B$ does not, primarily because of the pronounced distortion produced by the absorption at the head of the Balmer series (Fig. 43), and because further distortions are produced by the

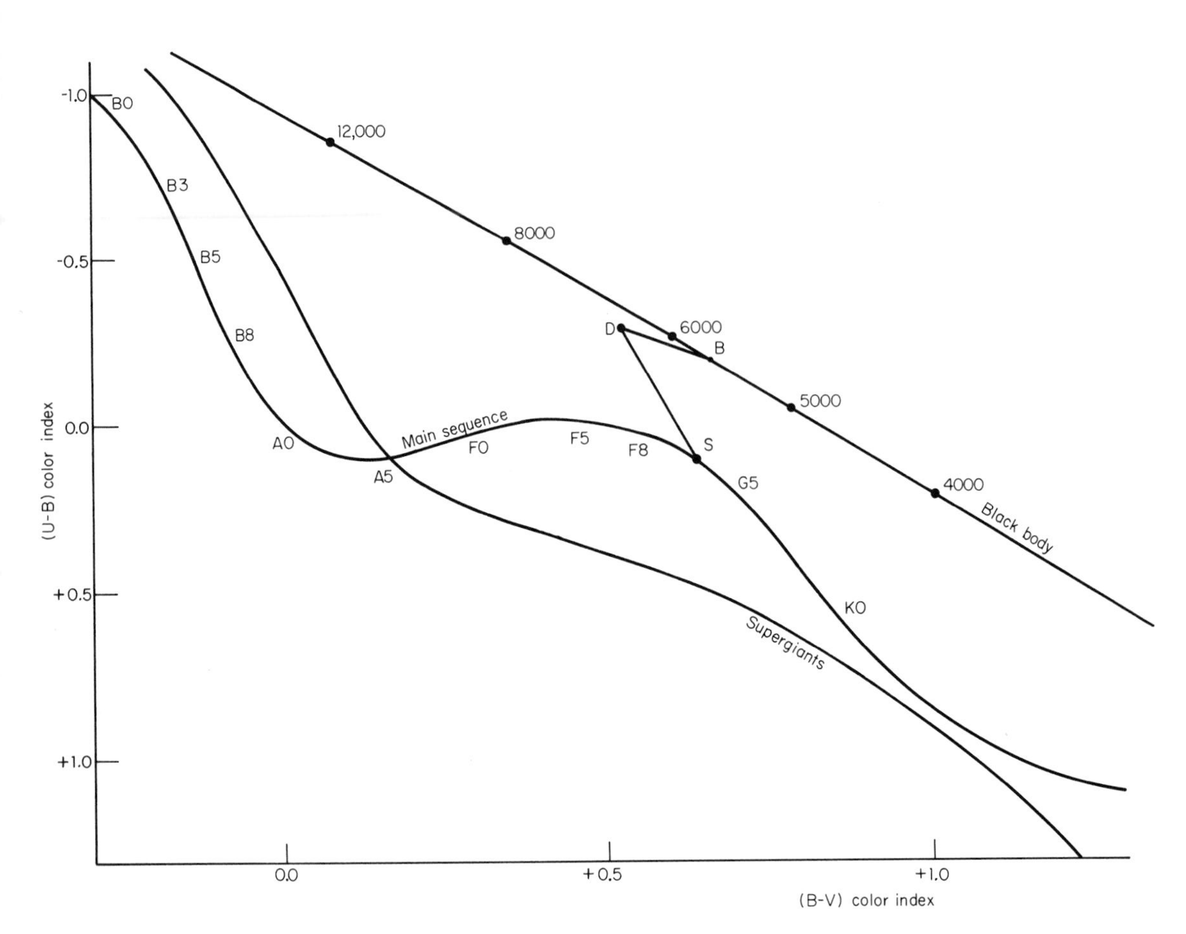

Fig. E–2. The relation between $(U - B)$ and $(B - V)$ color indices. The mean relation derived by Harold Johnson between $(U - B)$ and $(B - V)$ color indices is plotted for main-sequence stars and supergiants. The corresponding curve as calculated by H. C. Arp for a sequence of black bodies is also included. Compare the smooth slope of the black-body relation with the *S*-shaped curve for the main-sequence stars. The supergiant curve shows a less marked curvature because the Balmer jump is less prominent in supergiant F, A, and B stars than in dwarfs. A sequence of pure hydrogen stars would show a considerable departure from a black-body curve but absorption by strong metallic lines produces marked effects also. *S* denotes the position of the sun, *B* the position of a black body of the same effective temperature as the sun, and *D* the position of a metal-deficient star of the same temperature as the sun. It is displaced from *B* by the amount *DB* because the negative hydrogen ion is a nongray absorbing agent (see Fig. 45). The displacement *DS* corresponds to the influence of strong metallic absorption in the sun. (Adapted from H. C. Arp, *Astrophysical Journal 133* (1961), 878, 880 copyright University of Chicago Press.)

influence of strong absorption lines. The two-color graphs for metal-deficient stars differ appreciably from those for normal stars. Supergiant stars show a less pronounced kink because the Balmer jump is less important for stars near class A0.

Another very important use of the two-color $(U - B)$–$(B - V)$ graph is to determine space absorption. Interstellar reddening affects the two-color indices by different amounts, so by comparing the observed $(U - B)$–$(B - V)$ graph for a star cluster with the standard graph one can determine the extra coloring or color excess $\Delta(B - V)$ produced by space absorption. Magnitudes and colors corrected for space absorption are usually denoted by V_0, $(B - V)_0$, $(U - V)_0$. See Appendix F.

The various kinds of magnitudes that we have been discussing, U, B, V, and infrared magnitudes, utilize radiation over a limited wavelength range. For many problems, such as those of stellar evolution, we want to compare the luminosities of two stars with reference to the radiation summed over all wavelengths.

We express the total luminosity of a star in terms of its so-called bolometric magnitude. The difference between the bolometric and V (essentially visual) magnitudes is called the *bolometric correction:*

$$\text{B.C.} = m_{\text{bol}} - m_v.$$

The system of bolometric magnitudes is so adjusted that the corrections vanish for a class F0 star and are small for the sun. They become very large for hot stars, where most of the energy is in the far ultraviolet, and for very cool stars, which radiate most of their energy in the infrared.

These corrections may be established reasonably well for stars similar to the sun; they are uncertain for very hot stars, where we have to rely on predictions of model-atmosphere theory with occasional checks from rocket and satellite observations. They are likewise uncertain for cool stars whose spectra are strongly distorted by molecular-band absorption and most of whose energy is extinguished by absorption by water vapor in the earth's atmosphere.

Table E–1 summarizes our data on temperatures, intrinsic, that is, $(B - V)_0$, colors, and bolometric corrections for various types of stars. We discussed the temperature scale in Chapter 4. The $B - V$ colors are taken primarily from the work of Harold Johnson. The bolometric corrections are derived from a number of sources, notably model-atmosphere studies by Mihalas, by Strom, and by Morton and their associates for the hotter stars, from Davis and Webb for stars of intermediate temperature (7,000–12,000°K), and from the work of Popper for yet cooler stars. For the

Table E-1. Temperatures, $(B - V)$ colors, and bolometric corrections for normal stars.

Main sequence								Giants and supergiants			
Spectral class	T(°K)	$B - V$	Bol. corr.	Spectral class	T(°K)	$B - V$	Bol. corr.	Spectral class	T(°K)	$B - V$	Bol. corr.
O6V	40,000	−0.31	−3.70	F2V	7100	+0.36	0.00		*Supergiants*		
O7V	35,000	− .31	−3.37	F5V	6470	+ .46	− .02	B0I	21,100	−0.23	−1.75
O8V	32,000	− .30	−3.10	F8V	6120	+ .55	− .05	B8I	11,200	− .03	−0.36
O9V	31,000	− .29	−3.00	G0V	5970	+ .59	− .06	A2I	9,200	+ .10	− .08
B0V	28,000	− .28	−2.62	G2V	5780	+ .64	− .08	F0I	7,510	+ .15	.00
B1V	23,000	− .24	−2.06	G5V	5570	+ .70	− .11		*Giants*		
B2V	20,800	− .22	−1.76	G8V	5330	+ .80	− .16	G2III	5,300	+0.89	−0.18
B3V	17,200	− .19	−1.27	K0V	5150	+ .86	− .21	G5III	5,100	+ .91	− .23
B5V	15,000	− .16	−0.94	K2V	4840	+ .98	− .31	G8III	4,840	+ .96	− .30
B7V	13,000	− .12	− .64	K5V	4370	+1.13	− .55	K0III	4,680	+1.03	− .38
B8V	12,000	− .09	− .50	K7V	4000	+1.30	− .80	K2III	4,400	+1.10	− .52
B9V	11,400	− .06	− .38	M0V	3680	+1.44	−1.12	K5III	3,800	+1.50	−1.00
A0V	10,200	.00	− .22	M2V	3400	+1.52	−1.58	M0III	3,580	1.55	−1.24
A1V	9,700	+ .03	− .16	M4V	3180	+1.57	−2.10	M2III	3,300	1.58	−1.80
A2V	9,300	+ .04	− .09	M5V	3070	+1.60	−2.40	M4III	3,050	1.59	−2.45
A3V	9,000	+ .11	− .06	M6V	2960	+1.70	−2.70	M6III	2,800	1.59	−3.25
A5V	8,750	+ .14	− .03	M7V	2820	+1.88	−3.20	M8III	2,650	1.59	−4.25
A7V	8,100	+ .22	.00	M8V	2700	+2.0	−3.70				
F0V	7,450	+ .30	.00								

coolest stars, the radiometric observations of Pettit and Nicholson and the infrared measurements of Harold Johnson supply information on the bolometric correction.

Earlier than Class K5 the supergiants may be taken as 200°K cooler than giants of the same spectral class. At each spectral class from G2 to M8, the $B - V$ color indices for supergiants are about 0.09 redder than for corresponding giants. At spectral classes later than $K7$–$M1$, giants and supergiants tend to have similar temperatures and colors. Notice that bolometric corrections are always negative, so the star is always brighter bolometrically than visually (except for spectral classes near $F0$, where the two are the same). That is, the bolometric magnitude is equal to or less than the visual magnitude.

Example. The magnitude of ϵ Eridani is $V = 3.74$; its spectral and luminosity class is K2V. The corresponding bolometric correction from Table E–1 is −0.31, whence $m_{\text{bol}} = 3.74 - 0.31 = 3.43$.

Appendix F Some Uses of Color and Luminosity Measurements

1. Relation Between Absolute Magnitude, Temperature, and Radius of a Star

If we know the absolute brightness of a star and its size we can find its temperature. Alternatively, if we know the temperature and the true brightness of a star we can find its size.

The relation between the absolute magnitude of a star M_λ (as measured in a magnitude system of effective wavelength λ_{eff}), its radius R, and its temperature T (assuming that it radiates approximately like a black body of temperature T) is

$$M_\lambda = C_\lambda - 5 \log R + \frac{1.561}{\lambda_{\text{eff}} T} + X_\lambda;$$

(see Russell, Dugan, and Stewart, *Astronomy* (Ginn, Boston, 2nd ed., 1938), 2:733; L. H. Aller, *Astrophysics I: Atmospheres of the Sun and Stars* (Ronald Press, New York, ed. 2, 1963), p. 288), where C_λ is a constant depending on wavelength and X_λ is a small correction factor which may be important at high temperatures:

$\frac{1.561}{\lambda_{\text{eff}} T}$	5.0	4.0	3.0	2.0	1.0
X_λ	−0.01	−0.03	−0.07	−0.19	−0.55

For V magnitudes, let us adopt

$$\lambda_{\text{eff}} = 5480 \text{ Å} = 5.48 \times 10^{-5} \text{ cm}.$$

To evaluate C_λ we note that a black body of the same size and effective temperature as the sun would have approximately the absolute V magnitude 4.84. Since R is measured in terms of the solar radius, $R = 1.0$, $\log R = 0.00$. Hence, if the effective temperature of the sun is 5776°K,

$$4.84 = C_\lambda + 4.94 - 0.01$$

or

$$C_\lambda = -0.09.$$

The relation between visual magnitude, radius, and effective temperature is

$$M_V = -0.09 - 5 \log R + \frac{28{,}500}{T}.$$

For B magnitudes, we might take $\lambda_{\text{eff}} = 4400$. The absolute B magnitude of the sun is $M_B = 5.46$ (Stebbins and Kron, 1957). Hence a similar calculation gives $C_B = -0.66$, so

$$M_B = -0.66 - 5 \log R + \frac{35{,}500}{T}.$$

Eliminating R between these two equations and noting that $M_B - M_V = B - V$, we find

$$T = \frac{7000}{B - V + 0.57},$$

which yields a crude conversion from $B - V$ color indices to temperature for stars whose temperatures do not differ very much from that of the sun.

Example. The V magnitude of Wolf 359 is 13.66. Its parallax is 0″.425. The corresponding absolute magnitude is 16.80. The temperature corresponding to its spectral class is 2960°K. What is its radius in terms of that of the sun? The equation for M_V gives

$$16.80 = -0.09 - 5 \log R + 9.63,$$

or

$$\log R = -1.45,$$

whence

$$R = 0.035.$$

Example. The radius of a B3V component of an eclipsing binary system is 4.23 in terms of the sun. What is the absolute magnitude of the star? The temperature of a B3V star (Table E–1) is 17,200°K. Then (compare Fig. F–4),

$$M_V = -0.09 - 5 \times 0.627 + 1.65 = -1.57.$$

The assumption that stars radiate like black bodies is only a crude approximation, particularly rough for Class A stars, which show a strong continuous absorption due to hydrogen, and for cool stars, which show strong molecular bands. Over narrow spectral ranges, the black-body approximation may sometimes be a fairly good one. The $B - V$ color index, which avoids the Balmer jump, yields reasonable temperature estimates, but a $V - R$ or $B - R$ index might have been better. The $U - B$ or $U - V$ indices would have given nonsense.

Of course bolometric magnitudes and effective temperatures permit one to determine stellar radii readily. The relation is an extremely simple one. Since

$$L = 4\pi R^2 \sigma T_{\text{eff}}^4,$$

if we measure L and R in terms of the corresponding solar values, $L_\odot$ and $R_\odot$, we obtain

$$\frac{L}{L_\odot} = \left(\frac{R_*}{R_\odot}\right)^2 \left(\frac{T_*}{T_\odot}\right)^4,$$

or in terms of bolometric magnitudes

$$M_\odot - M_* = 5 \log R + 20 \log(T/5800).$$

2. Estimation of Space Absorption

Since the interstellar material reddens the light of the stars it dims, it acts to increase their color indices. Suppose one observes the main-sequence stars of a distant cluster on the U, B, V system. If one plots the $U - B$ color indices against the $B - V$ color indices, the standard S-shaped curve will be displaced (Fig. F–1). To fit the standard $(U - B)_0 - (B - V)_0$ curve to the observations, one shifts it along a line whose slope is given by

$$\frac{(U - B)_{\text{color excess}}}{(B - V)_{\text{color excess}}} \sim 0.7$$

until coincidence is obtained. In the example illustrated, $\Delta(B - V) = 0.20$, $\Delta(U - B) = 0.14$. The total space absorption is then

$$A_V = 3.0\ \Delta(B{-}V) = 0.6 \text{ magnitude},$$

although some observers, notably Harold Johnson, have concluded that the coefficient may vary from point to point in the galaxy. In some regions, 3 may have to be replaced by 5 or 6.

This method is strictly applicable for stars of "normal" chemical composition and could not be applied for a cluster of metal-deficient stars; supplementary spectroscopic data would then be needed.

3. Analysis of a Color-Magnitude Diagram for a Star Cluster

The detailed procedures differ for galactic and globular clusters. For the former, one often can obtain spectroscopic as well as photometric observations. Galactic-cluster main sequences usually extend to early types, F, A, or even B or O, whereas main sequences of globular clusters do not

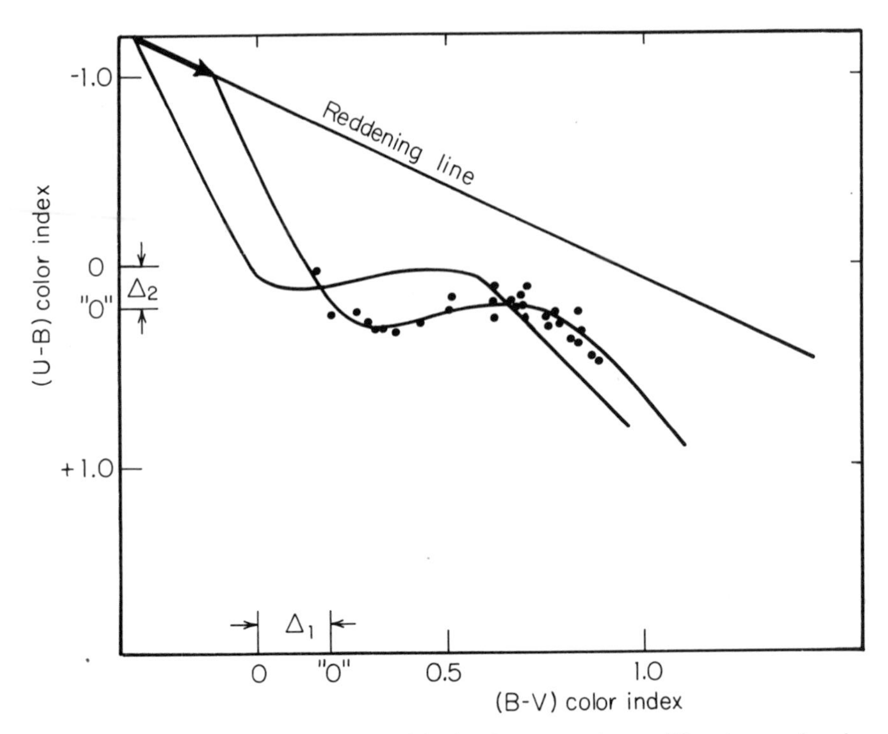

Fig. F-1. Determination of space reddening for a star cluster. The observed main sequence is indicated by the points to which the standard $(U - B)$ vs. $(B - V)$ curve is fitted by a shift, whose amount is indicated by the heavy arrow. The displacements $\Delta(B - V) = \Delta_1$ and $\Delta(U - B) = \Delta_2$ determine the color excess. The ratio $\Delta_2/\Delta_1 = 0.7$ defines the slope of the "reddening line" along which the curve is to be shifted.

extend much earlier than F8 or G0. Spectroscopic observations of individual stars in globular clusters require the very largest telescopes; hence there is heavy reliance on color-magnitude diagrams.

The steps in the procedure are as follows:

(1) Construct the best possible color-magnitude diagram by using both photoelectric and photographic observations. The photoelectric photometer is used to establish certain stars as standards of color and brightness and the photographic observations are used to interpolate magnitudes and colors of other stars.

(2) Establish the space absorption by using the method described in the previous section or some variation thereof. Then one can convert V magnitudes and $B - V$ colors to corrected apparent V_0 magnitudes and true $(B - V)_0$ colors.

(3) Next compare the V_0 vs. $(B - V)_0$ diagram thus obtained with the standard Hertzsprung-Russell diagram, which gives $M_V - (B - V)_0$, to obtain the distance modulus $y = V - M_V = 5 \log r - 5$, and hence the distance r of the cluster and the absolute magnitude of its members.

The resulting H–R diagram can be compared with published data to decide whether the cluster is a normal metal-rich one or a metal-deficient one.

We may also transform the color-magnitude diagram to a bolometric-magnitude–effective-temperature diagram for comparison with predictions of stellar evolution theory.

Consider the color-magnitude diagram for the old galactic cluster M 67 (Fig. F–2), which is taken from the work of Sandage. Apparent magnitude V is given on the right-hand side and absolute magnitude M_V on the left-hand side. The quantities V, M_V, and $B - V$ are all corrected for space absorption. Proceeding from fainter to brighter stars, the main sequence departs increasingly from the zero-age main sequence and finally breaks off to run continuously up into the giant region.

Using the data of Table E–1, we may transform M_V vs. $(B - V)_0$ measurements to M_{bol} vs. T_{eff} values. Thus

M_V	$(B - V)_0$	B.C.	M_{bol}	T_{eff}
0.0	1.31	0.80	−0.80	4060
0.5	1.22	0.66	−0.16	4200
1.0	1.14	0.55	+0.45	4320

In this manner we construct Fig. F–3, which can be compared directly with theoretical predictions. A similar transformation has been carried out to go from Fig. 60 to Fig. 61.

4. Interpretation of Luminosity Classes

Figure F–4, which is based primarily on the work of W. W. Morgan, gives the relation between luminosity and spectral classes and absolute visual magnitudes. The calibration is accurate for the main sequence (except perhaps for the very brightest stars), but the supergiant data are more uncertain, for two reasons:

(1) Since these stars are very distant, determinations of their distances are intrinsically difficult because they depend on association with star clusters and other uncertainties;

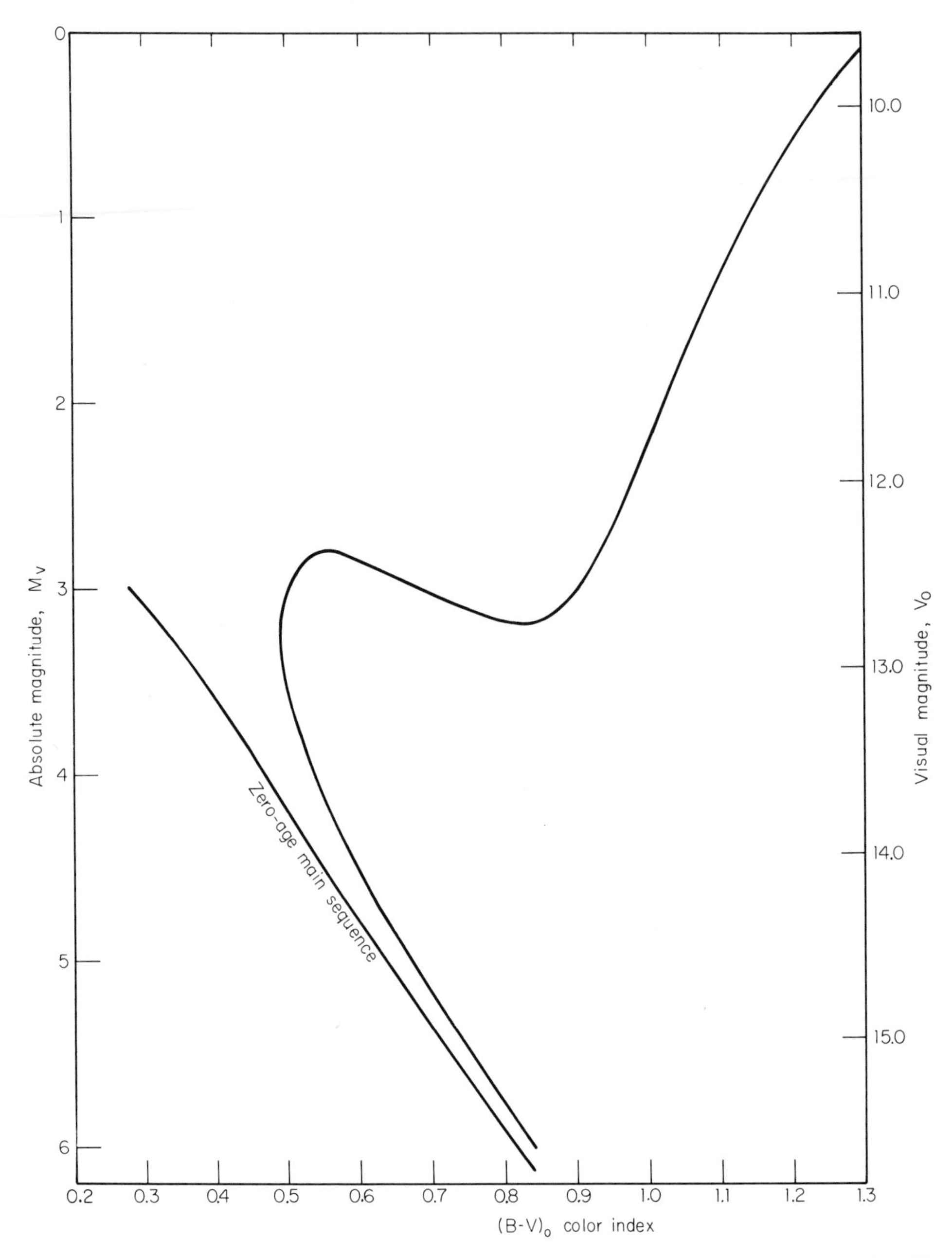

Fig. F–2. The color-luminosity relations for the galactic cluster M 67. "Visual" magnitudes V_0 are plotted against $(B - V)_0$ colors for mean points for this cluster. The left-hand ordinates give the absolute magnitude M_v obtained by fitting the curve to a standard Hertzsprung-Russell diagram. The zero-age main sequence is included. (After A. R. Sandage, *Astrophysical Journal 135* (1962), 349 copyright University of Chicago Press.)

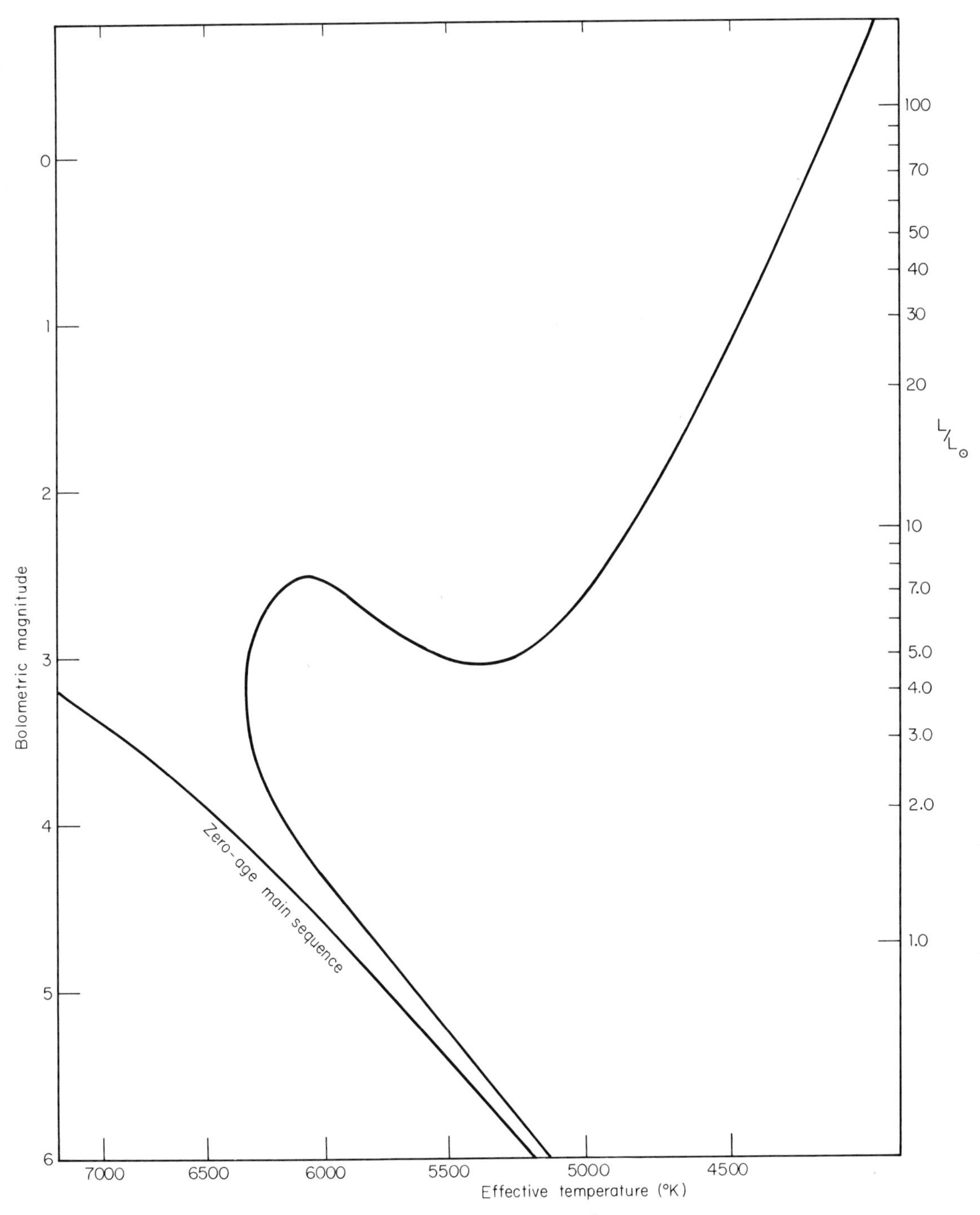

Fig. F–3. The luminosity–effective-temperature relation for M 67. By use of the data of Table E–1, Fig. F.2 has been transformed from a graph of M_v vs. $(B-V)_0$ to one giving bolometric magnitude vs. effective temperature. The right-hand scale gives the luminosity in terms of that of the sun. This curve differs from that given by Sandage, *Astrophysical Journal 135* (1962), 349, because a different color-temperature and bolometric-correction–temperature relation has been used.

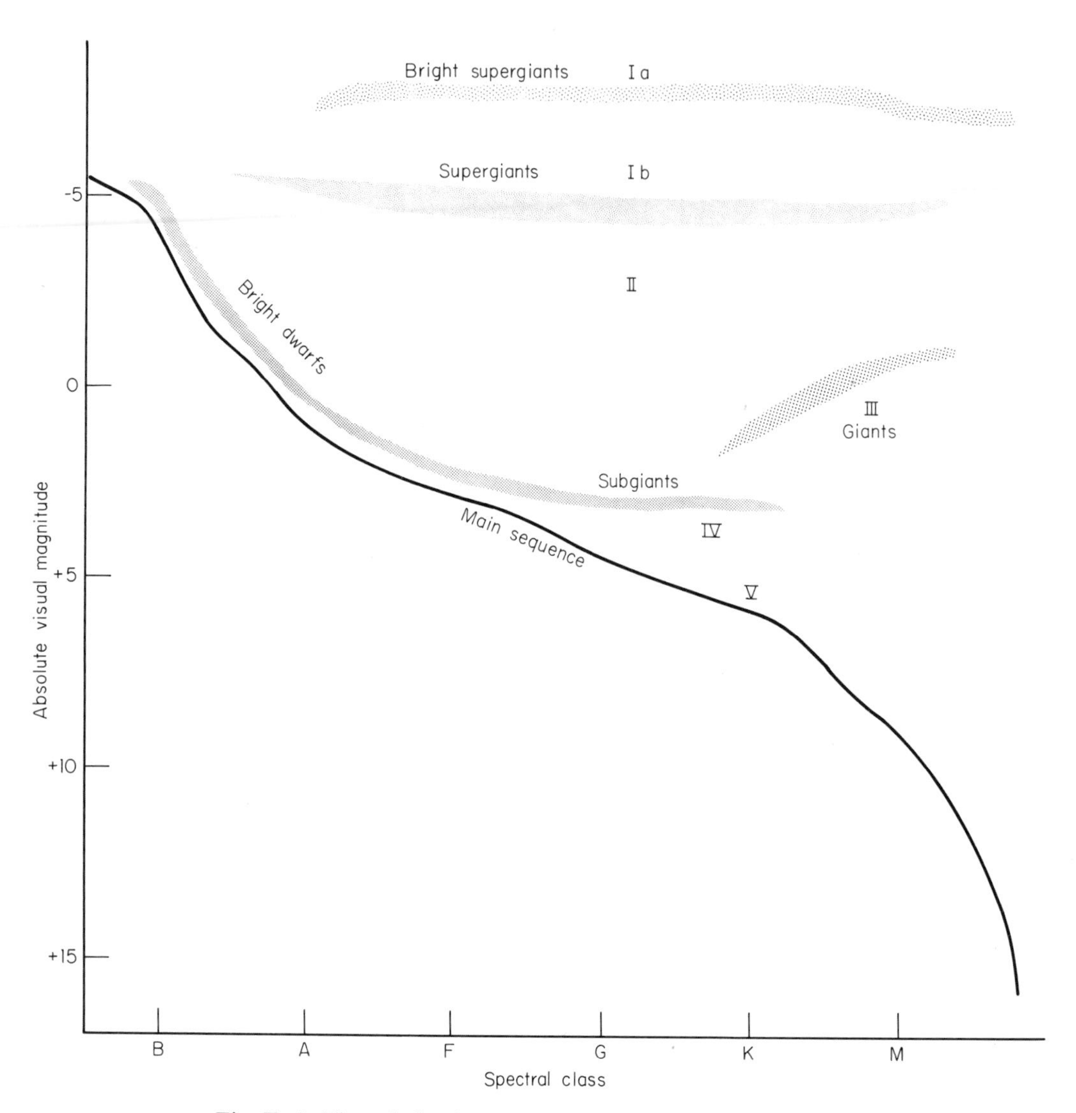

Fig. F–4. The relation between luminosity class, absolute magnitude, and spectrum. From the appearance of the spectrum, one can estimate not only spectral class but also a luminosity class, Ia, Ib, II, III, IV, or V. These have to be calibrated in terms of absolute magnitude, as has been done by W. W. Morgan and others. For supergiants, and even for giants, there is a considerable spread in intrinsic brightness for each luminosity class. Conventional giants constitute luminosity class III. Notice that in the later spectral classes luminosity class IV refers to subgiants, whereas in earlier classes it refers to bright dwarfs—in each instance stars that have evolved away from the main sequence, but by differing amounts.

(2) There is a large intrinsic spread in the brightnesses of these stars; that is, although main-sequence stars, particularly young main-sequence stars, tend to hug the dwarf sequence, giant and supergiant stars both show a huge range in intrinsic brightness, depending on the mass and chemical composition of the main-sequence star from which they evolved. Accordingly, giant and supergiant luminosity classes have been indicated by broad bands rather than by a narrow line as for main-sequence dwarfs.

Appendix G The Ionization and Excitation Formulas

As we saw in Chapter 4, the theory of ionization explains the great changes exhibited by the spectra of the stars as we proceed along the spectral sequence from a hot O star to a cool M dwarf or giant. In the hotter stars the metals are ionized and no longer absorb radiation in the spectral ranges where we can observe them; the spectra of cooler stars are jammed full of metallic lines. We shall devote some attention to both the excitation and the ionization formulas.

1. The Meaning of Thermal Equilibrium

Returning to the discussion of Chapter 4, let us fix our attention upon the excitations and ionizations of a group of atoms in our hypothetical box whose walls are maintained at some temperature T. If the temperature is sufficiently high, say 4000 or 5000°K, atoms will dash wildly about, absorb and emit energy, collide with one another, and lose and regain electrons.

In Figure G–1, let A be the ground level, and $B, C, D, \ldots$ excited

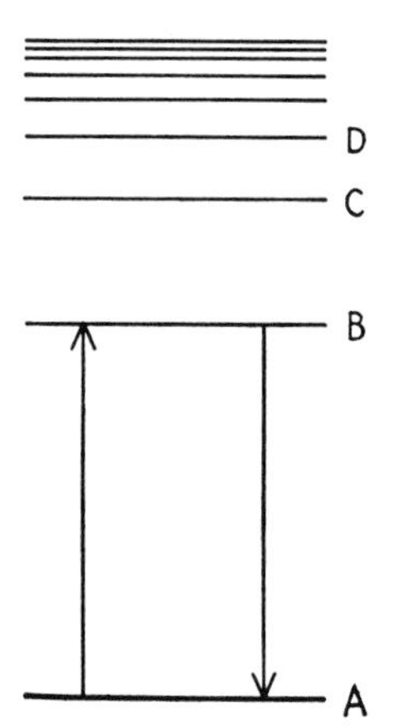

Fig. G–1. Schematic energy-level diagram.

levels. A fast electron may hit an atom in level A, lift it to level B, and then go away with less energy. Similarly, another electron may hit an atom in level B, de-excite it to the ground level A, and bounce off with increased energy. In an enclosure, these two processeswill exactly balance. Likewise, the collisional excitations of level C will exactly equal the collisional de-excitations. A similar situation obtains for the emission and absorption of radiant energy:

$$\text{Number of absorptions } A \to B = \text{number of emissions } B \to A;$$
$$\text{Number of absorptions } A \to C = \text{number of emissions } C \to A.$$

The number of ionizations from the level B will exactly equal the number of recaptures by the ion of electrons upon level B, and so on. Thus every process is exactly balanced by its inverse process. Under such conditions, the assemblage of atoms is said to be in *thermodynamic equilibrium*.

2. The Excitation Equation

Under conditions of thermodynamic equilibrium, the relative numbers of atoms in two levels A and B is given by the Boltzmann equation:

$$\frac{N_B}{N_A} = \frac{g_B}{g_A} e^{-\chi_{AB}/kT},$$

where k is Boltzmann's constant, 1.380×10^{-16} erg/deg, e is the base of natural logarithms, 2.718, T is the absolute temperature, and χ_{AB} is the energy necessary to excite the atom from level A to level B. The factors g_A and g_B are constants depending on the level involved; they are called *statistical weights*. In each instance they are numerically equal to the number of Zeeman states into which the level is split when the atom is placed in a magnetic field, and may easily be computed from atomic theory. If ν_{AB} is the frequency of the line emitted in the transition from B to A, then

$$\chi_{AB} = h\nu_{AB}.$$

Generally, it is more convenient for numerical computations to have this formula in another form. If we take logarithms to the base 10 and express χ_{AB} in electron volts, then

$$\log \frac{N_B}{N_A} = -\frac{5040}{T} \chi_{AB} + \log \frac{g_B}{g_A}.$$

For a derivation of the Boltzmann equation and the ionization equation (Sec. 3) see, for example, L. H. Aller, *Astrophysics I: Atmospheres of the Sun and Stars* (Ronald Press, New York, ed. 2, 1963), chap. 3.

In many cases g_A and g_B are small numbers of about the same size and for a qualitative notion of the ratio N_B/N_A we may omit them and write simply:

$$\log \frac{N_B}{N_A} \sim -\frac{5040}{T} \chi_{AB}.$$

Example. If A is the ground level of the O III ion (in this case A is actually a group of three levels close together, but we may treat them as one level for the present problem; see Fig. 72) and the excitation potential of level B is 2.48 volts, $g_A = 9$, $g_B = 5$, what will be the relative numbers of atoms in level B in thermodynamic equilibrium at a temperature of 10,000°?

$$\log \frac{N_B}{N_A} = -1.25 - 0.25 = -1.50,$$

$$N_B = 0.032 N_A.$$

Example. What would be the fraction of hydrogen atoms excited to the second energy level in the sun, if the excitation temperature is 5800°?

$$\chi_{AB} = 10.16 \text{ volts}, \qquad g_B = 4, \qquad g_A = 1;$$

$$\log \frac{N_B}{N_A} = -\frac{5040 \times 10.16}{5800} + \log 4 = -8.83 + 0.60 = -8.23,$$

$$N_B = 5.9 \times 10^{-9} N_A,$$

that is, under these conditions, about 6 atoms in every 1000 million are excited to the second level and thus become capable of absorbing the Balmer lines.

3. The Ionization Equation

Under conditions of thermodynamic equilibrium, the relative numbers of ionized and neutral atoms are given by the Saha equation:

$$\log \frac{N_1}{N_0} P_\epsilon = -\frac{5040}{T} I + 2.5 \log T - 6.48 + \log \frac{2B_1(T)}{B_0(T)}$$

where N_1 is the number of ionized atoms, N_0 the number of neutral atoms, P_ϵ (atmos) the electron pressure, I (volts) the ionization potential, and T (°K) the temperature. The correction term $\log [2B_1(T)/B_0(T)]$ is a function of the temperature for any given atom. It depends on the number and kind of energy states and may be computed from atomic theory. For

Table G-1. Data for calculation of ionization equilibrium.

Atom		Atomic No.	χ_0	χ_1	χ_2	log $2B_{r+1}/B_r$ $r = 0$ $T = 5800$	log $2B_{r+1}/B_r$ $r = 1$ $T = 10,000$
Hydrogen	H	1	13.60	—	—	0.00	
Helium	He	2	24.58	54.40	—	.60	0.00
Carbon	C	6	11.26	24.38	47.87	.11	− .48
Nitrogen	N	7	14.53	29.59	47.43	.65	.07
Oxygen	O	8	13.61	35.11	54.89	− .05	.63
Neon	Ne	10	21.56	41.07	63.50	1.08	.48
Sodium	Na	11	5.14	47.29	71.65	−0.03	1.08
Magnesium	Mg	12	7.64	15.03	80.12	.58	0.00
Aluminum	Al	13	5.98	18.82	28.44	− .47	.58
Silicon	Si	14	8.15	16.34	33.46	.08	− .47
Sulfur	S	16	10.36	23.40	35.00	− .03	.50
Argon	A	18	15.76	27.62	40.90	1.03	.51
Potassium	K	19	4.34	31.81	46.00	−0.14	1.05
Calcium	Ca	20	6.11	11.87	51.21	.54	−0.25
Titanium	Ti	22	6.82	13.57	27.47	.48	− .20:
Vanadium	V	23	6.74	14.65	29.31	.27	.15:
Chromium	Cr	24	6.76	16.49	30.95	.16	.50:
Manganese	Mn	25	7.43	15.64	33.69	.37	.23:
Iron	Fe	26	7.87	16.18	30.64	.49	− .07
Cobalt	Co	27	7.86	17.05	33.49	.25	.08
Nickel	Ni	28	7.63	18.15	35.16	− .12	.31
Strontium	Sr	38	5.69	11.03	—	.50	− .24
Barium	Ba	56	5.21	10.00	—	.43	− .55

many practical purposes it is legitimate to replace this term by log $(2b_1/b_0)$, where b_1 and b_0 are atomic constants depending on the kind of ground energy levels in each ion.

Table G–1 gives the ionization potentials for a number of the most abundant elements of astrophysical interest, together with values of the term log $(2B_{r+1}/B_r)$ for $r = 0$ at $T = 5800$°K and for $r = 1$ at 10,000°K. These numbers are chosen to illustrate the effects in the neighborhood of the solar temperature where neutral atoms are becoming singly ionized and near a Class A0 star where singly ionized atoms are tending to become doubly ionized. For most elements the ratio changes slowly with temperature.

Example. What are the relative proportions of neutral and ionized sodium in the sun if $T = 5800$°K and $P_\epsilon = 10^{-5}$ atmosphere? The ionization potential of sodium is 5.14 electron volts, $\log (2B_1/B_0) = -0.08$, $2.5 \log T = 9.41$, and $5040I/T = 4.46$. Hence we find that

$$\log (N_1/N_0) = -4.46 + 9.41 - 6.48 - 0.08 + 5.0 = 3.39,$$

that is, $N_1/N_0 = 2460$, and only 0.041 percent of the sodium in the atmosphere of the sun is neutral.

Example. What is the relative amount of singly ionized iron (Fe II) in the atmosphere of Sirius, for which we take $T = 10{,}000°K$ and $P_\epsilon = 3 \times 10^{-4}$ atmosphere (see Table G–2)? The first ionization potential of Fe I is 7.87 eV; we take $\log (2B_1/B_0) = 0.49$. Then $5040\ I/T = 3.96$. $2.5 \log T = 10.000$, $\log (N_1/N_0) = -3.96 + 10 - 6.48 + 0.49 + 3.52 = 3.57$, or $N_1/N_0 = 3720$, that is, iron is almost completely singly ionized. Is it appreciably doubly ionized? Apply the ionization equation again with the following data: second ionization potential of iron is 16.18 eV, $\log (2B_2/B_1) = -0.07$. Then we find $\log (N_2/N_1) = -1.18$, $N_2/N_1 = 0.066$, $N_2/(N_1 + N_2) = 0.062$. Hence about 6 percent of the iron is already doubly ionized and 94 percent is singly ionized.

In applications of the ionization formula we must know the electron pressure as well as the temperature. For most types of calculation it suffices to use a mean value of the electron pressure as well as the temperature. Table G–2 gives representative mean values of the electron

Table G-2. Electron pressure and mass above photosphere, representative values.

		Electron pressure (10^{-6} atm)			Mass "above photosphere" (gm/cm^2)		
T(°K)	$\frac{5040}{T}$	Dwarfs	Giants	Super-giants	Dwarfs	Giants	Super-giants
10080	0.5	320	—	33	0.071	—	0.37
8400	.6	190	—	23	.25	—	.71
7200	.7	70	—	13.6	0.76	—	2.5
6300	.8	23	10	4.0	1.7	3.0	10
5600	.9	7.9	2.7	1.0	3.0	10.0	26
5040	1.0	4.4	.76	0.22	3.7	20	60
4582	1.1	2.8	.35	.10	3.9	28	76
4200	1.2	1.8	.19	.052	3.9	40	89
3800	1.3	1.2	.10	.025	4.0	50	112
3600	1.4	0.81	.04	.012	4.1	76	162
3360	1.5	.57	.018	.004	4.1	100	200

pressure for giant and supergiant stars as well as data for main-sequence objects.

Finally, to compare the numbers of, say, neutral atoms above the photosphere of the sun with those above the photosphere of some other star, we must have some knowledge of the relative transparencies of the two atmospheres, that is, the opacity variation. If N_s is the number of atoms above the photosphere of a star of temperature T, and $N_\odot$ is the number above the photosphere of the sun, we may tabulate the ratio of N_s to $N_\odot$ for different values of T for dwarfs, giants, and supergiants. Alternatively, we may give the number of grams per square centimeter above the photosphere for different temperatures and types of stars (Table G–2).

4. An Illustration of the Curve of Growth

In actual use of the curve of growth it is necessary to combine line-intensity data from different spectral regions. Exact theory shows that one should use log (W/λ) rather than log W as ordinate and, say, log $Nf(\lambda/5000)$, with λ in angstroms, rather than log Nf as abscissa. Figure G–2 shows the theoretical curve of growth for the sun. Table G–3 lists data for four sodium lines, all of which arise from the same ground level. The second column gives the equivalent width, in thousandths of an angstrom, as taken from the *Revised Rowland Table* and other published sources. The third column gives the f-value as adopted from Brian Warner's (1968) discussion.

Our empirical curve of growth, Fig. G–3, consists of a graph of log (W/λ) against log $f(\lambda/5000)$, which is then fitted to the theoretical curve, Fig. G–4. The scale at the bottom is that of the theoretical curve; the scale at the top is that of the empirical curve. From a comparison of the two scales, we find that

$$\log N = 14.98,$$

that is, the number of neutral sodium atoms in the ground level above the photosphere of the sun is 0.95×10^{15}. An application of Boltzmann's formula will show that nearly all sodium atoms are in the ground level. With sufficient accuracy we can take the number of neutral sodium atoms above the solar photosphere to be 1×10^{15}. If $P_\epsilon = 10^{-5}$ atmosphere and $T = 5800°K$, $N(\text{neutral Na})/N(\text{total Na}) = 4.1 \times 10^{-4}$, so the total number of sodium atoms above the photosphere is 2.46×10^{18}. Since the mass of the sodium atom is $23 \times 1.66 \times 10^{-24} = 3.8 \times 10^{-23}$gm, the total mass of

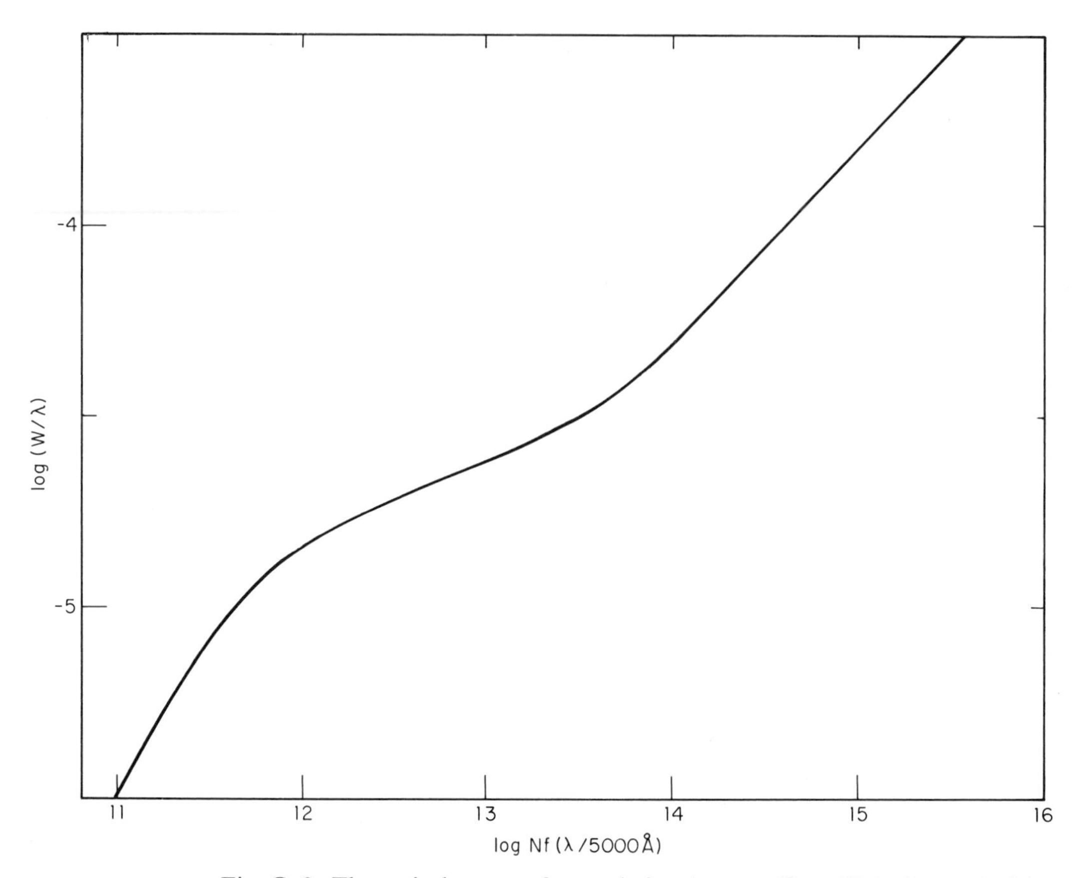

Fig. G–2. Theoretical curve of growth for the sun. Here W is the equivalent width and λ the wavelength (both in angstrom units) of a given spectral line. N is the number of atoms "above the photosphere" in the lower level of the transition involved, capable of absorbing the line in question, and f is the oscillator strength. (See C. W. Allen, *Astrophysical quantities* (Oxford University Press, New York; ed. 2, 1963), p. 165; also *Astrophysics I*, p. 292.)

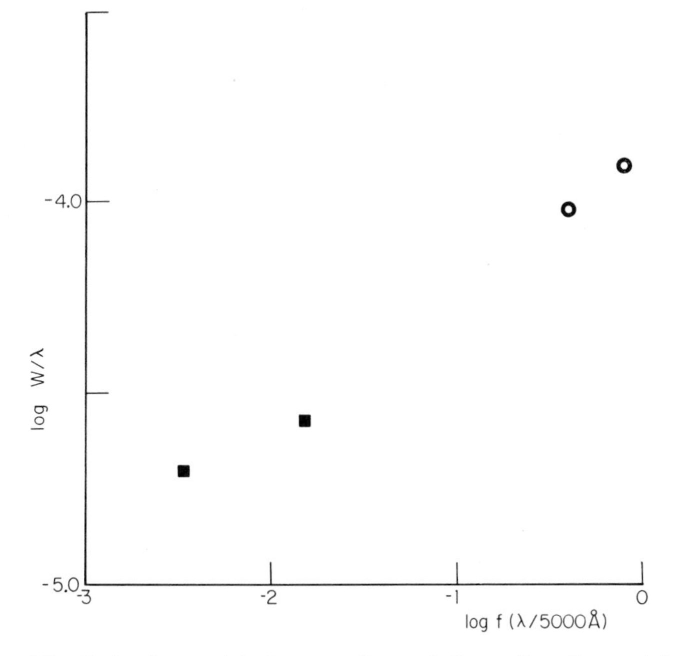

Fig. G–3. An empirical curve of growth for sodium lines arising from the ground level.

Table G-3. Data for solar sodium lines.

λ	W (mÅ)	f	$\log W/\lambda$	$\log f\lambda/5000$
3302.38	88	0.0214	−4.58	−1.85
3302.98	67	.0049	−4.70	−2.49
5889.97	730	.645	−3.90	−0.12
5895.94	560	.325	−4.02	− .415

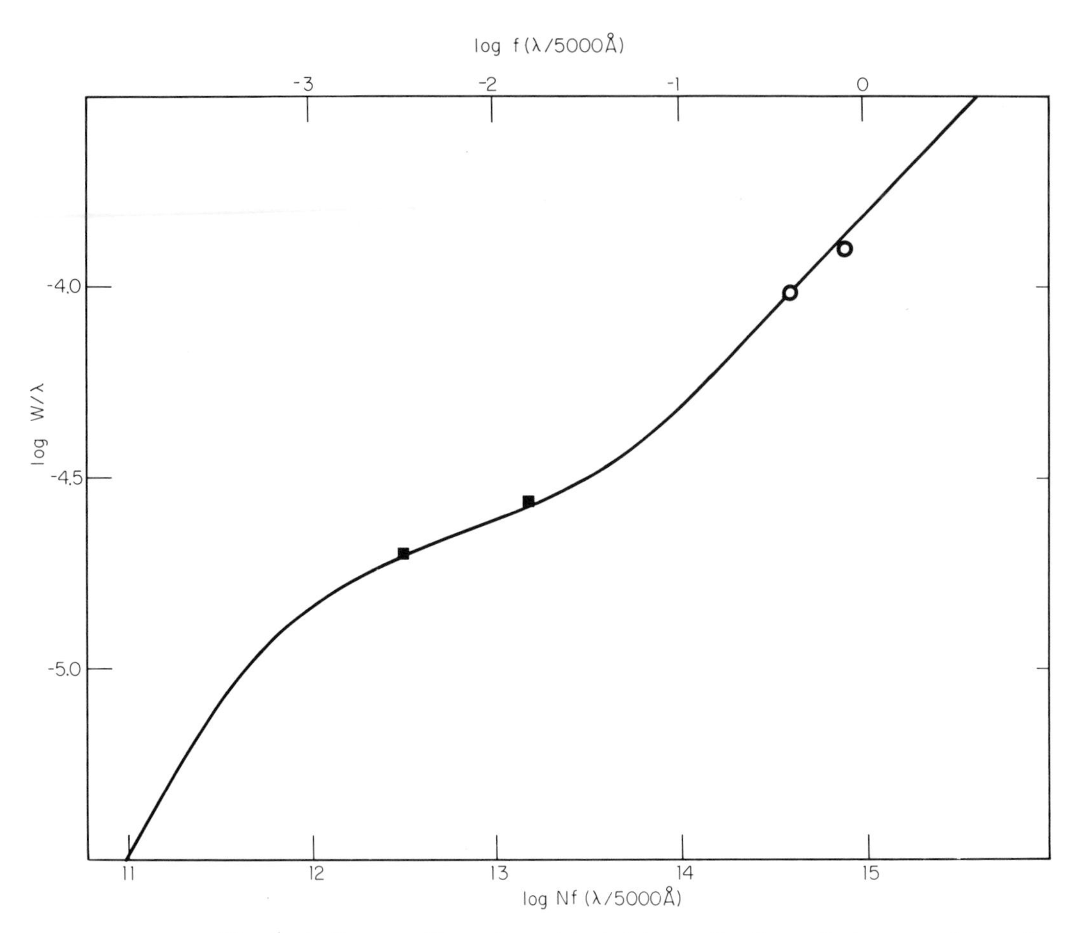

Fig. G–4. Superposition of empirical data on the theoretical curve of growth. The abscissae are (*top*) log $f(\lambda/5000\ \text{Å})$ from the empirical curve in Fig. G–3 and (*bottom*) log $Nf(\lambda/5000\ \text{Å})$. The zero of the empirical curve falls at 14.98 of the theoretical curve.

sodium above the photosphere is 9.4×10^{-5} gm = 0.094 mg. Similar analyses can be carried out for other metals, so their relative abundances can be found. Hydrogen produces most of the continuous absorption and its abundance controls the "depth of the photosphere." Detailed applications of the curve of growth may be found in *Astrophysics I*, chap. 7.

Appendix H The Determination of Stellar Masses

If two stars in a binary system are separated by an average distance a and move about one another with a period P, Kepler's third law says that the sum of the masses, $M_1 + M_2$, is given by

$$M_1 + M_2 = \frac{a^3}{P^2}.$$

Here P is measured in years, a in astronomical units (mean distance from earth to sun), and M_1 and M_2 in units of the sun's mass.

Suppose we know the velocities of the two stars and that they move in circular orbits. Then

$$2\pi a_1 = V_1 P, \qquad 2\pi a_2 = V_2 P,$$

and

$$M_1 a_1 = M_2 a_2,$$

so we can solve for the individual masses.

For a visual binary where the velocities are not directly measured but where the distance is known,

$$a = a'' r,$$

where a'' is the mean separation and r is the distance in parsecs. In terms of parallax p $(= 1/r)$,

$$M_1 + M_2 = \frac{a''^3}{P^2 p^3}.$$

We cannot determine the masses of the individual stars unless their motions are determined with respect to the background, that is, with respect to a fixed reference frame.

Appendix I. Interstellar Molecules

Table I-1 gives for each molecule its chemical formula, date of discovery, discoverers or early contributors to the investigation, and the institutions involved. Note that isotopes of carbon and oxygen, C^{13} and O^{18}, are involved in some of these molecules. Molecular hydrogen, H_2, was discovered from its ultraviolet lines observed with a spectrophotometer flown above the earth's atmosphere. The observed lines often show unusual intensity ratios which change with time and cannot be interpreted as normal radiation from a heated gas. Peculiar, highly selective excitation (perhaps similar to action of a maser) is often invoked for OH and H_2O. Formaldehyde shows a curious "refrigeration" effect. The population of one of its lowest levels is actually less than that calculated for a temperature of 3°K (as computed by Boltzmann's equation).

The structures of some of these molecules are indicated in Fig. I-1.

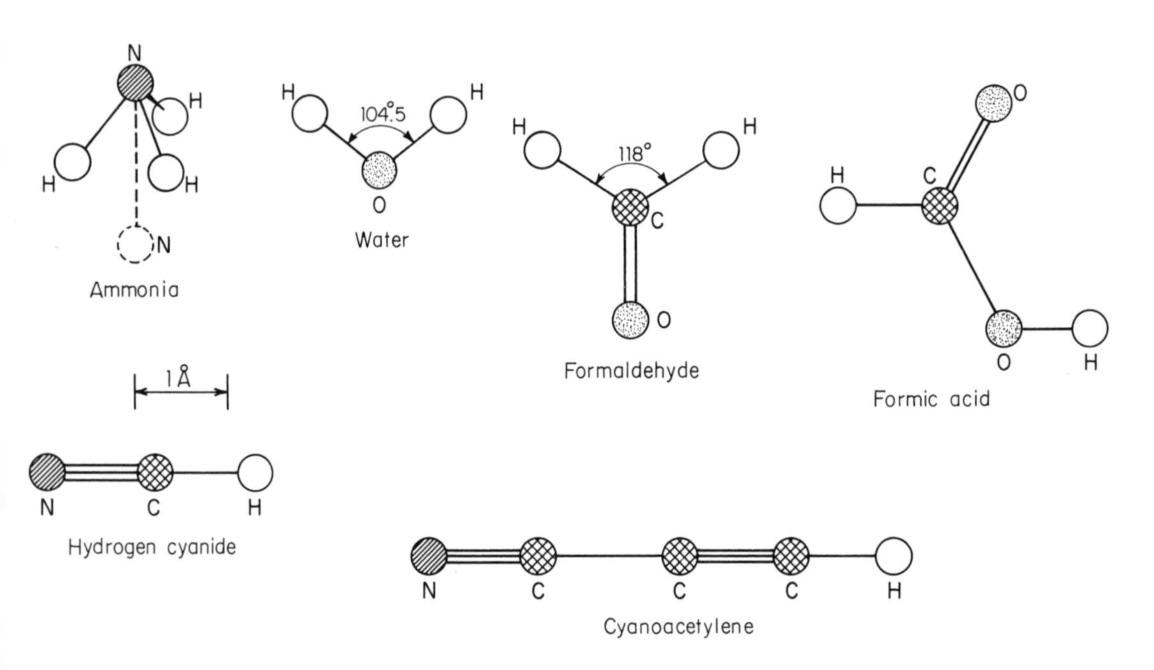

Fig. I-1. Structures of some interstellar molecules.

Table I-1. A list of recently discovered interstellar molecules.

Molecule		Year of discovery	Reference
Hydroxyl	$O^{16}H$	1963	1M, 2C, 3H
	$O^{18}H$	1966	4M
Ammonia	$N^{14}H_3$	1968	5C
Water	H_2O^{16}	1969	5C
Formaldehyde	$H_2C^{12}O^{16}$	1969	6R
	$H_2C^{13}O^{16}$	1969	6R
Hydrogen	H_2	1970	7S
Hydrogen cyanide	$HC^{12}N^{14}$	1970	6R
	$HC^{13}N^{14}$	1970	6R
Cyanoacetylene	HC_3N	1970	6R
Wood alcohol	CH_3OH	1970	8H
Carbon monoxide	CO	1970	9R
Formic acid	HCOOH	1970	10R

[1] S. Weinreb, A. H. Barrett, M. L. Weeks, and J. C. Henry (1963)
[2] N. H. Dieter, H. Weaver, and D. Williams (1966)
[3] E. J. Gunderman, S. J. Goldstein, and A. E. Lilley (1966)
[4] A. Barrett and A. Rogers
[5] A. C. Cheong, D. M. Rank, C. H. Townes, D. Thornton, and W. J. Welch
[6] L. E. Snyder, D. Buhl, B. Zuckerman, and P. Palmer
[7] G. Carruthers
[8] C. A. Gottlieb, J. A. Ball, A. E. Lilley, and H. E. Radford
[9] R. Wilson, K. Jefferts, and A. Penzias
[10] B. Zuckerman, J. A. Ball, C. A. Gottlieb, and H. E. Radford

M, Massachusetts Institute of Technology; C, University of California Hat Creek Radio Observatory; R, National Radio Astronomical Observatory; H, Harvard College Observatory; S. Hulburt Center for Space Research.

Index